T0327707

SECURITY AND PROFIT IN CHINA'S ENERGY POLICY

Contemporary Asia in the World

CONTEMPORARY ASIA IN THE WORLD
David C. Kang and Victor D. Cha, Editors

This series aims to address a gap in the public-policy and scholarly discussion of Asia. It seeks to promote books and studies that are on the cutting edge of their respective disciplines or in the promotion of multidisciplinary or interdisciplinary research but that are also accessible to a wider readership. The editors seek to showcase the best scholarly and public-policy arguments on Asia from any field, including politics, history, economics, and cultural studies.

ØYSTEIN TUNSJØ

SECURITY AND PROFIT IN CHINA'S ENERGY POLICY

Hedging Against Risk

Columbia University Press / New York

Columbia University Press
Publishers Since 1893
New York   Chichester, West Sussex
cup.columbia.edu

Copyright © 2013 Columbia University Press

Library of Congress Cataloging-in-Publication Data
Tunsjø, Øystein.
    Security and profit in China's energy policy : hedging against risk /
Oystein Tunsjo.
        pages cm. — (Contemporary Asia in the world)
    Includes bibliographical references and index.
    ISBN 978-0-231-16508-2 (cloth : alk. paper) — ISBN 978-0-231-53543-4
(e-book)
    1. Energy policy—China. 2. Energy security—China. 3. Energy resources
development—China. I. Title.

    HD9502.C62T86 2013
    333.790951—dc23
                                                    2013006636

Columbia University Press books are printed on permanent and durable
acid-free paper.
Printed in the United States of America

c 10 9 8 7 6 5 4 3 2

JACKET DESIGN: Thomas Stvan

To my father

CONTENTS

CONTENTS

LIST OF MAPS

ACKNOWLEDGMENTS

In writing this book I have had the advice and assistance of many individuals. I benefited greatly from my interviews with experts and analysts in the petroleum industry, the shipping sector, and commodity research in China, the United States, and Norway.

Academics following Chinese energy security policy and international relations scholars have been generous with their time and helpful in passing on suggestions and comments. In China, I am indebted to Wang Haibin, Zhao Hongtu, Zhang Tuosheng, Zha Daojiong, Xue Li, Wang Yizhou, Wu Baiyi, Lu Dehong, Ouyang Wei, Xu Weidi, Kate Westgarth, and Linda Jakobson. I also gained valuable experience and contacts when I participated in co-organizing with the China Foundation for International and Security Studies and the Chinese Academy of Social Sciences the Energy Security in Asia conference in Beijing in May 2009, which brought together leading Chinese and international energy experts. In the United States, I am grateful to Bud Cole, Mikkal Herberg, Edward Chow, Bo Kong, David Pumphrey, Daniel Rosen, Gabriel Collins, Andrew Erickson, Peter Dutton, Chin-hao Huang, and Edward Steinfeld. In Norway, I want to thank Jo Inge Bekkevold, Dag Harald Claes, Stein Tønnesson, Ingerid Opdahl, Tor Vidar Mykland, and Jonas Grätz. In Japan, I am grateful to Tatsuo Masuda and Yoji Koda.

I am especially indebted to the following for providing invaluable feedback throughout the book project:

To Robert S. Ross, for providing constructive criticism on my work and for guiding me in my academic career. To be a recipient of Bob's knowledge, collegiality, and friendship have been a privilege.

To Erica Downs, for many stimulating discussions, for reading part of the manuscript, and for encouraging me to develop the hedging framework.

To Johannes Rø, for always finding time to discuss and read my work and for pushing my thinking on hedging beyond what I could have achieved alone.

While writing the book manuscript, I had the support of Nancy Hearst, who proofread the entire manuscript and made valuable suggestions. Jan Bølstad at the Norwegian Military Geographic Service was very helpful in creating maps. I want to thank my editor Anne Routon at Columbia University Press, whose interest in this project has been most helpful.

The Norwegian Embassy in Beijing, the Norwegian Embassy in Tokyo, the Norwegian Embassy in Washington, D.C., the Norwegian Ministry of Petroleum and Energy, and the Norwegian Ministry of Foreign Affairs have all been forthcoming and helpful during my research. In particular, I would like to thank the Norwegian Ministry of Defense for generously providing economic funding for this research. I also benefited from a Fulbright scholarship that allowed me to stay as a visiting scholar at the Fairbank Center for China Studies at Harvard University in 2010. I am grateful to my colleagues at the Norwegian Institute for Defense Studies (IFS) and, in particular, Director Sven Holtsmark and former Director Rolf Tamnes, for providing an excellent environment that allows research to thrive and academic ambitions to prosper. Needless to say, the shortcomings of this book are my own.

My biggest debt is to my family and my wife, Hege. Without Hege's love, continuing support, and understanding, this book would never been completed. Axel and Felix have reminded me of the critical balance between academic life and the other rewards that life has to offer. I have dedicated this book to my father, Jack Tunsjø, to whom I owe more than I can ever repay.

ABBREVIATIONS

APEC	Asia Pacific Economic Cooperation
ARF	Association of Southeast Asian Nations Regional Forum
ASCM	antiship cruise missile
ASEAN	Association of Southeast Asian Nations
ASC	antiship capability
ASW	antisubmarine weapon
AU	African Union
bpd	barrels per day
CSCAP	Council on Security Cooperation in the Asian Pacific Region
CASS	Chinese Academy of Social Sciences
CBM	coal-bed methane
CCP	Chinese Communist Party
CDB	China Development Bank
CGNPC	China Guangdong Nuclear Power Group
CIC	China Investment Corporation
CICIR	China Institute of Contemporary International Relations
CIIS	China Institute of International Studies
CIMA	China Institute for Marine Affairs
CIS	Commonwealth of Independent States
CITIC	China International Trust and Investment Corporation

CLCS   UN Commission on the Limits of the Continental Shelf
CMSNC   China Merchant Steam Navigation Cooperation
CNNC   China National Nuclear Corporation
CNOOC   China's National Offshore Oil Company
CNPC   China National Petroleum Company
COSCO   China Ocean Shipping Company
CSCO   China Shipping Company
CSI   container security initiative
DOE   U.S. Department of Energy
E&P   exploration and production
EEZ   Exclusive Economic Zone
EIA   Energy Information Administration
EIBC   Export and Import Bank of China
ESPO   East Siberian-Pacific Ocean (pipeline)
GAIL   Gas Authority of India Ltd.
GICNT   Global Initiative to Combat Nuclear Terrorism
GMD   guided missile destroyer
GMF   guided missile frigate
GNPOC   Greater Nile Petroleum Operating Company
GW   gigawatts
HSE   health, safety, and environment
HNOC   hybrid national oil companies
IAITO   International Association of Independent Tanker Owners
ICG   International Crisis Group
IEA   International Energy Agency
IMB   International Maritime Bureau
IMO   International Maritime Organization
IOC   international oil company
IR   international relations
ISPFSC   International Ship and Port Facility Security Code
KMG   KazMunaiGaz (Kazakhstan)
KMT   Guomindang (Kuomintang)
LACM   land-attack cruise missile
LNG   liquefied natural gas

mbd	million barrels per day
MFA	Ministry of Foreign Affairs
MMG	MangistauMunaiGaz (Kazakhstan)
MOGE	Myanmar Oil and Gas Enterprise
MOP	Ministry of Personnel
MOOTW	Military Operations Other Than War
MOU	memorandum of understanding
MRBM	medium-range ballistic missile
NCP	National Congress Party
NDRC	National Development and Reform Commission
NDU	National Defense University
NEAB	National Energy Administration Bureau
NEC	National Energy Commission
NIOC	National Iranian Oil Company
NOC	national oil company
ONGC	Oil and Natural Gas Corporation Ltd. (India)
PDVSA	Petroléos de Venezuela
PLA	People's Liberation Army
PLAAF	People's Liberation Army Air Force
PLAN	People's Liberation Army Navy
PRC	People's Republic of China
PSA	production-sharing agreement
ReCAAP	Regional Cooperation Agreement on Combating Piracy and Armed Robbery against Ships in Asia
SASAC	State Asset Supervision and Administration Commission
SCO	Shanghai Cooperation Organization
SDPC	State Development and Planning Commission
SETC	State Economic and Trade Commission
SIPRI	Stockholm International Peace Research Institute
SLOCs	sea lines of communications
SOA	State Oceanic Administration
SOE	state-owned enterprise
SPLM	Southern People's Liberation Movement
SPR	strategic petroleum reserve

SRBM	short-range ballistic missile
tU	tons of uranium
ULCC	ultralarge crude carrier
UNCLOS	United Nations Convention of the Law of the Sea
UNPKO	United Nations peacekeeping operations
UNSC	United Nations Security Council
USGS	United States Geological Survey
USNS	United States Naval Ship
VLCC	very large crude carrier
WMD	weapons of mass destruction
YOY	year over year

GLOSSARY

dangwei	party committee
dangzu	party group
fangzhen	policy direction
luxian	policy line
zhengce	policy

SECURITY AND PROFIT
IN CHINA'S ENERGY POLICY

1

INTRODUCTION

C HINA'S RISE AS a global power is likely to be one of the most significant developments in international affairs in the twenty-first century. One dimension of China's growing influence is its rapid emergence as a major force in world energy markets, and energy security is undoubtedly influencing China's diplomatic and strategic calculations. This book examines how China is adapting to its increasing reliance on imported crude oil and petroleum products and discusses the strategies that China has developed to strengthen its energy security by reducing its exposure to potential supply disruptions and sudden price rises.

The book focuses on oil because China needs to import oil in substantive quantities and hence needs to "secure" its supply. Much has been written about China's energy security policy, but there are not many books that comprehensively study China's domestic, global, maritime, and continental petroleum strategies and policies.[1] In contrast to much of the existing literature, this book seeks to demonstrate how strategic and market elements interact, show that China has developed hedging strategies to insure against its growing net oil import gap and oil supply disruptions, and explain that the behavior of China and its national oil companies (NOCs) fit a hedging pattern. The most important argument of this book is that hedging is a central feature of China's energy security policy and that a hedging framework captures how government[2] concerns about *security of supply* and energy companies' *search for profits*[3] are linked. As an investor seeking profits but also security, China ends up hedging its energy security bets.[4]

By combining security and profit considerations, the Chinese government has developed a number of strategies to minimize the risk that its increasing reliance on imported crude oil, exceedingly high oil prices, and oil supply disruptions will negatively affect China's economic growth and domestic stability. China's leaders have also developed policies to manage and minimize the risk that the commercial interests of China's NOCs might undermine China's diplomatic and national interests. It is contended that these strategies and policies are best understood as hedging.

Hedging is often a preferred strategy when confronted with uncertainty and the objective of managing risk.[5] Hedging in international politics can be defined as a strategy aiming to reconcile conciliation and confrontation in order to remain reasonably well positioned regardless of future developments. China remains uncertain as to whether market or strategic strategies best enhance its energy security, and it recognizes that risks to its oil supply security cannot be eliminated. Hence, there are strong incentives for China to combine market and strategic consideration and to manage energy security risks by developing hedging strategies.

It is unlikely that China will eliminate the threats and strategic vulnerabilities accompanying its increasing reliance on imported oil by strengthening its army's ability to safeguard China's NOCs' overseas production, pipelines, and important sea lines of communications (SLOCs) or by developing its own world-class, globally active NOCs. Instead, the potential disruption of oil supplies to China will remain a risk that it will manage through hedging. Consequently, China's energy security cannot be secured, but it can be insured by the decision makers' continued ability to process and manage risks through hedging strategies and policies.

A number of analysts and observers have debated whether China will address its increasing reliance on overseas oil supplies and associated vulnerabilities by taking strategic steps that might lead to conflict or by accommodating market mechanisms. Other experts reject such a simplistic and dichotomized debate between market and strategic approaches, pointing out that energy security combines aspects of both. Indeed, most states, including China, mix strategic and market strategies to secure energy supplies. This is often referred to as a "comprehensive" approach. Nonetheless, loosely mixing market and strategic aspects and pointing to a comprehensive approach is insufficient. This book contends that the concept of hedging needs to be added to traditional approaches and linked to risk management in order to provide a more rigorous theoretical framework for understanding China's energy security, thus enabling researchers to explore the balance and the interrelationship between market and strategy.

This argument goes beyond the semantics of simply replacing the word "comprehensive" with "hedging." By broadening our analytical tools and incorporating hedging into the analysis, we are able to capture the complexity that states face when developing energy security policies and to address important nuances in China's. In chapters 3 through 6, I emphasize the new perspectives that a hedging framework brings to the table when analyzing China's energy security policy.

Some studies focusing on the overseas investments in crude oil production and equity deals of China's NOCs argue that these strategies either are driven by commercial interests and the pursuit of profits or represent a neomercantilist strategy to lock up oil reserves and ship more oil back to China. Others argue that the picture is mixed and that security and profit considerations are both important. The hedging framework shows how profit and security concerns are linked and how the Chinese NOCs' preoccupation with profits under normal market conditions also provides a hedge against periods of crisis in the international petroleum market that might threaten China's oil supply security and against the very high crude oil prices that might negatively affect the Chinese economy.

Similarly, the hedging framework demonstrates how boosting China's refinery capacity and the nation's ambition to develop a large state-owned tanker fleet both provide profit opportunities for state-owned tanker companies and NOCs and improve China's energy security. The hedging framework also illustrates how the buildup of a strategic petroleum reserve (SPR) and the construction of a number of cross-border pipelines enhance China's risk management capability and provide important oil supply safeguards. The hedging strategies are identified by examining the domestic, global, maritime, and continental aspects of China's energy security policy and by tracing the hedging strategies back to the five-year and long-term plans developed by the central government in the early 2000s as Chinese leaders became more attentive to energy security issues.

The development of a hedging strategy will not insulate China from a major crisis in the international petroleum market. But in such a case China is likely to be better off than other major petroleum importers, and its economy might too be reasonably well off, minimizing a potential economic downturn and the risk of social instability, which is the core objective of China's energy security policy. Pursuing a hedging strategy to manage and minimize energy security risks is a strategy other states may favor, but there are important differences among major oil importers, and a comparative analysis of other countries' energy policies is beyond the scope of this study.

This book draws a distinction between hedging *strategies* reflected in the five-year plans, which have included building an SPR, developing a large state-owned tanker fleet, developing cross-border pipelines, and promoting three flagship NOCs that operate and produce petroleum worldwide, and hedging *policies*, an outcome of a policy-making process in sensitive areas, such as Sudan and Iran, whereby Chinese leaders strive for consensus and balance between actors that promote profit/commercial interests (NOCs and, at times, the PLA) and security/national interests (the Ministry of Foreign Affairs, the State Council, the PLA, and the National Development Reform Commission).

CORE FINDINGS

This study makes a number of important contributions to the existing and rapidly growing literature on the subject. Both theoretically and conceptually, it provides a framework to analyze China's energy security policy that explains and develops a better understanding of its energy security behavior. This hedging framework fills a theoretical void in the literature on China's energy security policy by systematically combining the traditional market and strategic approaches and by linking hedging to risk management in the study of energy security.

Although hedging with future contracts in order to protect against price risks and volatility has been normal practice in the oil market since the 1980s, and even though some scholars have referred to China's energy security policy as hedging, few analysts have elaborated on the actual meaning of hedging.[6] There are few studies that provide in-depth analysis of how hedging translates into the study of security and how it can be used to explain China's energy security policy.[7] I contend that China's leaders have the authority to control the NOCs and the means to develop a hedging strategy and that a hedging strategy has been developed and implemented in the domestic, global, maritime, and continental sectors.

Much has been written about China's energy security policy, with many publications focusing on an "either/or" perspective. China's energy security policy is *either* guided by strategic and mercantilist ambitions *or* it is shaped by market mechanisms. It is *either* organized and controlled by the government *or* it is manipulated by powerful NOCs pursuing their own corporate and commercial interests.[8] In contrast, this book emphasizes a "more or less" perspective. Both market and strategic considerations shape China's energy security policy. In some cases, the pursuit of profit may be the most important factor; in others, the government prioritizes security of supply.

This "more or less" perspective is an integral part of the domestic, global, maritime, and continental dimensions of China's energy security policy. China is pursuing both commercial and strategic interests through its global search for petroleum deals, its development of a state-owned tanker fleet, the building of pipelines, and its domestic energy sector. Similar to an investor who hedges by combining "longs" and "shorts" in his portfolio, China's energy security policy effectively combines strategic and security goals with market and profit considerations.

With respect to the important relationship between the Chinese government and the Chinese NOCs, the government has the authority to instruct and control the NOCs and the state-owned tanker companies during any crisis that might threaten China's petroleum supply. We find two opposing views on this issue within the existing literature. On the one hand, a mercantilist approach emphasizes coordinated and state-directed efforts that drive China's energy security policy. On the other hand, a more commercial approach focuses on the role of ineffective institutions and powerful NOCs that pursue corporate interests even at the expense of national interests.[9] This book maintains that Chinese NOCs are largely autonomous and competitive actors pursuing profits in their daily activities both domestically and internationally. Nonetheless, their corporate interests cannot undermine the overarching government objective of providing energy supply security, which is not only essential to China's economic growth and development but also closely tied to the CCP's political legitimacy and stability.

Even though the NOCs are influential in China's strategy of overseas expansion, or "going abroad," they depend on support from the Chinese government in order to be successful. More importantly, although the NOCs are largely independent and pursue their own commercial interests on a daily basis, when corporate and national interests clash in sensitive and strategically important cases, notably in Sudan and Iran or with respect to cross-border pipelines, the buildup of SPRs, or domestic petroleum prices, the Chinese government intervenes to maintain and advance national interests.

This book draws important distinctions between threats and risks, between peacetime and wartime contingencies, and between pipeline and seaborne supply routes. This is not new, but too many studies do not pay adequate attention to these important distinctions. More attention to the differences between threats and risks and the linkage between hedging and risk management will clarify and alert us to the vulnerability of China's energy security. In fact, Chinese energy security is a problem of risk management. Risk can be understood as the possibility that a potential threat will materialize. In theory, threats can be eliminated, but risks cannot. Instead, risks need to be managed.[10] One way to manage risks is through hedging.

Moreover, too much of the existing literature fails to make the important distinction between wartime and peacetime scenarios. For example, if China were to find itself at war with the United States over Taiwan, all bets would be off, and China's seaborne oil supply would likely be disrupted. The PLA would find it difficult to protect China's overseas oil production, and China's seaborne oil supply and state-owned tanker fleet would face wartime contingencies. In short, during a major war China would not have the military capability to withstand a blockade by the United States.

Nevertheless, a war between the United States and China is unlikely. Because the costs would be extremely high, the probability of such a conflict remains low. But even in the event of a conflict or war with the United States, China's seaborne oil imports only constitute about 10 percent of its total energy mix, and it is unlikely that the United States could completely cut off the flow of oil to China. This suggests that China would be able to withstand a blockade for a number of years, even though there are few substitutes for oil in the transportation sector and though China's economy and domestic stability would likely suffer. China produces domestically more than 4 mbd (million barrels per day) and in a crisis or war could produce synthetic oil from its large coal production. Despite China's record growth in energy demand, it remains about 90 percent self-reliant. China has not taken sufficiently strong measures to reduce the environmental effect of coal use, reflecting both a preference for relying on cost-effective and abundant coal reserves and a short-term strategy that emphasizes security and self-reliance. This is the critical domestic part of China's hedge against a disruption to its oil supply.

Contrary to conventional and popular wisdom that emphasize the so-called Malacca dilemma in China's energy security thinking, short of a major war China is not vulnerable to interruptions in its seaborne oil supply. First, a blockade of China's oil supply would most likely lead to war, and few states are willing to go to war with China. China can withstand a blockade during a war because of its domestic energy production. A blockade would have political and economic effects, but in wartime everything becomes critical, and the blockade would likely not top the list. Second, a distant or limited peacetime blockade of China's oil supplies would not prevent the arrival of oil shipments to China because tankers can manage exchanges of oil at sea. Third, a close blockade can quickly escalate into an act of war and potentially undermine the limited objectives of the blockade. Fourth, a blockade by convoy would logistically and operationally overwhelm the blockading states, including the powerful U.S. Navy. Fifth, a distant blockade of China's oil supply would cause significant disruptions in the international petroleum market, and the blockading power would rapidly lose diplomatic support while achieving only limited objectives. Sixth, precision strikes on Chinese oil installations and infrastructure would be an act of war and thus unthinkable short of initiating a major war.[11] Hence, instead of focusing my analysis on the threat of war, it is more useful to explore peacetime risks to China's petroleum supply security.

Risks, such as a war that does not involve China, environmental disasters, terrorism, piracy, and accidents at sea, that may cause a shock to the international petroleum market and result in a significant rise in the prices of crude oil and petroleum cannot be eliminated. Nevertheless, China can hedge and has been hedging against these risks through domestic production of 90 percent of China's energy; the development of overseas equity oil deals by the NOCs; the buildup of a state-owned tanker fleet; the development of cross-border pipeline projects; the buildup of SPRs; future and long-term contracts; and diversification of its sources, routes, and energy mix.

Finally, it is important to distinguish between seaborne and pipeline petroleum supplies. The United States may be reluctant to escalate a conflict, for example, in the Taiwan Strait, by attacking mainland China or neighboring countries in order to destroy cross-border pipelines that supply China with petroleum.[12] If the United States decided to attack China's domestic pipelines and oil fields, it would face the risk of escalation,[13] and it can also be assumed that China would be more capable of defending its own territory than defending sea lanes in the Indian Ocean, the South China Sea, or the western Pacific.

At the same time, pipelines are more vulnerable during peacetime because they are less flexible. That is, oil pipelines do not change hands like oil on tankers. The capacity of a pipeline is fixed, though more oil could be bought and shipped to China through the international petroleum market. And pipelines can be disrupted by an attack on the pipeline or on the pumping stations, but tankers can circumnavigate a distant or limited blockade. When it comes to pipelines, buyers and sellers are locked in an interdependent relationship that does not exist in the international petroleum market. During peacetime, it is easier for terrorists, insurgent groups, or even thieves to attack the thousands of miles of pipelines than the tankers on the high seas.

In sum, in the case of China's energy security, pipelines are more vulnerable to peacetime risks but safer from wartime threats, whereas seaborne energy supplies are safer from peacetime risks and more vulnerable to wartime threats. Thus the security difference between inflexible pipelines and flexible seaborne supply routes; the distinction between a pipeline "market" and the international petroleum market; and the diplomatic, strategic, and domestic spillover effects from constructing cross-border pipelines have led China to develop a hedging strategy by combining pipeline and seaborne supply routes to protect against disruptions of supply.

SOURCES AND OUTLINE OF THE BOOK

This volume does not engage the Chinese literature and debates regarding energy security because of the author's lack of Chinese-language skills. Although this is a drawback, it is somewhat compensated for by drawing on several interviews with energy experts in China. Interviews were conducted with representatives from the government NDRC, the NOCs, the PLA, the shipping industry, and a number of researchers at universities and think tanks during eight visits to China between 2007 and 2011.[14] Secondary literature in English by Chinese energy experts is used throughout the study, in addition to English-language long-term strategy and government papers related to energy policy.

I also conducted interviews with experts on China's energy security policy during several visits to the United States and during a spring 2010 visit to the Fairbank Center for Chinese Studies at Harvard University as a Fulbright scholar. I also interviewed some European energy experts and top executives from international oil companies (IOCs) and shipping and insurance companies operating in China.

Few researchers have access to top-level policy makers in China or to in-depth knowledge of the calculations of various stakeholders in China's energy policy. In many ways, China's energy security policy making remains a black box. Despite extensive interviews in China over a four-year period, I have found no silver bullet explaining China's energy security policy or the crucial relationship between the government and the NOCs. My experience, although far from conclusive, is that top officials from the government emphasize their control over the NOCs, and the representatives from the NOCs remind you how powerful they are. Interviews with representatives from IOCs, who interact with both the Chinese government and the NOCs, have been valuable for a better understanding of the relationship between the government and the NOCs. Conversations with Chinese and Western energy experts and at various think tanks have also been helpful. This study argues that the NOCs are largely independent and powerful actors in their daily business activities and that their search for profit may sometimes undermine national interests. But the government can control the NOCs during a crisis and force them to undertake measures that safeguard China's broader national interests.

Neither is there a smoking gun in the enormous literature on the subject, even in the literature that consults Chinese sources. Accordingly, this study is based on interpreting and observing the behavior of the Chinese government and important stakeholders in China's energy security policy. I contend that policy planning, such as the recommendations in the five-year plans, and the implementation of these objectives represent a hedging strategy to preserve China's energy interests. Lacking comprehensive insight into the thinking of China's top leaders on energy security, I can provide no conclusive evidence as to whether these strategies were formulated with foresight and awareness of contingencies and with any assumption of independence for the NOCs or other players in China's energy and foreign policy.

Chapter 2 provides theoretical and conceptual guidance for studying China's energy security policy. It draws on the conventional strategic and market approaches but argues that it is unsatisfactory to combine them into a comprehensive approach. The chapter details the tensions and interrelationships between the strategic and market approaches based on a hedging framework and risk management theory.

Sources, transportation routes, and the energy mix, in addition to physical availability and economic affordability, are all integral parts of energy security. By distinguishing between securing supply and constraining demand and differentiating between external and domestic aspects of China's energy security policy, chapter 3 focuses on the crucial internal measures taken to enhance energy security. Oil and gas account for roughly 21 and 4 percent, respectively, of China's energy mix.[15] Given China's abundant coal reserves, it is estimated that about 70 percent of China's energy mix is dominated by coal. There is little evidence to suggest that this heavy reliance on coal will change in the coming years.

Chapter 3 first examines the important relationship between the central government and the NOCs. It argues that the central government has a number of mechanisms to control the NOCs, thereby providing the government with the means to implement a hedging strategy and to pressure the NOCs into following it if the security of China's oil supply is endangered. Chapter 3 then shows that maintaining close to 90 percent of China's energy mix from domestic sources has been an important "short" strategy in its hedging. The chapter briefly analyzes China's energy mix and the policies adopted to improve energy efficiency and conservation, including securing alternative energy supplies, researching nuclear energy, and implementing initiatives to develop new and more infrastructure to distribute energy sources, strengthen domestic oil development, restructure energy companies and institutions, and reform taxes, fees, and the pricing of domestic oil products. Finally, I point out that building an SPR is a classic hedging strategy against the consequences of an oil disruption. Increasing China's refinery capacity also provides a hedging tool to manage and minimize the risk of a disruption or a crisis in the international petroleum market.

Chapter 4 explores China's strategies to gain access to overseas sources of oil. China's exceptional economic growth was an important factor in its becoming a net importer of oil in 1993. Since then, the Chinese government has facilitated overseas investments by its NOCs, which are now involved in more than two hundred upstream projects in more than fifty countries throughout the world.

Acknowledging that many Chinese NOCs now often pursue their own profit goals independently, one objective of chapter 4 is to determine whether the corporate interests of Chinese NOCs in some cases undermine national interests and to assess the extent to which the yuan "follows the flag." As Chinese NOCs largely sell their (upstream/equity) oil locally or on the international oil market, little oil produced overseas is shipped back to China. So why should the Chinese government risk losing status and prestige in the international community and undermining its relationship with important actors such as the United States and the European Union by supporting energy deals with rogue states, when China, irrespective of its "going-abroad" strategy, buys its oil on the international market? By exploring two sensitive cases, Sudan and Iran, I argue that the Chinese government has actively sought to prevent corporate interests from undermining national interests and has developed a hedging policy to balance profit and security considerations.

The marginal oil supplies that China receives from a number of rogue states represent only a small percentage of China's total energy mix and have a limited effect on China's energy security. Additionally, this chapter discusses whether the goal of Chinese NOCs is to gain access to "off-limits" countries to profit from upstream production or whether the government's aim of diversifying China's energy supplies is the driving factor behind China's global pursuit of energy resources. I argue that although diversification, corporate interests, and the ad-hoc nature of China's energy-related decision-making process are important characteristics of China's goal of securing overseas energy deals, a hedging framework captures the mixed interests and the broader commercial, diplomatic, strategic, and military incentives that shape China's global energy security policy.

Finally, I note that there is currently limited cooperation among Chinese NOCs. Rather than "locking up" reserves and overseas oil production by shipping oil back to China in a mercantilist fashion, Chinese NOCs engaged in upstream overseas production prefer to sell their products on the international market to obtain the best price, whereas downstream companies in China favor buying oil on the international market in order to obtain the lowest price (produce and sell locally, buy and sell locally). This is the profit aspect of China's energy security policy. Nonetheless, by investing in overseas upstream projects and capacity building, the government has the ability to force its NOCs to transport more of their overseas oil production home during a potential crisis or during a war not directly involving China. But such a strategy would be difficult to implement during a major war in which China is involved. This is the security aspect of China's energy security policy. Taking them together, I consider these aspects of China's global search for petroleum important factors in China's hedging strategy, which seeks to manage peacetime risks, insure against high oil prices during periods of crisis in the international petroleum market, and minimize supply disruptions.

Chapter 5 examines the main transportation route for Chinese oil supplies: the maritime route. It contends that China is hedging its energy supplies by building a large state-owned tanker fleet. This allows the state-owned tanker companies both to profit from a more prominent position in the tanker market under normal market conditions and to enhance China's energy security; during potential crises or emergencies, the Chinese government can instruct the NOCs and state-owned tankers to ship oil produced overseas back to China on Chinese tankers and also use the state-owned tanker fleet to access oil terminals and operate in war-risk exclusion zones. However, this strategy would be counterproductive if China were at war with another major power because it will be easier to identify Chinese tankers. Accordingly, again, the distinction between peacetime and wartime contingencies is crucial to understanding how China hedges its energy security bets.

It may appear puzzling that China is building a large state-owned tanker fleet because this would make it easier for the United States, for example, to identify tankers going to China during a potential blockade. However, a blockade of Chinese oil supplies by another state would pose a threat to China's national interests and present a wartime contingency, which China would try to eliminate most likely through the use of military force. But hedging is both a strategy and a policy to deal with peacetime contingencies and risks. A large state-owned tanker fleet may be of strategic value if a conflict in which China is not involved were to erupt.

For example, if the turmoil in the Middle East were to spread to Saudi Arabia and Iran, or if war erupts between the United States and Iran, rising oil prices would hurt the Chinese economy and contribute to social instability. During such a crisis, the Chinese government could call on the cooperation of its own oil and shipping companies. Consequently, oil supplies to China would be insured through a hedging strategy that allows the country to manage the risks of a disruption to oil supplies or of rising prices and that can "secure" oil supplies at a lower price during a crisis (the NOCs can be pressured to ship the equity production to which they have marketing rights back to China) and with lower shipping costs (China's state-owned tanker fleet can be pressured to help the NOCs ship their overseas equity production back to China) than if China were to import all of its oil supplies through the international market during a conflict in an oil-rich region. Simultaneously, tankers not needed to safeguard the Chinese oil supply can profit in the global market.

Of course, Chinese NOCs do not produce sufficient oil overseas to meet Chinese demand.[16] But having a state-owned tanker fleet also makes it possible to operate with high risks in a conflict area in order to access oil terminals when other shipping companies are unwilling or unable to operate. Finally, by having its own tanker fleet, China can provide its own insurance and avoid the high insurance premiums that it would need to pay to shipping and oil companies operating in the international market when a war-risk exclusion zone has been declared by the major insurance companies or where sanctions have been enforced. Thus, a large tanker fleet becomes an important hedge that provides benefits during peacetime (shipbuilding, infrastructure for a naval buildup, competitive shipping companies, and profits from the shipping market), an important insurance during a potential crisis or conflict, and an opportunity to profit from a more dominant position in the tanker market during an emergency. This hedging strategy will not insulate China from a crisis in the international petroleum market; it would still have to pay high global prices for oil. Nonetheless, China would be relatively better off than other major petroleum importers, and its economy would be less affected by the crisis, like a financial hedger in a bear market who has insured his portfolio with longs and shorts. China will have the capacity to access oil terminals and ports where other ships will not go; China will be able to draw on its NOCs' equity overseas oil production; and China will be paying lower shipping and insurance costs, thus allowing it to manage better a potential economic downturn and the concomitant risk of social instability.

It is important to note a number of assumptions guiding this argument. First, and for the foreseeable future, China will have limited naval power to withstand a naval blockade by the United States during a major war. China can only expect to receive limited seaborne oil supplies during a war with the United States, as the U.S. Navy will control the SLOCs. Second, a more limited or distant blockade of oil shipments to China, for instance, an attempt to blockade the Strait of Malacca or the Strait of Hormuz, would have a limited chance of success or might easily escalate into a major war.[17] Third, China has sufficient domestic energy sources and would not need to depend on overseas oil supplies to wage a war, and in wartime contingencies China can make petroleum from its abundant coal resources.[18] Finally, both the United States and China are seeking to avoid war or conflict, and energy issues are not likely to trigger a war between the two countries.

This chapter also examines the link between China's energy security policy and its maritime strategy and growing naval power-projection capability by looking at the "pseudo-national-interest" argument developed within China that justifies and promotes the buildup of the navy under the banner of protecting vital seaborne energy supply routes.[19] In short, although energy security is likely to be an important part of wartime contingency plans, a wartime scenario may not be a fruitful starting point to assess China's energy security policy because there would not be much China can do about being cut off from seaborne energy supplies during a major war with the United States. We can gain a better understanding of China's energy security policy by examining its goal of developing a large tanker fleet as an important hedging strategy to deal with peacetime contingencies and risks.

Because China's imports of petroleum products, today and tomorrow, will not solely arrive via sea lanes, chapter 6 examines China's efforts to secure petroleum supplies through pipeline projects on the Asian continent. Again, diversification of transportation routes is a major factor in China's involvement in a number of pipeline projects. I examine whether a hedging strategy can be observed here as well, whereby energy deals are part of a broader economic, diplomatic, and strategic agenda, that is, a broader and more sophisticated strategy that encompasses more than diversification. Although conventional analysis shows that diversification is often linked to diplomatic, political, strategic, and economic objectives, diversification is only one aspect of a hedging strategy that not only diversifies a portfolio but also short-sells some stocks and currencies and relies on future contracts and derivative products.

I argue that such hedging mechanisms are observable in China's continental petroleum pipeline strategy. Not only does China diversify its petroleum transportation routes, but since its pipelines and sea lanes carry different security risks, the pipelines become a short strategy to insure against disruptions in longs (seaborne transportation routes). Further, the differences between a pipeline "market" and the international petroleum market mean that China's relationship with its continental petroleum suppliers can provide a hedge if there is a crisis in the international petroleum market. By examining the different contracts negotiated for pipeline deliveries of petroleum products, in particular by looking at the price and amount of petroleum that will be delivered and the duration of the contracts, I highlight how these long-term and partially fixed contracts resemble futures contracts in finance, which can be considered an important part of a hedging strategy, adding additional insurance to what traditionally has been considered China's petroleum diversification strategy.

China's continental diversification strategy not only provides China with influence in neighboring countries and important strategic areas but also prevents and insures against encroachments upon Chinese borders from Western or other major powers. China invests in expensive cross-border pipelines not necessarily to maximize profits or for long-term economic benefit; instead, it follows a hedging strategy of buying security/insurance and seeking to be reasonably well off in the future. Finally, the hedging strategy is not one dimensional or horizontal; rather, it combines both horizontal and vertical dimensions. Although China seeks to diversify its transportation routes by building long-distance pipelines to boost its energy security, these measures simultaneously give the central government an opportunity to maintain domestic control and to develop the remote provinces. Linking a hedging framework to risk management provides a theoretical tool to analyze these multifaceted aspects of China's energy security policy and goes beyond analyses that focus only on traditional diversification strategies.

Chapter 7 considers some implications of China's energy security policy. Globally, the chapter examines how China's NOCs compete and cooperate with IOCs and shape the international energy market. China's energy deals and ties with energy-producing countries throughout the world will be scrutinized, broadening our understanding of how China's energy deals relate to economic, strategic, and diplomatic initiatives both regionally and globally. China's growing involvement with international energy institutions such as the IEA will be assessed to determine whether energy institutions can mitigate rivalry and potential conflicts over energy resources. Finally, this first section asks how China will cope with growing global energy interests and responsibilities.

The second section in chapter 7 focuses on the maritime domain. First, it examines how the buildup of a large state-owned tanker fleet affects the international shipbuilding and tanker market. Then I consider China's effort to manage peacetime risks to energy transportation routes at sea, including piracy, terrorism, and accidents at vital maritime chokepoints. China has been active in promoting public goods in the global commons through its participation in protecting vital SLOCs and contributing to UN peacekeeping operations in regions where it is important to safeguard transportation routes and maintain stability for the exploration and production of oil. What are the drivers of change in China policy? What are the implications of China's more active role in carrying out such missions? What are China's broader goals, and how do they shape the international energy order?

Moreover, are there tensions between continental and maritime interests in China's strategic outreach and search for energy sources? Will the PLA, for example, be able to focus its interests and capabilities in both strategic directions, or will the maritime domain be increasingly prioritized, with energy security promoted as a justification for this strategic outlook? This section ends by exploring China's interests with respect to the melting Arctic ice cap and the opening of new sources of petroleum products and transportation routes. Will China turn its attention toward this new petroleum frontier in order to diversify sources and transportation routes in its energy security policy, and what are the implications of a more prominent Chinese role in this region?

Finally, chapter 7 explores how Russia's possible ambitions to develop oil and gas pipelines in eastern Siberia and the Far East affect Sino-Russian relations and regional developments in northeastern Asia. I argue that energy is a key factor that still provides Sino-Russian relations with a strategic rationale and that it can maintain and accommodate these relations as the "strategic partnership" becomes constrained by a lack of common goals, China's growing dominance in Asia, and weaker and more limited arms export.

Chapter 8 concludes the study by summarizing the main arguments and findings. The analysis focuses on the influence of the key actors in the energy-policy-making apparatus (the CCP, NOCs, governmental energy bodies, local governments, the MFA, the PLA, the media, and academics) and how their relationships evolve and shape strategy and policy. Returning to the alarmist scenarios about China's energy security policy and the potential for conflict between China and other major energy consumers over limited energy resources, the conclusion examines the possibility of avoiding classical security dilemmas and spiraling conflicts by engaging in hedging strategies that promote security sensibility and develop trust through the management of risk.

2

CHINA'S ENERGY SECURITY

A New Framework for Analysis

THIS BOOK OFFERS a new theoretical framework based on the concept of hedging to analyze China's energy security policy. This chapter seeks to flesh out the core features of that hedging framework. After briefly reviewing traditional market and strategic approaches to energy security, the main section of the chapter shows how a hedging framework can enhance our understanding and provide a better explanation of China's energy security policy. Thereafter, hedging is linked to the broader theory of risk management.

STRATEGIC AND MARKET APPROACHES

Strategic considerations and the importance of market mechanisms have traditionally been the main focus when devising an energy security policy. According to Andrews-Speed, the perceptions and priorities of the Chinese government underscore an adherence to the "strategic" approach.[1] "Most obvious is the case of China," argue the contributors to the *CAEC Task Force Report* on Asia-Europe energy cooperation, when energy security is described in strategic terms. The report maintains that the Chinese government favors a hands-on approach, be it in the form of a strong government monopoly over indigenous production and foreign investments or in enhancing relationships with producer countries. This is achieved through political and financial factors, diplomatic relations, and the sale of goods and important investments.[2]

With energy-consuming states competing and scrambling for limited energy resources to power their economic growth, a strategic approach often emphasizes zero-sum and relative-gains thinking.[3] Energy is often seen as a source of conflict between states: one state's gain is another state's loss. China, it is argued, is "unilaterally securing its future energy needs by a neo-mercantilistic 'lock up' supply strategy" and pursuing close diplomatic ties with countries shunned by most of the world as pariah states.[4]

Such a perspective often yields alarmist scenarios that see China's quest for overseas oil supplies leading to rivalry, discord, and confrontation. As a consequence, because of Beijing's quest for ties that forge energy security, China and the United States might find themselves on a "collision course."[5] This view is fueled by the historical analogies of the great-power quests for energy to facilitate their rising power. "Sixty-seven years ago, oil-starved Japan embarked on an aggressive expansionary policy," notes Gal Luft. "Today, another Asian power thirsts for oil: China."[6]

Promoters of this strategic and alarmist perspective are challenged by those who emphasize growing global interdependence and international cooperation to maintain stability in the world oil market. A market approach emphasizes that by liberalizing international and domestic energy markets states can prevent disruptions of oil supplies through improvements in the flow of information, the maintenance of strategic oil stockpiles, and the promotion of investments in new capacities.[7]

The Chinese government has emphasized the importance of international cooperation and has promoted investment in new capacities both at home and abroad.[8] Several experts point out that instead of taking oil off the market, Chinese oil companies are investing in the few remaining underdeveloped oil regions of the world.[9] By extracting oil from "off-limits" places such as Sudan and by operating with higher risk in "problem states," China's national oil companies (NOCs) are contributing to world oil supplies. For example, China National Petroleum Company (CNPC) estimates that only 10 percent of the thirty million tons of Chinese overseas oil production was shipped back to China in 2005; the remainder was sold to overseas markets.[10] It is acknowledged that data are difficult to access and verify, but an International Energy Agency (IEA) report from 2011 focusing on Chinese NOCs' overseas investments argues that only a portion of Chinese NOCs' equity production overseas is shipped back to China. Evidence suggests that most of the equity oil is instead sold to local or international markets.[11]

Other observers see the world energy market as one large pool of oil. If China seeks oil from Sudan, it buys less elsewhere, which means that there will be more oil available to other consumers.[12] In addition, approximately 80 to 85 percent of the world's oil reserves are controlled by state governments or state-controlled enterprises.[13] Therefore, China will inevitably have to rely on purchasing oil from mature producers in the international oil market.[14]

Finally, it is held that Chinese strategies to establish a foothold in the world energy market are similar to the strategies of the major Western companies.[15] In contrast to a strategic approach, a market approach highlights the importance of nonstate actors and sees international politics largely in terms of a positive-sum game as opposed to a zero-sum game, although it is recognized that the international oil market is characterized by strong competition. Whereas a strategic approach can be linked to a realist international relations (IR) tradition, a market approach draws inspiration from liberalist IR theory.[16]

There are a number of studies that provide a more nuanced analysis than that implied by the above dichotomized debate.[17] Much scholarship combines elements of the market and strategic approaches to assess China's energy security policy.[18] Clearly, states mix strategic and market strategies to secure energy supplies; hence the objective here is not to jettison traditional approaches. Nonetheless, traditional analysis does not provide the tools required to examine the tensions and interrelationships between the market and strategic approaches.

Therefore, the key theoretical objective of this study is to explore how hedging and risk management can explain some of the complexity that is lost in the gap between the strategic and market approaches and thereby provide a more complete understanding of China's energy security policy. Because few studies have utilized the explanatory power of hedging and, in particular, hedging as an analytical framework for understanding China's energy security, the following sections aim to clarify the meaning of hedging, link hedging to risk management, and illustrate how this theoretical framework explains Chinese energy security policy.

HEDGING, RISK MANAGEMENT, AND ENERGY SECURITY

Hedging has become a buzzword in international relations.[19] The concept of hedging is used by scholars to examine contemporary Chinese foreign policy, explore how Southeast Asian states maneuver between the United States and China, understand the European security environment, analyze how China affects transatlantic ties, and go beyond the U.S. debate about whether containment or engagement is the best way to deal with the rise of China.[20] Indeed, a strategy to "hedge against unknowns" can be found in several major official U.S. policy statements, and it has been an important aspect of the U.S. "engagement plus hedging" policy toward China.[21] However, despite its frequent occurrence, the term remains elusive and not fully understood.

As pointed out above, hedging can be defined as state strategies aiming to reconcile conciliation and confrontation in order to remain reasonably well positioned regardless of future developments. Great powers hedge by combining moderate balancing elements, which includes strategic partnership but not alliances, military buildups but not arms races, and cooperation as well as assertive policies but not armed conflict. It is also contended that the incentive for hedging is positively correlated to uncertainty about systemic power distribution.[22]

Hedging has also been defined as "a set of strategies aimed at avoiding (or planning for contingencies in) a situation in which states cannot decide upon more straightforward alternatives such as balancing, bandwagoning, or neutrality." In hedging, states "cultivate a middle position that forestalls or avoids having to choose one side [or one straightforward policy stance] at the obvious expense of another."[23] These definitions are important starting points for any conceptual clarification, but it is simultaneously acknowledged that "there is no satisfactory exposition of hedging as a strategy. What does hedging behaviour look like?"[24] Similarly, it has been noted that "our received theoretical guidance about the conditions under which states will hedge—with what instruments and with how much recklessness—remains nearly as contradictory as the history of international relations."[25] In other words, it is necessary to develop further the concept of hedging to realize its explanatory potential and determine how hedging can be applied to energy security.

One important first step is to take into account the analytically useful distinction between threats and risks. The difference between threats and risks hinges on the different degree to which it is possible to judge whether a certain scenario is heavy with peril: threats are easier to detect and compute; risks are more elusive. Whereas threats are already formed and can be identified, risks are dangers in the offing whose potential costs are unknown. States face both threats and risks, but there are different ways to cope with the two. In principle, threats can be eliminated, whereas risks must be managed.[26] For instance, China can use military force to eliminate an identified blockade of its oil supply by an adversary, but it cannot prevent or eliminate conflicts that may erupt in oil-rich regions and disrupt the Chinese oil supply. Neither can China avoid adverse long-term trends in the international petroleum market or a mismanagement of its domestic energy sector. Instead, these risks can be managed and minimized, and hedging is one way of doing so.

This key point is succinctly acknowledged by the IEA in its *2007 World Energy Outlook* report: "[no] energy system can be entirely secure . . . energy security, in practice, is best seen as a problem of risk management, that is reducing to an acceptable level the risks and consequences of disruptions and adverse long-term market trends."[27] This change of focus to risk management is a rational adjustment to an uncertain security environment and can account for why China has been prone to hedging in its energy security policy. The hedging framework developed in this book shifts the analysis from focusing on wartime threats to peacetime risks, from maximizing profit to remaining reasonably well off, from securing to insuring an adequate oil supply, and from controlling to managing any disruptions to petroleum transportation routes.

The desire to prevent dismal outcomes by forgoing potential benefits is an age-old strategic response to uncertainty. As Jon Elster points out: "when choosing among crops, farmers have to consider the likelihood of early frost in the fall, of too little rain in the spring and of too much in the summer. Often they hedge their bets, by choosing a crop that leaves them reasonably well off regardless of weather."[28] Managing China's energy supply security and preventing the affordability and availability of energy from disrupting China's economic growth are priorities for Chinese leaders. In both farming and state-craft, it seems, survival sometimes depends on a successful hedging strategy.

There are several studies that link hedging to systemic factors and provide a framework for how hedging can explain contemporary great-power relations.[29] However, linking China's energy security hedging strategies and policies to systemic factors are unconvincing and problematic for a number of reasons. First, Chinese energy security hedging strategies are not a result of systemic factors or a "unipolar deconcentrated system."[30] The international system became unipolar in the 1990s, but China did not develop its energy security hedging strategies until the 2000s. While states' hedging strategies are positively correlated to uncertainty about systemic power distribution, and while great powers are most likely to hedge in a unipolar system, less so under multipolarity, and least in a bipolar system,[31] China's energy security hedging strategies are largely shaped by unit- and actor-level factors. In short, polarity does not drive China's energy security policy. China did not develop its energy security hedging strategies because the international system shifted from bipolarity to unipolarity. Instead, China developed hedging strategies as a result of concerns about a growing net oil import gap triggered by China's exceptional economic growth.

Second, both the United States and China have developed grand hedging strategies in the post–Cold War period that combines moderate balancing and cooperation.[32] China and the United States may choose more confrontational strategies and balancing in the future, as the power distribution within the international system becomes more attuned to bipolarity or multipolarity. However, China is most likely to continue developing energy security hedging strategies and policies. In short, energy security strategies are not comparable to a grand strategy, and China's energy security policy can be guided by the aim of hedging against supply disruption irrespective of whether China pursues balancing, bandwagoning, soft balancing, hedging, or buck passing at a grand strategic level.

Third, although China is concerned about SLOC security and has benefitted from the world order maintained by the United States, there are reasons to doubt that there are strong linkages between the alleged decline of the United States and its inability to lead and provide public goods and China's energy security hedging strategies. Some have argued that China hedges against a future where "the absence or reduction of U.S. naval forces in Southeast Asia may increase the frequency of pirate attacks against shipping in places such as the Malacca Straits, which would undoubtedly affect oil imports bound for China."[33] Southeast Asian waters were prone to piracy in the early 2000s, but this did not disrupt oil shipment to China. Piracy in Southeast Asian waters has been effectively combated by the littoral states during the last few years, largely without any prominent U.S. role. The United States may provide stability in the Middle East, but its actions have also been disruptive and alarming to China and have fueled concerns about China's energy security. At the same time, though, China has maneuvered into a stronger position in a number of countries to fill the vacuum produced after U.S. sanctions or withdrawals, such as in Iran, Iraq, Libya, Sudan, and Saudi Arabia. Thus U.S. decline may provide China with more opportunities than challenges when it comes to energy security matters.

Fourth, China's energy security hedging strategies are not compatible with the so-called type A and type B hedging. By disregarding the important distinction between wartime threats and peacetime risks, the writers advocating a distinction between type A and type B hedging cannot sufficiently show how China's energy security supply can be safeguarded in an "armed confrontation with a strong naval power like the United States" (type A hedging) and how China's oil supply security is vulnerable to piracy, terrorism, or natural disasters (type B hedging). Some of the hedging strategies that China has developed, such as diversifying transportation routes, increasing overseas oil production, developing an SPR, and improving ties with important oil suppliers, has been noted within the literature, but without distinguishing between wartime threats and peacetime risks, many observers of China's energy security are unable to grasp how and why these strategies encompass hedging. In short, the "hedging strategies" put forward in the current literature are not that different from the strategies emphasized by traditional "strategic" approaches to China's energy security policy. Accordingly, the following chapters will empirically show how and why China is hedging against oil supply disruptions and seeks to minimize and manage the risk of exceedingly high oil prices.

HEDGING AGAINST UNCERTAINTY

States that hedge aim to manage uncertainty and limit risk. Sensing that rivalry and instability might ensue if policy signals are taken as evidence of a one-dimensional intent, a hedging state deliberately sends out mixed signals. Hedging is a contingent strategy that simultaneously combines elements of cooperation, competition, and rivalry. Uncertainty about systemic power distribution and threats, transition periods, contingent unipolarity, and the possession of nuclear weapons makes great powers more inclined to develop hedging strategies. Uncertainty and risks are profound features of the international petroleum market, and they drive China's energy security hedging strategies and policies. Accordingly, I argue that when Chinese leaders sense uncertainty about whether a market or a strategic approach— or what kind of mix of these two approaches—best enhances China's interests, they will hedge their bets rather than choose one strategy at the obvious expense of another.

How is a mixed hedging strategy developed to manage uncertainty and limit risk? To limit market risk and at the same time make a profit, financial investors hedge by keeping a percentage of their portfolio on the "short" side—that is, they bet that some stocks, currencies, or food prices will go down. Since "shorts" will make money in a bear market, they act as a protection, a hedge, when "longs" (assets that are believed will rise in value) perform poorly. One can never eliminate risk, but if a hedging strategy is thoroughly implemented, investments will not be wholly dependent on whether the market goes up or down, as in either case there will be a profit. Instead of following a strategy to maximize profits, whereby investors simply speculate as to whether stocks will go up or down, a hedger follows a strategy by investing in both "longs" and "shorts," forgoing maximum possible profit in order to manage and minimize risk.[34] It is this idea that has transfer value to international politics and energy security: no matter how the market develops and whether or not risks on the supply side materialize, a hedging strategy can leave China reasonably well off tomorrow. In short, hedging amounts to insurance.

One perspective analyzes how China invests in "longs" by joining multilateral institutions, international organizations, and regimes; by cooperating to stabilize the international energy market; and through acting as a responsible player in the international community: for example, through the UN Security Council, through participation in UN peacekeeping operations (UNPKO), and by taking part in efforts to secure sea lanes against piracy. Note that all of these behaviors could be considered "positive" ones. An important aim of investing in "longs" is to insure against the potential detrimental effects or costs of a strategic or "short" strategy that emphasizes close ties with oil-producing states, thereby insuring against the vulnerability of becoming too reliant on market mechanisms or on U.S. control of seaborne transportation routes.

As with hedging in finance, "longs" are associated with cooperation or positive developments, such as profiting from the market or preventing crises. "Shorts" are tied to strategic and security considerations. "Shorts" insure against the possibility that cooperation will not damage Chinese interests or that instability or the higher costs of petroleum in the international market will disrupt China's supply security by, for example, building up a strategic petroleum reserve (SPR),[35] signing long-term contracts, securing overseas equity in oil production, developing a state-owned tanker fleet, diversifying the energy mix, constraining demand, and improving efficiency.

Investors can develop a hedging portfolio whereby it is possible to profit from both "longs" and "shorts." Similarly, China can profit from both its "long" and "short" strategies. States often recognize the need to cooperate to deal with common problems, thereby stimulating "long" strategies. For example, China has developed closer ties with the IEA, has become an active member of a number of multinational institutions, and has participated in antipiracy operations in the Gulf of Aden. All these efforts can enhance China's energy security. However, the urgency of risks does not create transnational consensus. In dealing with energy security challenges, different interpretations, judgments, and scenarios about current risks and future threats lead to different calls for action. Should China become a free rider on efforts by the IEA to stabilize world energy markets and efforts by other states to secure SLOCs? Should China take advantage of sanctions against rogue petroleum-exporting states or instead cooperate with the international community in seeking to isolate these states?

Additionally, mistrust is difficult to transcend in an anarchic international system, and relative gains are often more important than absolute ones. Put differently, conditions of insecurity at the systemic level still compromise cooperation among states. Most states prefer win-win situations, but, as Waltz explains, even when all agree on the goals and when the actors have an equal interest in cooperation, "one cannot rely on others."[36] When faced with the possibility of cooperating for mutual gain, states that feel insecure must ask how the gain will be divided. They are compelled to ask not "'will both of us gain?' but 'Who will gain more?' . . . Even the prospect of large absolute gains for both parties does not elicit their cooperation so long as each fears how the other will use its increased capabilities."[37]

Accordingly, China does not completely trust that the international petroleum market, interstate cooperation, or other "longs" will safeguard China's energy interests. Hence, China has taken a number of steps to develop "short" strategies that seek to insure against supply disruptions, high prices, and instability in the international petroleum market. Like an investor seeking to hedge against a volatile market through derivatives, future contracts, and "short" selling, China bets against the international petroleum market or cooperation and takes insurance "outside" the petroleum market through broader diplomatic and strategic measures, which include more costly strategies such as maintaining an energy mix that emphasizes self-reliance, developing a large state-owned tanker fleet, investing in overseas equity oil production, and building a comprehensive SPR and network of cross-border and domestic pipelines. These are not sufficiently accounted for by conventional analysis and the emphasis on diversification.

It has been argued that the potential risks to China's growing net oil importer status and its effect on national security "will prompt measures to hedge against reliance on the market as a source of energy security." The more

"China feels threatened by the US, the more likely it is that China seeks to hedge dependence on oil supplies vulnerable to American power." Accordingly, China is considering a number of strategies "to hedge against interruptions in supplies" and to avoid "putting all their faith in market mechanisms."[38]

Although such analyses nuance our understanding of China's energy security policy, they miss the fact that China not only hedges against relying on market mechanisms or U.S. dominance of important SLOCs but also against downside risks to the measures advocated by a strategic approach. For example, China's oil production in Sudan can offer a hedge and be strategically important if a conflict erupts in the Persian Gulf, producing a crisis in the international petroleum market. At the same time, China knows that its equity oil investments in Sudan are controversial. China has initiated diplomatic steps to participate constructively in solving the conflict in Sudan. In addition, CNPC has been selling most of its Sudan oil on the international petroleum market, thereby bringing more oil to the market. To some extent, these measures insure against the adverse consequences of China's strategic objectives or its "short" strategy in Sudan. This study aims to show that China hedges against the adverse consequences of *both* strategic *and* market approaches, a fact often overlooked when assessing China's energy security policy. Hedging is not simply about relying on "shorts" or "longs"; the two need to be combined to become a hedging strategy. If an investor only relies on "shorts," then she is a speculator, not a hedger. Strategic approaches are not necessarily more "secure" or safer than market approaches. The risks inherent in both approaches need to be managed through hedging strategies and policies that combine "shorts" and "longs." Hence, the question is more one of "longs" and "shorts" than one of either strategic or market strategies.

A characteristic of a hedging strategy is the balancing of uncertainty and the risk of opportunity loss. To illustrate, to secure a mortgage on a home, one can often choose between a fixed or a floating interest rate. If one decides on a fixed rate, the risk is minimized, and the monthly payments are fixed. But if the floating rate is consistently lower than the fixed rate, this strategy will lead to continual opportunity loss. A central feature of hedging is judgment and interpretation: any financial manager needs to decide how much a rise in one asset could affect another asset.[39] Some may decide on a fixed or floating rate;

others may choose a hedging strategy that mixes floating and fixed rates. Using this example to guide our thinking about how to analyze Chinese energy security policy, we can also explore how China seeks to prevent opportunity losses in its energy security policy and what risks China is willing to take to "secure" its energy supplies.

A hedging strategy is contingent on the risk profiles of the analysts, firms, or state stakeholders: some are risk takers, and others are more risk averse. Strong advocates of a Chinese blue-water navy, which includes aircraft carriers, argue that China "does not possess the ability to safeguard its equal right to energy in the world" and that China needs stronger military capabilities to protect its overseas oil import routes.[40] However, the Chinese government has not been willing to risk much in investing in aircraft carriers or in developing power-projection capabilities to secure the SLOCs through which 85 percent of China's oil imports pass. Instead, the Chinese leadership has shown caution, sending mixed signals that balance between China as a responsible stakeholder patrolling the SLOCs in the interest of the public good and the importance of a strong navy as a symbol of the emergence of China's status as a great power. Thus far, energy security has not fueled China's ambition to maximize military power; rather, China has preferred to limit risks and to obtain moderate goals through hedging.

Recent events suggest that China has now adopted a more confrontational and assertive stand. The March 2009 incident in which Chinese fishing vessels harassed the USNS *Impeccable* surveillance ship marked a turning point. In addition, belligerent diplomacy has been seen among countries with sovereignty claims in the South China Sea; China adopted combative diplomacy toward Japan after the Japanese Coast Guard arrested a Chinese fishing vessel near disputed islands in the East China Sea; tensions reemerged on the Korean peninsula; there have been renewed tensions between China and India over border disputes; the United States has conducted naval exercises with its regional allies and sold arms to Taiwan; President Obama's meeting with the Dalai Lama provoked an assertive Chinese diplomatic response; and increased great-power tensions have added to the general unease by preventing cooperation within international organizations and multilateral forums.

Even though China's more assertive stand in the South and East China Sea and its more firm posture toward safeguarding important SLOCs may be driven by energy security objectives, it is more likely that other factors, such as domestic instability, protecting Chinese sovereignty claims, nationalism, and broader great-power competition and rivalry, are shaping Chinese behavior.[41] Nonetheless, it is important to point out that even if the foreign policy or grand strategy that China has pursued since the end of the Cold War is shifting from a hedging strategy to more traditional balancing, China's energy security policy is still best explained and understood through a hedging framework.

Hedging states attempt to preclude conflict and brinkmanship. Some have argued that China has taken steps to position itself along important sea routes in the Indian Ocean to facilitate Chinese power-projection capabilities that might protect Chinese oil supplies in the future.[42] Although China helped Pakistan build a port at Gwadar and provided technical and financial aid to expand the port into a naval base, and although some analysts have referred to China's "string of pearls" strategy in the Indian Ocean, China has not stationed troops or set up bases overseas.[43] China has shown a more assertive stance since 2009, and there are signs that neighboring states may balance against China by developing closer ties with the United States. Nonetheless, China has engaged in talks over disputed areas in the South China Sea and the East China Sea, which promise to be rich in petroleum resources. Although China has become a dominant power on the Asian continent, it has done so by successfully enhancing its power through a "peaceful rise" strategy that resolved a number of border disputes and avoided war with other states.[44]

An increasing need for energy supplies has accompanied China's rising power, but even though China has become more reliant on oil imports, its behavior can often be described as showing "security dilemma sensibility" by developing hedging strategies that do not seek to maximize Chinese power, outreach, or a strategic position that would exacerbate conflict.[45] That is, Chinese leaders have frequently shown an ability to understand the role its actions may play in provoking fear, mistrust, and instability. Despite China's more assertive behavior over the past several years and increased regional tension, China has reemerged as a great power over the last three decades and has become a dominant power in Asia without waging war with its neighbors. Never in the history of mankind has another great power enhanced its relative and absolute power as formidably as China has done through its "peaceful rise" strategy over the past thirty years. This illustrates that the Chinese government has been careful not to take steps so provocative that they might lead to conflict and that it has sought to avoid brinkmanship.

The Chinese government has also showed a willingness not to risk too much political capital, diplomatic influence, or prestige over energy security issues, because more than oil is at stake in China's foreign policy. Hedging strategies can facilitate and guide decision makers with a reasonably comfortable attitude toward risks, for example, by balancing the corporate interests of Chinese energy companies, which largely focus on maximizing profits, against national interests. Indeed, as will be discussed, the Chinese government has taken steps to insure that its limited energy interests in discredited regimes, notably Sudan and Iran, do not undermine China's international reputation, status, or prestige.

Clearly, there is no consensus among Chinese researchers, government officials, and energy experts when it comes to assessing what increasing reliance and dependence on foreign oil will mean for Chinese security.[46] Accordingly, there is a tension between those arguing that China's increasing dependence on imported oil supplies presents a strategic threat to China's national security and those arguing that China's increasing oil imports should be treated as a normal byproduct of its growing interdependence with the rest of the world.[47] Uncertainty related to energy security and tensions between those favoring market or strategic solutions make hedging a favorable option both in terms of strategic planning, as is evident in the five-year plans and other long-term planning, and in terms of the policy-making process when Chinese leaders balance corporate and national interests on sensitive issues such as the cases of Sudan and Iran.

Emphasizing interdependency, commercial elements, and market mechanisms, Professor Zha Daojiong of Peking University argues that "China's dependence on maritime energy transportation is a natural state of affairs that must be managed."[48] Zha's observation is an important reference point in taking a further step to link hedging and risk management. The argument in this book is that hedging incorporates more scope for limiting and managing risk and for loosely mixing strategic and market elements in a comprehensive approach than do traditional diversification strategies.

Diversification, in terms of sources of oil, types of fuels consumed, and how energy is delivered to the consumer, has traditionally been the overarching principle behind energy security.[49] Hedging encompasses more than the traditional strategy of diversification, since hedging involves broader financial, political, military, and diplomatic tools. Diversification is only one important aspect of the insurance mechanism embedded in a hedging strategy. For example, a currency hedger will not only diversify by investing in different currencies but simultaneously will short sell a number of currencies in order to limit market risks. Which currencies and how much of each is "shorted," in addition to what short-selling strategy is preferred, will depend on the investor's risk profile. These measures provide a hedger with more tools than an investor who merely diversifies to limit and manage risk.

As will be illustrated in chapter 5, China's emphasis on cross-border pipelines has been an important diversification strategy. However, China's energy security policy encompasses more than the traditional diversification of sources, routes, and mix. In contrast to the international petroleum market, a pipeline market is less flexible and more political, and it often locks buyers and sellers into a mutually dependent or interdependent relationship. When developing cross-border pipelines, long-term contracts are signed, and prices are often partially fixed and sometimes rewritten through buyer-seller negotiations. The differences between transportation routes (seaborne versus pipeline) carry different security implications. Thus pipelines must be considered more than simply diversification, since they provide insurance "outside" the international petroleum market and carry different security risks than seaborne energy supplies. Although linked to developments in the international petroleum market, pipelines insure against supply disruptions and allow China to hedge its bets through a "short" strategy. In addition, the construction of pipelines provides broader opportunities to stimulate development and maintain stability in the remote and underdeveloped Chinese provinces and to enhance China's influence in neighboring states.

Such a "short" strategy trades profit for security. Instead of maximizing its search for profit by relying on cheaper seaborne petroleum supplies and profiting commercially in the international petroleum market, the Chinese government has decided to buy security through insurance in pipelines. Finally, hedging through the development of pipelines also carries broader spillover effects, such as developing and consolidating diplomatic ties, enhancing strategic objectives, maintaining domestic stability, and stimulating domestic development, all effects not accounted for by traditional diversification.

LINKING HEDGING AND RISK MANAGEMENT

Exploring the link between hedging and risk management[50] is an important refinement to the use of hedging in IR and to explaining China's energy security policy. A number of studies, borrowing eclectically from Beck's risk society, Giddens's reflexive modernity, and Foucault's emphasis on governmentality, use the concept of risk to analyze new security challenges such as terrorism, immigration, health, climate change, and the spread of weapons of mass destruction.[51] Additionally, the risk framework has recently entered the traditional security sphere.[52] I argue here that energy security also fits this analytical framework.

To date, few scholars have coupled the concept of hedging to risk or have explored how such an analysis can shed light on China's energy security policy. Linking hedging to risk management requires attention to the distinction between threats and risks. Risk analysis, which is preoccupied with the uncertainty of tomorrow and evolving dangers, differs from threat assessment, which predominantly focuses on a specific danger that can be identified and measured on the basis of capabilities plus intent.[53] Risk can be understood as the possibility that a potential threat will materialize. "The openness of the future and the always present need to decide in time precludes *the avoidance of risk by managing risks*."[54] Risk management quantifies and minimizes risk, and energy security is largely a problem of risk management.[55]

At present, the worldwide supply of oil is adequate and does not represent a threat to China. History shows that oil has generally been available on the market, that the key to market supply has been price, and that China currently is in an advantageous financial situation, with enough cash to accommodate high oil prices. In peacetime risk scenarios, China can buy the energy it needs, but states are concerned about their relative gains, and the fact that countries or investors are rich is not a disincentive for hedging.[56] China also seeks to reduce the risk that oil disruption will interrupt its economic growth and trigger social instability. In a wartime scenario involving China and the United States, the seaborne oil supply is likely to be disrupted, but one may question whether energy security is really a security issue when oil imports only account for roughly 10 percent of China's energy mix. Many commentators often confuse energy security during wartime and peacetime. As argued above, drawing a distinction between energy security in wartime and in peacetime is crucial for a more nuanced analysis.

The threat that a major power will attempt to disrupt Chinese energy supplies constitutes a wartime scenario. If China is faced with a blockade or sabotage of its energy supplies in the future, it will no longer hedge but rather will deal with it through extraordinary measures, including the use of military force. In fiscal year 2004, the U.S. military, fighting wars in Iraq and Afghanistan while also maintaining normal operations, used approximately 395,000 bpd of oil.[57] China produces roughly 4.1 mbd. This is ten times more oil than the United States used to wage war and conduct military operations around the world. China can also turn to synthetic oil during a war, making petroleum from its large coal reserves.[58] Thus, oil supply is not a security issue in a wartime scenario, both because China has sufficient domestic oil production to wage war and because everything becomes a security issue in a wartime contingency.

Hedging is about peacetime risk scenarios. This is crucial to understand how China's growing oil import reliance and supply (in)security are energy security issues. A war in an oil-rich region where China is not directly involved will affect the availability and affordability of petroleum on the international petroleum market. Very high oil prices or supply disruptions will affect Chinese economic growth, but China has developed "short" strategies that make it possible to take strategic/security measures "outside" of normal market mechanisms, either by Chinese NOCs shipping home more overseas oil production on Chinese state-owned tankers or by self-insuring and forcing state-owned tankers to pick up oil in a war exclusion zone if the price of oil becomes too high. China can also rely on its SPR, its increased refinery capacity, and its cross-border pipelines. Such "short" strategies do not insulate China from the risk of supply disruptions or high oil prices, but it will be reasonably well off in a "down market" (high oil prices, less availability, and supply insecurity)—and probably better off than other major oil consumers—similar to how an investor who has hedged safeguards a portfolio in a bear market.

Piracy and accidents at sea, among other nontraditional security threats, represent other risks during peacetime that might disrupt China's oil supply. It is doubtful whether such risks will lead to conflict between states. Indeed, the current antipiracy and escort missions being carried out in the Gulf of Aden show that all major energy-consuming states are taking important steps to cooperate in preventing and managing piracy and potential threats to the supply of oil. Peacetime scenarios cannot be completely eliminated; hence they need to be managed and hedged against so that they do not materialize into immediate threats.

Accordingly, China's current energy security policy deals with risks, however unlikely, and future threats. Hedging in security affairs can be thought of as a strategy to deal with current risks in order to prevent future threats. As we will see in the following chapters, Chinese energy security cannot be "secured," but it can be "insured" by the continued ability of decision makers to process risk and develop hedging strategies.

3

CHINA'S DOMESTIC ENERGY SECTOR

OIL IS THE only energy source China imports in substantial quantities. For this reason, this book focuses on oil; it remains the fuel that China is most concerned about "securing." The analysis generally examines China's measures directed at securing or, more precisely, *insuring* an adequate and affordable supply of oil. I differentiate between the external and domestic aspects of China's energy security policy and emphasize how China has developed global, maritime, and continental hedging strategies to manage both supply disruptions and high oil prices.

The domestic energy sector remains a crucial element in China's energy security policy. Energy, of course, includes oil, coal, gas, hydropower, nuclear power, and renewable energy. Given China's abundant coal reserves, its energy mix will be dominated by coal for the foreseeable future. China has attempted to constrain demand, improve energy efficiency, and develop a domestic energy infrastructure, which is clearly linked to the security of its energy supply and the objective of shaping China's energy mix. At the same time, China has not taken sufficiently strong measures to reduce the environmental effects of coal use, reflecting a preference to rely on a short strategy that emphasizes security and self-reliance. China has also developed domestic hedging strategies, such as building strategic petroleum reserves (SPRs) and expanding China's refinery capacity, to insure against supply disruptions and to enhance

China's ability to manage high oil prices and volatility in the international petroleum market. Finally, as an insurance policy against disruption rather than a large-scale substitution, China can make synthetic oil from coal as part of a self-sufficiency strategy.

China's energy mix and its efforts to maintain almost 90 percent of China's energy demand from domestic sources is a major factor in enhancing China's energy security. The central government has initiated policies to construct SPRs and hydroelectric dams, and it has constrained demand and improved energy infrastructure and efficiency. Not only does the government oversee an energy sector facing historically unprecedented growth in demand, but it has also developed and implemented hedging strategies, both domestically and externally, thereby improving China's ability to manage energy security risks.

In this chapter, I first examine the important relationship between the central government and the national oil companies (NOCs) and governance of China's energy sector. It is argued that China's leaders have the authority to control the NOCs and therefore the means to develop and implement a hedging strategy that provides opportunities to manage a potential crisis in the international petroleum market and minimizes risks to China's energy security. I then analyze China's energy mix and the drivers behind its demand for oil. Finally, I explore how domestic hedging strategies have been developed and implemented by focusing on the construction of SPRs and on developing a refinery capacity.

GOVERNING THE ENERGY SECTOR

To develop strategies and policies for the energy sector, the central government interacts with the NOCs, local governments, and a number of other actors. But how the government governs and who is in control is not straightforward. This chapter argues that the authority of the central government is upheld and that the long-term strategies put forward by the central government to manage China's growing reliance on imported oil are successfully developed and implemented. Since the critical relationship between the government and the NOCs is an important part of the argument throughout this book, it is necessary to examine this relationship in more detail.

THE IMPORTANCE OF THE NOCS

It is often argued that China lacks overarching guidelines for its energy security policy, especially in the coal and mining industries, and that the government strategy to secure imports of oil is characterized by ad hoc, reactive, and pluralist factors.[1] Domestically, policy is obstructed by special interests either within local governments or state-owned enterprises (SOEs), by rivalry between parallel lines of government, and by shortages of manpower in the relevant departments at all levels of government.[2] Externally, it is maintained that China does not yet have a government agency to coordinate implementation of its "going-abroad" strategies, despite the impression by foreign analysts that Chinese policy making in this area is coherent. Many writers point out that the apparently mercantilist character of the going-abroad strategy is much less state-directed, coherent, and strategic than is generally assumed.[3]

The three leading NOCs, China National Petroleum Company (CNPC), Sinopec, and China's National Offshore Oil Company (CNOOC), grew out of government ministries and were set up as state-owned companies in the 1980s. CNPC acquired administrative power over onshore exploration and production (E&P), CNOOC primarily focused on upstream offshore E&P, and Sinopec was set up to supervise and conduct downstream businesses. CNPC and Sinopec inherited both staff and an organizational culture from the former Ministry of Petroleum Industry. All three are vice-ministerial-level companies, and the general managers of the major NOCs hold the rank of vice minister. CNPC's general manager Jiang Jiemin and the former general manager of Sinopec, Su Shilin (in 2011, Fu Chengyu, the former chairman of CNOOC, became the head of Sinopec), are also both alternate members of the Central Committee of the Chinese Communist Party, and both present and past influential and prominent leaders of the CCP, such as Zhou Yongkang and Li Peng, have backgrounds in the energy sector.[4]

The energy sector underwent institutional reforms during the 2000s to improve the dysfunctional system after the line ministries were dissolved in the late 1990s and state-owned companies were established in each energy subsector. The National Development and Reform Commission (NDRC), replacing the State Development and Planning Commission (SDPC) and the State Economic and Trade Commission (SETC), was created in 2003 when bureaucratic consolidation sought to centralize policy making at the national level. In 2005, the government established a National Energy Leading Group, an advisory and coordination body under the State Council, headed by former premier Wen Jiabao. However, the energy sector still lacks a single institution, such as an energy ministry, to coordinate the interests of the various stakeholders and more effectively implement energy policy.[5] The January 22, 2010, announcement by the State Council to set up a National Energy Commission (NEC), headed by former premier Wen Jiabao and composed of twenty-three ministerial-ranked interministry members, could all turn out to be "old wine in a new bottle."[6]

Not only does China lack an energy ministry to coordinate policy, but the decision-making process has been undermined by "ineffective institutions and powerful firms."[7] The NOCs are better staffed than the energy institutions, and they often have more in-depth expertise and greater resources, thanks to the profitability of their internationally listed subsidiaries. The Chinese government also depends on the energy companies for strategy and policy implementation.[8] At times, profit considerations have undermined the national interest, and the NOCs often compete domestically and internationally despite government initiatives to prevent competition, with the result that "each soldier is fighting his own war."[9] This is a very sensitive issue in Beijing, notes another observer.[10] In summary, it is contended that

> observers of China's energy sector have long been drawn to a common set of conclusions relating to energy policy-making and implementation: that the agencies of government are fragmented and uncoordinated; that they lack the capacity for well-informed policy-making; that special interest groups, particularly state-owned energy companies, have undue influence over policy content; and that, as a result, the policies that emerge are inadequately formulated, fragmentary, contradictory, and tend to focus on ambitions and often unrealistic quantitative targets.[11]

MECHANISMS OF CONTROL

Although it is debated whether China's energy security policy is highly coordinated, with clearly defined strategic objectives in pursuit of national interests, the government controls the NOCs in a number of ways. The CCP appoints and removes the top leaders, managers, and members of the boards of directors of the NOCs and their subsidiaries. This is known as the *nomenklatura* system, whereby the CCP's Department of Organization and the Ministry of Personnel (MOP) are in charge of all top-level appointments, promotions, and dismissals at the NOCs and whose decisions are ratified by the Politburo Standing Committee and implemented by the MOP. The importance and authority of the CCP Organization Department was highlighted in interviews with IOCs representatives in China.[12] Not only does authority rest with the CCP, but positions within the NOCs are often one step up the career ladder in the party, thereby resulting in the leaders of the NOCs complying with and acting in accordance with CCP goals.[13] The CCP has set up "party groups" (*dangzu*) and "party committees" (*dangwei*) to "ensure that its broad policy lines (*luxian*), policy direction (*fangzhen*), and specific policies (*zhengce*) are implemented." These party groups decide on the most important issues ranging from investment strategy and corporate development to personnel selection.[14] The government and energy institutions are lawmakers, regulators, and law enforcers. This provides authority and instruments for controlling, overseeing, and supervising both the NOCs and the energy sector.[15] The government exerts strategic control through the State Asset Supervision and Administration Commission (SASAC), which oversees and controls the state-owned assets in the hands of centrally owned SOEs.[16] Before 2008, China's NOCs retained most of their profits and earnings, but beginning in 2008 they were required by the SASAC to pay dividends to the government.[17]

There is a linkage to international and market prices, but the central government or, more specifically, the NDRC sets domestic oil product prices and directly or indirectly compensates the NOCs for some of their losses in the domestic downstream sector.[18] The central and local governments also set the tax rate. Despite the longstanding opposition from local governments and the NOCs, the government passed a fuel tax in January 2009 and in previous years also enforced windfall taxes.[19] The government approves the overseas investments of the NOCs. Investments exceeding $30 million are signed off by the NDRC, and those over $200 million must be reviewed by the NDRC and then submitted to the State Council for approval.[20] The Chinese government also seeks to prevent more than one Chinese oil company from bidding on the same project. If there are conflicts of interest, the NDRC or the top leaders will decide which bid will be allowed to go forward.[21] For example, in January 2013, CNOOC failed to receive Beijing's approval for its planned purchase of coal-bed methane (CBM) assets in Queensland from Australia's Exoma Energy.

The government maintains leverage through credit from China's state-owned banks,[22] and the government controls the NOCs by setting petroleum production targets; health, safety, and environmental (HSE) standards; and social and economic targets.[23] The government approves the overall development plan for the energy sector, together with exploration and license rights.[24] Although the NOCs are powerful and at times autonomous actors that provide important inputs to the energy-policy-making process and have access to top decision makers, the government possesses tools to control the NOCs and the energy sector.

In assessing the relationship between the central government and the NOCs, it is important to examine the key strategy papers published by the government in the early 2000s, which spell out the various goals of China's energy security policy.[25] Some point out that the five-year plans and other long-term strategy reports might take the form of targets for the individual line ministries and the state-owned companies, thereby resembling a summation of individual industry strategies rather than a coherent package of policies.[26] For China's NOCs to operate effectively and become internationally competitive oil companies, they must be given autonomy.[27] The NOCs have participated in and provided input to the process of developing the five-year plans and thereby have given shape to the long-term planning for the energy sector; to some extent, the five-year plans for oil reflect the five-year plans that the NOCs have written for themselves.[28]

Nevertheless, some of the strategies developed and implemented have not always been supportive of NOC interests or the pursuit of profits. For example, the NOCs were reluctant to support and implement the decision to develop SPRs.[29] Furthermore, the NOCs, because of high costs and limited profitability, initially did not favor plans for cross-border pipeline projects. Nonetheless, SPRs and cross-border pipelines have become important parts of China's strategy to enhance energy security. Thus, it can be argued that when concerns about the security of supplies became a top central government consideration in the early 2000s, the government initiated some energy security strategies and policies that partly contradicted the business strategies of the NOCs.

IMPLEMENTING ENERGY SECURITY STRATEGIES

Former premier Li Peng advocated in the 1990s that China should "vigorously" cooperate with foreign governments and energy companies for exploration and development of petroleum resources abroad. Li strongly called for using lobbying and financial aid to boost China's goal of diversifying the sources of energy supply, which initially stimulated cooperation between China and Central Asian states.[30]

The Tenth Five-Year Plan of the PRC for National Economic and Social Development (2001–2005) specifies a number of strategies, including diversifying petroleum imports; raising natural gas consumption, exploration, development, and utilization; building SPRs; taking measures to conserve petroleum consumption; and stepping up exploration within China, offshore, and abroad.[31] For the first time, a "Special Energy Development Plan" was added to the Tenth Five-Year Plan, highlighting the central government's concern about energy security. The Special Energy Development Plan not only elaborated on and reiterated the policy in the Tenth Five-Year Plan but also provided an authoritative interpretation of Chinese strategies to safeguard energy security, stating:

> Energy security constitutes an important part of China's economic security. . . . China must base its energy consumption on domestic supply, with coal continuing to be [the] mainstay. . . . China should actively implement the going-abroad strategy, attach great importance to establish SPRs commensurate with China's petroleum productive capacity, diversify sources of

petroleum imports, and develop petroleum substitution and conservation technology so as to guarantee the country's oil and gas supply.[32]

Promoted by former president Jiang Zemin and former premier Zhu Rongji's call for petroleum strategies, a joint proposal by the SETC and the SDPC (the predecessor of the NDRC) added six additional strategies to the Tenth Five-Year Plan and the Special Energy Development Plan in November 2002: first, to set up a state petroleum fund, construct a petroleum financing system, and participate in the global petroleum futures market (hedging); second, to reinstitute the State Energy Commission; third, to develop a domestic tanker fleet and a blue-water navy; fourth, to speed up development of a petroleum conservation plan; fifth, to build three flagship oil companies; and sixth, to restructure China's energy consumption system and establish a system based on "clean coal" and natural gas.[33]

In 2003, the NDRC also put forward a national petroleum strategy that stressed five additional strategies: first, to conduct petroleum diplomacy; second, to advance the development of renewable energy; third, to hedge against price risks to China's petroleum imports by establishing future contracts; fourth, to improve petroleum efficiency and reduce petroleum intensity; and fifth, to strengthen the collective statistical capacity of the petrochemical industry and improve the timeliness and accuracy of petrochemical statistics.[34]

Astonishingly, almost all of the strategies put forward by the government and the energy institutions in early 2000 have been advanced and implemented. China has diversified its imports of petroleum products and has invested in large cross-border pipeline projects, it has built three flagship oil companies that operate worldwide; it has started building SPRs, it has improved energy efficiency and conservation; it is a major investor in alternative and renewable energy, it has diversified its energy mix and increased the use of natural gas and nuclear energy, it has pursued an active energy diplomacy, it has hedged against price risks by establishing future contracts, and it has built a large state-owned tanker fleet. As pointed out above, the interests of the NOCs may have infiltrated the energy policy planning, and the decisions are often basically in line with what the NOCs generally favor, but such remarkable results lead one to wonder whether there really has been a "policy paralysis within the energy bureaucracy" and whether the "activism of China's state-owned energy companies" and market mechanisms alone can account for developments in China energy security policy.[35]

The existence of long-term energy strategies for more than a decade and the evolution of China's energy security policy in accordance with these strategies challenge the argument that "the central government in Beijing today has neither a coherent national energy strategy nor much capacity to monitor, support, or impede the actions of state-owned energy companies."[36] Moreover, it questions the view that China's energy institutions have "failed to produce focused energy policy" and are "crippled" in their "ability to establish and carry out a national energy strategy."[37] In contrast to an emphasis on the ad hoc, fragmented, and poorly coordinated nature of China's energy security policy, together with the role of local entrepreneurship and provincial actors and the importance and activism of China's NOCs, it is difficult to see that the Chinese government and its energy institutions have not played a vital role, when long-term energy strategies that address many of the challenges to China's energy security identified in the early 2000s have been successfully advanced and implemented.

Although some of the central government's long-term strategies in the early 2000s endorsed what the NOCs had already been doing in the 1990s, after the Chinese government put forward its goals and ambitions for the energy sector, the government become more active. It is unlikely that there would have been a push toward nuclear energy, natural gas, and renewable and alternative energy if the NOCs and local entrepreneurs, rather than the central government, had been the major players in China's energy sector. There probably would not have been a SPR buildup without the call for it by the government. Chinese NOCs would have had more difficulties in establishing themselves as leading international oil companies if the Chinese government had not encouraged this process and had not facilitated it through its energy diplomacy and financial support. It is unlikely that there would have been a major expansion in the buildup of a large state-owned tanker fleet if the government had not been concerned about China's shipbuilding industry and foreign tankers being largely responsible for supplying China with imported oil. It would have been difficult to develop costly pipeline projects, especially cross-border projects, without government support and concern about diversification and the government goals of maintaining stability, control, and developing the remote provinces. As will be seen below, governmental subsidies and the provision of cheap loans from state-owned banks have been an important factor allowing the NOCs to thrive and expand, for shipping and shipbuilding companies to succeed in a bearish tanker market, in sustaining a competitive shipbuilding industry, and in order to develop pipelines, which can be costly and often less cost effective than other means of transport. Finally, the NOCs would not be running huge losses in the downstream sector in China, as a result of the disparity between domestic and international crude oil and oil product prices, if the government was no longer able to control the NOCs and to set domestic ex-refinery prices.

As in other countries, there are still problems with China's energy sector and with implementing government policy.[38] The central government does not have sufficient resources to control the entire energy sector—an industry growing at an unprecedented pace.[39] There are many "illegal" power plants, the central government is not able to impose mandatory pollution emission fees on all power plants, diesel-fired generators are common, and a self-help approach is pursued to meet power needs as a result of the "fragmented system of governance" and "blurred lines of governance and accountability."[40] But these challenges should be viewed with some perspective and understanding: controlling the energy sector is a difficult task in any country.

How would any leaders cope with the rise in demand for energy by nearly 1.4 billion people, coupled with China's weak energy infrastructure? Remarkably, and probably unprecedentedly in a historical context, China's energy consumption grew about fivefold from 1978 to 2009, and China moved from being a net exporter of oil in 1992 to being the world's second-largest importer less than twenty years later. China has surpassed the United States to become the largest oil importer and energy consumer in the world. "Between 2005 and 2010, China's total electricity capacity doubled. It is thought the country built in just half a decade a new electricity system of identical size to the system in place in 2005!"[41] Nonetheless, almost 90 percent of China's energy demand is still satisfied by domestic supplies, indicating that domestic reforms and policies have been adequate to meet China's record demand for energy consumption. Even though the government cannot control everything and lacks the capacity to achieve all its goals, it cannot be concluded from such findings that the government has lost control of the energy sector.

This book largely focuses on China's oil supply security and the external dimensions of China's energy security policy. Clearly, these are linked to developments in China's domestic energy sector, but it should be noted that there are many fewer actors within the petroleum sector than within the electricity or coal sectors, thereby enhancing the government's ability to control the key actors in the former. Finally, frictions between central and local governments are not as prominent in the oil sector as they are in other energy sectors, and in most cases the central government does not need to go through the local and provincial governments when it comes to securing overseas oil supplies and controlling the overseas activities of Chinese NOCs.[42] Hence, it is reasonable to believe that it is easier for the Chinese government to oversee the major players in the petroleum sector—the NOCs—and advance the government's goal of petroleum supply security than it is to control the domestic electricity and coal sectors.

HEDGING STRATEGIES AND POLICIES

China's energy security policy has evolved through a multifaceted process with many actors. These are, largely in order of their influence, the CCP Politburo, the State Council, the National People's Congress, the Chinese oil companies, the government energy institutions, provincial and local governments, the Ministry of Foreign Affairs (MFA), the People's Liberation Army (PLA), the media, research institutions and academia, and civil society and the public. China's energy security policy has been characterized by a struggle between economic liberalists and economic nationalists, that is, a struggle between market and strategic forces.[43] Petroleum policy making under a "dual governance" or "cogovernance" structure is another way of characterizing this interactive process or relationship between the "principal and the agents."[44]

The argument developed in this book draws on these findings but holds that after the government came to see energy as a security concern in the early 2000s it sought to control policy making and other processes considered to be of strategic importance. Concerned about a number of risks to China's energy security, mainly how the security, affordability, availability, stability, and sustainability of supply would affect economic growth and domestic stability, the Chinese government became more actively involved but remained uncertain as to whether the strategic or market approaches would best accommodate these risks. As demonstrated in the long-term strategy plans discussed above, to manage the emerging risks and uncertainties the central government developed a number of hedging strategies. These strategies included, for example, promoting cross-border petroleum pipelines, constructing SPRs, developing a large state-owned tanker fleet, facilitating NOC access to equity oil production, and arranging long-term loan-for-oil deals.

The debates and interactions between the stakeholders in China's pluralist energy security policy-making apparatus sometimes led to policy outcomes in some sensitive cases whereby the central government, pulled between the impulses of the security (strategic) and the profit (market) camps, ended up hedging its energy security bets. Decision making in China, as in most countries, is characterized by bargaining, consensus building, and balancing among different interest groups.[45] For example, in chapter 4 I maintain that China has hedged its energy security policy in the sensitive cases of Sudan and Iran as a result of a policy-making process whereby China's position (as determined by the Politburo and the State Council) is shaped by diplomatic interests and largely promoted by the MFA, corporate interests promoted by the NOCs, and strategic and military considerations promoted by the PLA.

It is not feasible for—nor is it an objective of—the central government to control the daily business activities of the NCOs and their pursuit of profit. Instead, government authority to control the NOCs and other sectors of the energy industry is important during times of emergency or crisis that threaten China's oil supply and national security interests and when it comes to implementing long-term plans and strategies so that the government better can manage risks. If an oil or energy crisis were to erupt, it is argued, the government can impose more cooperation among the NOCs (which remain fierce competitors in their day-to-day business), enforce collaboration between the NOCs and the Chinese state-owned shipping companies (and even the non-state-owned Chinese shipping firms), instruct the shipping sector to aid the domestic market by shipping crude oil and petroleum products back home, and force the energy sector to maintain reasonable prices for domestic consumers.[46] In short, the government can provide for the security of petroleum supplies. This hedging strategy will not insulate China from a major crisis in the international petroleum market, but China is likely to be relatively better off than other major petroleum importers, and its economy might remain reasonably well off, thereby minimizing a potential economic downturn and the risk of social instability, which essentially is the core aim of China's energy security policy.

Finally, this book argues that the Chinese government largely focuses its energy security policy on overall and long-term strategy. This includes those areas that are considered important to China's national interests and sensitive issues that may affect China's security, diplomatic standing, and not necessarily the daily business activities of the NOCs. As pointed out above, the central government's long-term strategies with respect to China's energy security policy, as stated in the early 2000s, have been pursued and implemented both domestically and internationally. Although there is a danger that corporate interests may undermine national interests, the Chinese government has taken steps to ensure that the tail is not wagging the dog on issues considered strategically and diplomatically important, while simultaneously allowing the NOCs to continue their daily pursuit of profits.

PATH DEPENDENCY AND SELF-RELIANCE

Past experiences shape governance of the energy sector and influence perceptions about whether the strategic and market approaches best enhance China's energy security. Energy security in China cannot be understood without acknowledging how history, tradition, and culture have shaped energy policy.[47] China's historical wealth and abundance of natural resources created long-held ideas and beliefs that determine a degree of path dependency and a preference for self-reliance, in addition to asserting mastery over nature and an ideological commitment to social equity.[48] Content with the variety of domestic resources, Chinese dynasties were often reluctant to engage in foreign trade. After taking power in 1949, the Chinese Communists were forced to maintain a high degree of self-reliance as a result of the Western embargo, diplomatic isolation, and the split with the Soviet Union in the 1960s.[49] The stress on self-reliance was accompanied by a preference for direct government involvement in natural resource projects, especially if the projects were critical to the interests of the state. The rapid development of the Daqing oil field in Heilongjiang province during the 1960s was a major achievement and became a prominent example of the self-reliance strategy. China's distrust of the outside world and the oil crisis of the 1970s and early 1980s reinforced and confirmed the need for self-reliance in China's energy security policy.[50]

When China's energy sector liberalized in the 1980s and 1990s, the self-reliance strategy, which promoted competence building in Chinese onshore oil and the coal and hydro sectors, restricted foreign-company involvement in the domestic energy sector except in projects where foreign technologies and skills were absolutely necessary. These projects included offshore oil exploration, offshore and onshore gas production, nuclear power generation, liquefied natural gas importation, and the manufacturing of large turbines.[51]

The open-door policy in the 1980s increased tensions between the long-held preference for self-reliance and the increased economic liberalization. Strategic, market, and environmental aspects have shaped China's energy security policy.[52] Although concerns about the environment and the sustainability of China's growth model became more emphasized in the Twelfth Five-Year Plan, environmental factors have largely been secondary to concerns about energy security and commercial considerations. This book draws on a hedging framework to examine how these strategic and market drivers, or security and profits, are linked and mixed in China's energy security policy. In establishing a hierarchy or prioritizing between strategic and market approaches, it is argued that concerns about the security of supply override commercial interests.

It is important to note that the government's concern about security of supply relates to long-term strategic planning and crisis situations where energy supplies may be disrupted. The Chinese government is not heavily involved in the daily activities and business strategies of the NOCs. Nor does it have the capacity, expertise, or resources to play such a role. As Chinese NOCs have expanded domestically and internationally, becoming involved in more than two hundred overseas projects and undertaking a number of joint ventures and M&As, the Chinese government may find it even more difficult to control their daily activities. Although the government is represented on the boards and in the decision-making apparatus of the NOCs, government representatives may lack the expertise, technical competence, and resources to control NOC projects and activities.

Therefore, it is often contended that the oil companies have become powerful new actors in China's foreign policy.[53] This may be a result of what some researchers have described as "strong firms and weak government." Nonetheless, the government has maintained ownership, and whether the government is "weak" in its capacity and lacks the resources to manage the energy sector should not be conflated with its authority to direct, control, and implement China's energy security policy.[54] In short, the government may be weak in overseeing the commercial activities of the oil companies and the energy sector, but the government is not weak in terms of its authority and in directing the NOCs to preserve China's oil supply security and to minimize risks to oil supply disruptions.

CHINA'S ENERGY MIX: KEY DEVELOPMENTS

COAL

China relies on coal for about 68 percent of its energy needs. Coal is the dominant source for electricity production, representing roughly 80 percent of the supply and more than 70 percent of the capacity.[55] The possibility of creating synthetic oil from China's abundant coal reserves provides important insurance in times of emergency or war. The main production base for coal is the northwestern provinces, where about 70 percent of China's coal resources are located. Most power plants are located in Shandong and Heilongjiang provinces.[56] China's electric power sector, largely fueled by coal, has roughly doubled in size since 2000.[57]

Coal has generally been distributed by a national allocation system managed by the state, along with an ad hoc and market-like system of transactions between mines and power plants. Coal prices have traditionally been controlled and subsidized by the government, providing few incentives to switch to cleaner alternative fuels or to use technologies that burn coal more efficiently. Since 2002, when coal demand began to surge, the market channel has become far more important, and domestic coal prices have been rising, effectively converging toward world levels.[58] China has traditionally been a net exporter of coal, delivering about 9 percent of internationally traded coal to world markets in 2001. In 2009, China was importing roughly 3.5 percent of its annual coal consumption.[59]

OIL

Oil is China's second-largest source of energy, accounting for about 20 percent of China's total energy consumption. In January 2011, the National Bureau of Statistics announced that China's crude oil production in December 2010 stood at 4.13 mbd.[60] In 2010, China was the world's fifth-largest producer of oil (after Saudi Arabia, Russia, the United States, and Iran), accounting for about 5 percent of global production.[61] It is projected that China's oil production over the next decade will be close to 4 mbd. According to the U.S. Energy Information Administration (EIA), approximately 85 percent of China's domestic oil production is located onshore. Much of China's offshore production is located close to the shore and at water depths of only half a meter.[62] Daqing in the northeast and Shengli southeast of Beijing have been major oil fields since the 1960s, but production has now peaked, and China's oil companies are moving west, seeking exploration and production opportunities in interior provinces such as Xinjiang, Sichuan, Gansu, and Inner Mongolia. Offshore exploration and production have been concentrated in the Bohai Bay and the South China Sea, with more recent steps taken in the East China Sea.

China has witnessed unprecedented economic growth: approximately 10 percent during the last three decades. This astonishing economic growth has fueled energy consumption and demand for oil surpassing most, if not all, forecasts, projections, and estimates.[63] For example, in 2004 the International Energy Agency (IEA) projected an average annual rate of increase in demand between 2002 and 2010 of 4.0 percent. The actual rate of increase during the

2002–2009 period was more than 10 percent per year. Both the IEA and the Chinese government widely missed the mark in their predictions about the growth of energy demand. Energy consumption grew four times faster than predicted, totaling 16 percent of global demand in 2006.

China became a net importer of oil in 1993. Its oil consumption now stands at close to 10 mbd, and the degree of foreign dependence surpassed 50 percent at some point between 2007 and 2009. China's net oil imports stood at 5.2 mbd in April 2011.[64] Although roughly 50 percent of China's oil imports continue to come from the Middle East, given China's record growth in oil demand, China has managed to diversify away from the Middle East and find new suppliers, increasing its imports from Africa, Russia, Central Asia, and Latin America. In 2009, Saudi Arabia (21 percent), Angola (16 percent), Iran (11 percent), Russia (8 percent), and Sudan (6 percent) were China's top suppliers, but in 2010, Brunei, Nigeria, Iraq, Chad, Kazakhstan, and Canada were the fastest growing Chinese suppliers. However, China is still more than twice as dependent on Middle Eastern oil supplies than the United States.

It is anticipated that China's growth in demand will continue to represent a major portion of the world's projected growth in oil demand.[65] Two researchers at the NDRC estimate that China's import dependence will reach 60 percent before 2020, while the IEA projects that China will import 79 percent of the oil it consumes in 2030, accounting for 22 percent of world demand, up from 17 percent today.[66] The 60 percent mark is likely to be surpassed before 2020 because China's oil imports already account for about 55 percent of demand.[67] Historically, the forecasts and projections have tended to underestimate China's energy consumption and demand for oil, but China's weaker economic growth in 2011–2012 reveals a sharp deceleration in demand, with preliminary estimates depicting YOY growth close to zero.[68]

HYDROPOWER

Hydropower accounts for roughly 6 percent of China's energy mix, and hydropower production is mainly concentrated in the central and southern regions. China is the world's largest producer of hydroelectric power, and its capacity has risen more than 10 percent in the period from 2008–2011. Its total hydropower capacity stood at 197 GW in 2009, and the target, according to the National Energy Administration (NEA), is that the capacity will reach 380 GW by 2020, an average increase of 16.5 GW.[69] Because of a priority for clean energy, the government has sought to boost hydropower through a number of large-scale hydroelectric projects planned or under construction and to increase the share of hydropower in the energy mix. By 2000, it was estimated that about 22,000 of the world's 45,000 large dams spanned Chinese rivers, and in the last decade there has been a large increase in the number of dams built and under construction, including twenty-six projects on the Lancang, the headwater of the Mekong; thirteen on the Nu, headwater of the Salween; and twenty-eight on the Yarlung Tsangpo, the headwater of the Brahmaputra.[70] Despite the large investments and the significant increase in capacity, hydropower as a share of the supply of electricity has declined from nearly 20 percent in 1990 to 16 percent between 2006 and 2009, and the government faces a range of obstacles in its aim to reverse the declining trend in the relative contribution of hydroelectricity.[71]

GAS

Natural gas stands at close to 4 percent of China's energy mix, but its role in that mix is increasing. Gas might comprise as much as 8 percent of the energy mix by 2015, and the government seeks to increase usage even further. China became a net importer of gas in 2007, although domestic production has risen substantially. Industry accounts for the majority of gas consumption, but the commercial sector will be the driver of future demand.[72] Sichuan, the Tarim basin in Xinjiang, the Ordos basin in northern China, and the South China Sea are expected to be major domestic supply production sites.

Piped gas from Central Asia, Myanmar, and potentially from Russia will also be large drivers of supply.[73] Gas imports through pipelines and liquefied natural gas (LNG) diversify China's sources, routes, and energy mix. Although there are differences in the security of the various pipelines,[74] they can provide a hedge during peacetime risk and wartime threat scenarios, as will be pointed out in chapter 6. The pipeline "market" differs from the international petroleum market because it is less flexible and locks suppliers and consumers into increased interdependency. Pipeline petroleum and LNG contracts are often longer term, which means that they can provide a hedge against the international spot oil market. In the early 2000s, China's NOCs' first LNG contracts were low and involved relatively fixed prices because the competition to sell LNG was so intense. More recent contracts have had to come to terms with normal LNG contract arrangements, similar to the basic Japan Crude Cocktail, which links LNG to oil prices.[75] LNG supplies are expected to increase substantially, and both CNOOC and CNPC have built LNG terminals in Guangdong, Fujian, and Shanghai. Eight more LNG terminals are either nearing completion, under construction, or being planned. By 2010, the volume of LNG contracted on a long-term basis by Chinese companies from the year 2014 exceeded 30 billion cubic meters.[76]

There is much uncertainty when it comes to estimating the rate of China's gas imports. Gas imports from Central Asia will progressively grow, a new pipeline from Myanmar is expected to be operational by 2012, and a gas pipeline from Russia to China may be built in the near future. With the melting of the Arctic ice cap, shipping LNG from Russia to China through the Northeast Passage might be another alternative. Of course, there remain a number of uncertainties related to developments in the high north, and Russian infrastructure needs to be developed in this area, but thanks to the shale gas "revolution" in North America and the limited demand in Europe, there are strong incentives for shipping Russian gas to the markets in East Asia. If Russia has enough reserves and is willing to ship more of its gas east, then LNG shipment may not prevent the development of a Russia-China gas pipeline, if China is willing to pay for an additional and more costly pipeline option. Indeed, developing both a gas pipeline and purchasing LNG from Russia would be characteristic of a Chinese hedging strategy. The pricing formula has been the key obstacle to a Russian-Chinese gas pipeline, together with the facts that Gazprom has signed long-term contracts for gas deliveries to the European market and that the Russian energy infrastructure is geared toward the West.

China has now abandoned its original bargaining position of tying gas import prices to coal, agreeing to market prices for LNG and recently agreeing to link a pricing formula on gas imports from Russia to oil prices. This could convince Russia and Gazprom to develop an eastward infrastructure, which eventually could facilitate the construction of a gas pipeline to China. In 2011, the gap between the border price for Russian gas of about $250 per thousand cubic meters that the Chinese is seeking, based on city gate prices and LNG from its older contracts, and the price the Russians are seeking of close to $500, which is roughly the present price or slightly above the price of Russian gas sold to Europe, remains an obstacle that might be too large to overcome.[77] Additionally, China is increasingly finding alternative sources of gas both overseas and domestically, and Russia does not appear to be in a hurry to conclude a gas sales agreement with China.[78] Unconventional gas resources such as CBM and shale gas may become a game changer for the role of gas in China's energy mix, bringing additional uncertainty to forecasts concerning China's total gas imports. One report estimates that domestic CBM production could reach 25 percent of China's total gas production by 2020.[79] If China successfully develops indigenous CBM production and manages to construct and improve its networks of domestic gas pipelines, then China's petroleum imports could be significantly reduced.[80]

Despite these challenges, it should be remembered that China and Russia eventually managed to overcome the impediments and difficulties in building an oil pipeline from Russia to China. Hence, we cannot rule out that a gas pipeline will be built in the years ahead.

NUCLEAR

Nuclear power accounts for approximately 1 percent of China's energy mix, but China has the most ambitious, and among the fastest, program for nuclear power construction in the world. According to some reports, about 40 percent of global nuclear power capacity is now under construction in China.[81] China is currently operating thirteen reactors in three provinces (Guangdong, Zhejiang, and Jiangsu), with a total capacity of more than 10 GW.[82] In 2010, twenty-four reactors were under construction, representing a combined capacity of 27.08 GW.[83] The long-term plan is to expand China's current installed production from the present 10.8 GW to 86 GW in 2020 and to increase nuclear power to 5 percent of China's energy output, with longer-term targets of 200 GW in 2030 and 400 to 500 GW in 2050. The Twelfth Five-Year Plan anticipates a large buildup of nuclear power plants and 43 GW of reactor capacity in operation at the end of 2015.[84]

China National Nuclear Corporation (CNNC) and China Guangdong Nuclear Power Group (CGNPC) dominate China's nuclear power industry, but neither of these two companies are among China's top five largest utilities, which together represent half of the country's installed electricity capacity.[85] It is reported that China has more than two hundred proven uranium deposits, accounting for 171,400 tU, but Chinese planners believe there are far more—between 1.2 and 1.7 million tU of potential conventional uranium reserves. Northern Inner Mongolia has 31,000 tU of proven reserves, which is 18 percent of the country's total, and Liaoning, Yunnan, Shaanxi, Jiangxi, Guangdong, Guangxi, and Hunan provinces account for 56 percent of China's known uranium resources.[86]

In the aftermath of Japan's nuclear crisis and the Fukushima disaster in March 2011, it was first announced that China would not alter its plans for nuclear power projects.[87] However, on March 16, 2011, after a meeting of the State Council, former premier Wen Jiabao stated that nuclear safety would be a top priority and that approval of new plans would be suspended until new safety rules were in place.[88] Tighter safety regulations may be enforced, but many experts believe that China's ambitious nuclear energy plans will nevertheless continue so as to reduce dependence on coal and energy imports.[89]

RENEWABLES

Renewables account for about 1 percent of China's energy consumption mix, but after passage of the Renewable Energy Law in 2005 China has become the world's top investor in renewable energy projects, which is indicative of the determination to enhance the role of renewables in China's energy mix.[90] China is currently among the world's largest wind power producers, and the aim is to increase wind capacity to 100 GW by 2020. The increase in capacity is remarkable. In 2005, total installed wind power capacity was about 1 GW. By 2008, total capacity reached 12 GW, and by the end of 2009 China was set to reach 20 GW, making China's wind power capacity the fourth largest in the world, behind only the United States, Germany, and Spain.[91] Since then, the dramatic growth of wind power has continued. In January 2011, it was reported that U.S. wind capacity, which stood at 40,180 GW, had fallen to second, behind China's 41,800 GW, making China the largest wind energy provider worldwide.[92] China has become a major producer of solar panels, but nearly all Chinese production is exported. Although solar water heating is used extensively, China has only installed about 80 MW of photovoltaics, which represents merely 0.01 percent of the nation's power generation capacity.[93]

WHAT IS DRIVING DEMAND?

In 2007, the industrial sector accounted for 70 percent of final energy consumption in China, whereas the residential, commercial, and transportation sectors accounted for 10, 2, and 7 percent, respectively.[94] Because of its dominance in the economy, the industrial sector will remain a significant contributor to the growth in total oil consumption, but the relative importance of industry to oil consumption is dropping.[95] The share of industrial oil demand declined from 59.2 percent in 1990 to 42.1 percent in 2007, which shows that industrial oil demand is growing more slowly than most other sectors.[96] In this respect, it is important to note that the Chinese government developed a strategy to limit industrial oil consumption in the Tenth Five-Year Plan (2001–2005) by setting fuel oil consumption targets for the power, chemical, ferrous metals, and nonmetallic minerals industries.[97]

Contrary to reports that highlight consumption-led growth, for instance air conditioning and automobiles, the main source of growth has been in energy-intensive heavy industry, but consumption-led energy demand, already significant in absolute terms, will be the major driver in the future.[98] Since China's current and future oil imports today reflect automobile use, this is not a significant energy security vulnerability: personal automobile use can be curtailed, as it was during the 2008 Summer Olympics in Beijing and as it already has been in large cities.

Historically, there is a strong correlation between the share of the transportation sectors in total energy consumption and per capita GDP. A number of studies show that when per capita GDP surpasses US$1,000, the ownership of vehicles takes off, and when countries reach $5,000 per capita GDP, the commercial and transportation sectors begin to surpass industry as drivers of energy demand.[99] Statistics from the IMF, the World Bank, and the CIA *World Fact Book* show that China's per capita GDP has surpassed $7,000. More importantly, the per capita GDP of hundreds of millions of Chinese citizens in the coastal provinces has already far surpassed $7,000, and they have become the main drivers of consumption-led energy demand in China. Accordingly, if the manufacture of steel, glass, and other materials to build office buildings, shopping malls, and automobiles drives China's energy demand today, lighting, heating, and cooling these malls and offices and the emerging consumption-oriented middle class that can afford energy-intensive consumer goods like air conditioners and automobiles will drive China's energy demand in the future.[100]

By the end of 2009, and largely supported by the government's pro-automobile policy, China became the world's largest auto market, with sales of cars, buses, and trucks reaching 13.6 million.[101] The numbers for 2010 showed that 11.5 million cars were sold in the United States, while China had reached 17 million. By 2020, sales in China could reach 30 million—and keep going.[102] In addition, air transportation, continuing urbanization, infrastructure expansion, the growing military force, and structural changes in the economy affecting the commercial and household sectors will continue to fuel Chinese oil demand.[103]

With a shortfall in domestic energy supplies, particularly electricity, in 2004 energy savings, conservation, and efficiency, in addition to environmental issues, were identified as areas of priority by the government. Energy policies to meet these challenges were laid out in long-term strategy reports by the State Council, the NDRC, and were also embedded in the 2006–2010 Five-Year Plan.[104] In his address to the Seventeenth Chinese Communist Party Congress in October 2007, President Hu encouraged the formation of an "energy and resource-efficient and environmental-friendly structure of industries."[105] Emphasis was placed on the industry and transportation sectors, on buildings, and on the pricing of energy. The key strategic measures were stimulated by significant financial support to investments in energy efficiency and backed up by policy and legal documents.[106] It still remains unlikely that China will sacrifice energy production from coal to be greener. The growth of renewables in China is significant, but its share of the energy mix remains small, and the annual increase of coal power capacity surpasses that of renewables by a wide margin.[107]

An important goal of the Energy Conservation Plan is to reduce energy intensity by 20 percent. Although per capita consumption of energy in China is much lower than in most large economies, China's energy intensity, when GDP is measured at market exchange rates, purchasing power parity, or per unit of GDP, is one of the highest in the world.[108] Emphasizing how much more energy intensive China is than, say, Japan, ignores the important issue of what the country makes. China's high energy intensity partly reflects the major role of the industrial sector; Japan has lowered its energy intensity in part by migrating its energy-intensive sectors to China; another rising power, India, has taken a more service-heavy approach.[109] Put differently, the United States has outsourced part of its energy consumption to China.[110] Media reports suggest that the government has achieved its goal of reducing energy intensity by 20 percent from 2006 to 2010, an impressive achievement that has required changes in the structure of the economy but without slowing China's economic growth.[111]

Whether the Chinese government in fact reached its goal of reducing energy intensity by 20 percent in 2010 is difficult to verify. Nonetheless, energy conservation and efficiency have improved, showing that the government is capable of introducing changes and implementing its policies in the energy sector. The difference with the Western countries is often striking when one thinks of how the Chinese government enforces its energy conservation plans in a large city such as Beijing. How would the citizens of a large American city react if the government were to ban driving cars once a week or if it prevented citizens from purchasing new cars? Moreover, the goals in 2011 of the Twelfth Five-Year Plan are at least as ambitious as those of previous years. The emphasis on nuclear energy and renewables will continue to focus on non-fossil fuels to account for 11.4 percent of primary energy consumption. Energy consumption per unit of GDP is to be cut by 16 percent, and carbon dioxide emissions per unit of GDP is to be cut by 17 percent.[112]

Finally, while China is actively embracing renewable energy, reducing oil in China's fuel mix is only a secondary objective because oil does not dominate China's energy mix. Whereas the United States wants to reduce oil use and enhance energy security by using more renewable energy, China is developing renewable energy to lower its dependence on coal, the largest single source of air pollution and energy inefficiency in China.[113]

DOMESTIC HEDGING MEASURES

Some aspects of China's domestic energy security policy can be considered to be hedging strategies. The construction of SPRs and increasing refinery capacity provide China with a hedge against supply disruptions and insure a reliable and adequate supply of oil at a reasonable price, thereby supporting the objective of the central government to sustain economic growth and domestic stability. China's strategies and ambitions are not so different from those of other governments that seek to minimize risks through the development of SPRs. Improving China's refinery capacity may not be as straightforward a hedging strategy as the buildup of SPRs against oil disruptions and their consequences. China's refinery expansion plans are clearly driven by competition among China's NOCs in the domestic downstream market, the growth in oil imports, and changes in China's crude import suppliers and types of imports. Nonetheless, China's increased refinery capacity provides insurance if China is forced to seek more oil, for example from Venezuela, if another conflict were to erupt in the Middle East. Finally, and as pointed out previously, making synthetic oil from coal provides an insurance against oil disruption.

THE BUILDUP OF SPRS

The construction of SPRs is a traditional way of hedging and insuring against oil supply disruptions and price volatility. SPRs can act as a "short" strategy to insure against risks in "longs" in the international petroleum market. Developing SPRs can reduce the effects of potential crude oil supply disruptions and minimize the influence of fluctuating global oil prices on China's domestic market. China officially announced its SPR program in the Tenth Five-Year Plan in 2001, corresponding to the increasing concern about China's energy security and growing dependence on imported oil. This decision put to rest a longstanding debate going back to 1993 over the merits of building SPRs.[114] The government and the NDRC have been in charge of the planning for SPRs through reserve base construction, procurement of oil, and execution of use and turnovers. The NOCs have been instructed to implement the strategies embedded in the five-year plans, such as the building of storage facilities, and they have been given responsibility for policies related to facility maintenance and site management.[115]

The construction of the first four sites in Zhenhai and Zhoushan in Zhejiang province, Qingdao in Shandong province, and Dalian in Liaoning province did not begin until June 2004. The combined storage capacity of the four sites during the first phase was about 100 million bbl, or 15 million tons. This would support 13 to 14 days of oil consumption, or 25 to 26 days of net oil imports based on 2008 levels. However, because of the high level of international oil prices the facilities were not filled until crude oil prices fell during the summer of 2008, with Zhoushan being filled as of June 2007 (capacity 33 million bbl), Zhenhai being filled as of December 2007 (capacity 33 million bbl), and Dalian (capacity 19 million bbl) and Qingdao (capacity 19 million bbl) being filled as of April 2009. Filling the tanks is price sensitive and challenging because it is difficult to avoid price increases caused by imported barrels for stockpiling. The final authority of China's fill rate and the decision to purchase crude for the reserves lies with the State Council. In June 2009, it was reported that the average crude price for the SPR phase 1 purchases was $58 per bbl, implying that the government approached crude purchases strategically and purchased only when it perceived prices to be low.[116]

China is currently building its second phase. The NDRC plans to build eight strategic stockpiling bases during phase 2, with a total storage capacity of roughly 207 million barrels. Two sites were completed in 2011, and a third, the Tianjin site, is reportedly set to be completed in 2012. The remaining sites are expected to become operational by 2013. By the end of 2011, China reportedly held a storage capacity of around 140 million barrels.[117] A third phase will

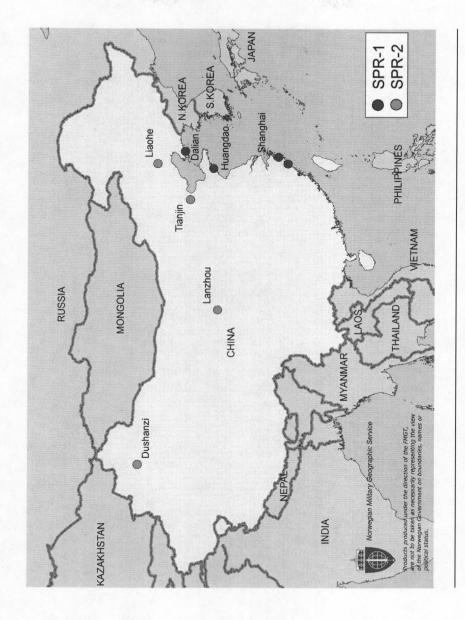

MAP 3.1 Locations of Strategic Petroleum Preserves, China

add a further 28 million tons, bringing total SPR capacity to about 70 to 80 million tons, or 500 million barrels, amounting to about six to seven weeks of total oil supply and roughly 125 days of net oil imports.

The government has encouraged the domestic oil companies to develop commercial reserves, and China plans to set up a refined oil reserve. Additionally, if its seaborne oil supply is lost, China can draw on its large domestic oil production, import oil through pipelines and railways, and convert coal to synthetic fuel. Thus, the SPR is only one aspect of China's hedging strategy. In contingencies whereby oil imports from, for example, Iran, Sudan, Angola, Venezuela, and the Persian Gulf are lost, China is likely to have the financial strength to secure oil on the international petroleum market, and it has insured against such emergencies by developing a large state-owned tanker fleet and overseas equity oil production.

China has not only sought to avoid filling its SPR tanks when international oil prices are high; it also has intended to minimize the effects of its SPRs on world oil markets. It has been pointed out that China might seek to fill its SPRs with either its overseas equity oil production or with oil from joint ventures or domestic production. In theory, the net effect on Chinese demand and overall world oil demand would be the same.[118] In other words, if filling China's SPRs adds 100,000 to 200,000 barrels per day to China's oil demand, it will affect an already tight oil market, and, in such a case, China will not be able to escape the fact that the global oil market is one big pool of oil. However, the current sanctions against Iran show that this may not be as straightforward as it might seem. If China and its NOCs have the will and the capacity to pick up oil from a state that is sanctioned or embargoed and where most other nations and IOCs cannot trade, then the global oil market would no longer be one big pool of oil.

One of the important differences between phase 1 and 2 is that all four sites in phase 1 are located close to coastal ports and aboveground, whereas some of the sites selected for phase 2 are located inland and are planned to be built in underground facilities. It is likely that this shift reflects an ambition to have SPR sites closer to important oil fields and to link the SPRs to various pipelines that will be pumping oil to China from Central Asia, Russia, and Myanmar. This does not detract from the fact that when the government decided to build SPRs in the early 2000s to improve China's energy security, they developed storage tanks aboveground and close to the coast that would be easy targets in a potential wartime scenario. Managing peacetime risks probably trumped wartime threat scenarios when considering building SPRs.

There is also commercial storage by NOCs and by local governments or companies, in addition to the national SPRs. At the start of 2010, China had at least 49.5 million cubic meters of storage capacity for crude oil, of which Sinopec owned nearly half. CNPC-PetroChina owned about 43 percent, and CNOOC-Sinochem and others owned the last 7 percent. Because these tanks—old and new—are located all over the country, overall average utilization of the crude storage facilities is only about 65 percent.[119] In mid-May 2009, it was announced that the Chinese government planned to build strategic reserves for refined products, its first official plan to expand beyond crude oil reserves. The State Bureau of Material Reserves, a subsidiary of the NDRC, was to operate China's stocks of strategic refined products, and storage facilities were to hold up to 10 million tons of refined products by 2011.[120]

As we have seen, phase 2 of the SPR buildup involves plans to locate several tank sites closer to cross-border pipelines. The effectiveness of China's SPRs in absorbing supply disruptions and price fluctuations will largely depend on whether China will be able to release oil rapidly into a well-developed pipeline and maritime transportation system linked to substantial refining capacity, thereby successfully shifting oil supplies between major demand and import areas.[121] Partly because of this, the Eleventh Five-Year Plan emphasized expanding the country's pipeline network for oil and gas products. Developing SPRs and a pipeline network infrastructure is strategically important and will improve China's energy security. The development of SPRs is clearly driven by a mix of security and profit incentives, but commercial opportunities are important secondary considerations.

To improve energy access and competitiveness in the Chinese domestic downstream market, CNPC plans to double the length of pipelines laid during the Twelfth Five-Year Plan period (2011–2015). CNPC, China's leading pipeline operator, had laid 500,000 km of oil and gas pipeline by the end of 2010, accounting for 70 percent of China's total crude oil pipelines in terms of distance and 90 percent of its total gas pipelines.[122] CNPC has thus benefited from the strategic decision by the Chinese government to boost the supply of gas and to increase the use of gas in China's energy mix. Similarly, Sinopec, China's largest petroleum refinery, runs a number of major refineries next to SPR sites to increase its competitiveness, but the government has simultaneously pushed for improving pipeline and refinery infrastructure to utilize SPR oil during a crisis.

The Oil Stockpiling Office within the NDRC and the State Oil Stockpiling Center are directly responsible for the management and operation of China's SPRs. Still, companies play an important role in construction and maintenance of the sites. Although the nature of the organization remains unclear, there are close interactions between government agencies and NOCs when it comes to operating the sites. For example, stored crude has been leased, and the government has allowed crude from the initial four phase 1 sites to be drawn and refilled. In addition, in early 2010 it was reported that the government would rent crude storage facilities for two years at a rate of $2.50 per cubic meter per month from six companies to stockpile crude in Shandong and Zhejiang.[123]

It is still unclear how the Chinese authorities will govern the stockpile. When comparing the stockpiling strategies of the world's top oil importers—the United States, Japan, and South Korea—there are both similarities and differences in terms of SPR development, forward cover protection, drawdown policy, and economic efficiency.[124] Which model will China follow? Will commercial stocks go into the SPRs? Will the Chinese government only release oil from the reserves in the case of supply disruptions or when international oil prices are exceedingly high?

China will face a number of challenges during phases 2 and 3. When importing barrels for stockpiling, it will be difficult to avoid oil price increases because China will need to add 100,000 to 150,000 barrels a day to meet Chinese demand for the years to come. High oil prices may also prevent China from reaching its targets. One alternative is to take advantage of existing commercial storage capacity by requiring the NOCs, as a first line of defense, to hold forty days of forward cover in a mix of crude oil and refined products. With SPRs expanding to at least fifty days of reserve during phase 2, China may achieve the goal of the IEA-mandated ninety days' cover,[125] but the progress and governance of China's SPRs remain uncertain. Nonetheless, the buildup of China's SPRs, like many other strategic measures taken by the Chinese government to manage China's energy security, represents a hedging strategy that mixes security and profit interests.

DEVELOPING DOMESTIC REFINERY CAPACITY

A mix of security and profit considerations can also be seen in the goal to expand China's refining capacity. Enlarging China's ability to refine a wider variety of crude oil provides both an opportunity to improve the refineries' profitability and insurance in cases of supply disruptions, as China will be able to use more crude oil from a wider range of suppliers. This will provide China with greater strategic and economic flexibility.[126]

China's refinery capacity grew from 2.9 mbd in 1990 to 7.7 mbd in 2007—almost the same as that of the entire Middle East—making China the second-largest refiner, behind only the United States, in terms of capacity.[127] In 2010, a consultancy report showed that China's total refining capacity stood at roughly 10 mbd, exceeding China's total oil demand of 8.65 mbd as calculated by Reuters based on official data. It was further estimated that China will add about 3 mbd of new refining capacity between 2011 and 2015.[128]

Sinopec and CNPC are China's two leading refiners, but there are more than one hundred smaller individual refining enterprises. In 2008, Chinese refineries processed on average 52,000 b/d of crude oil, compared with the international average throughput of 114,400 b/d. This relative lack of consolidation undermines efficiency, as indicated by China's high energy consumption per ton of refined oil compared with the rest of the world.[129] Smaller refineries, or "teapot" refineries, have been important swing players when the supply of domestic oil products is tight. Nonetheless, the Chinese government is now aiming to phase out refineries that process 40,000 b/d or less, similar to how it forced thousands of small coal mines to shut down in the past. Judging from the experience in the coal sector, it is likely that the smaller refineries will take refuge in local politics and be protected by local governments, but the central government is determined to shut old, small, and uncompetitive plants and to promote large refineries with 200,000 b/d capacity.[130]

After China became a member of the WTO in 2001 and China's wholesale oil market was opened to foreign investors in 2007, given China's increasing demand, China's downstream sector offered opportunities to both domestic and foreign refiners. Foreign firms face several challenges when seeking
to access China's downstream market. Oil import and export licenses are difficult to acquire, and at least 1.25 mbd of storage capacity is required to engage
in the wholesale oil business in China. China's oil companies are becoming
experienced in engineering and in constructing refining projects. This makes
the Chinese downstream market more competitive. Chinese oil companies
are becoming less dependent on foreign expertise and technology, and Chinese oil companies are more willing and able to proceed with the construction
of refineries on their own. Sinopec, China's leading refiner, has completed
more than 450 refineries and petrochemical units, and more and more of the
refining capacity that is coming on line is Chinese built.[131]

Much of China's new refinery capacity is intended to refine heavy, sour,
and high-acid crude oils, which are more plentiful and often cheaper than
the light, sweet crudes that most of China's refineries have traditionally
handled.[132] Being able to process discounted, lower-quality crude oils provides an important economic incentive for Chinese refineries. The NOCs
can increase their crude acquisition options and use cheaper crude oil in the
downstream market. Sinopec reportedly was aggressively importing high-
sulfur oil because it was $1 a barrel cheaper than sweet crude. The added
cost to Sinopec of removing sulfur in the refining process to meet Chinese
national standards was only 10 cents a barrel. With rising oil prices, the
incentive to reduce oil acquisition costs by importing more lower-quality
crude was therefore evident and understandable.[133] Concurrently, the NOCs
are partially subsidized by the government for losses in the downstream sector as a result of the disparity between international oil prices and domestic prices for oil products. Thus, processing cheaper, heavy, sour, high-sulfur,
and acid oil can greatly affect the NOCs' profitability. Increasing the refining
capacity provides an opportunity for the NOCs to profit despite government
control over domestic oil product prices, and it simultaneously enhances
China's import flexibility and security of supply by expanding China's ability
to refine crude oil from more suppliers.

Foreign firms have struggled to access China's downstream market, complaining that China's domestic pricing policy for oil products undermines economic viability,[134] but supply-rich national oil companies have benefited. Oil-exporting countries, particularly in the Middle East, have helped China expand and upgrade its refining industry in order to enlarge their market shares in the Chinese downstream sector.[135] The process has often been lengthy. Saudi Aramco initiated negotiations for the construction of a new refinery in Qingdao, Shandong province, in 1993. This led to an agreement in 2005, but the refinery only came on stream in 2008.[136] Sinopec sealed a deal with Exxon[137] and Saudi Aramco in 2005 to develop a 230,000 b/d capacity refinery in Fujian province.[138] The agreement was confirmed in 2007, with Sinopec controlling 50 percent and Exxon Mobil and Saudi Aramco each holding 25 percent. According to Reuters, crude oil was being pumped into the refinery beginning in late April 2009.[139] Other joint ventures with foreign NOCs include a $6 billion deal between CNPC and Venezuela's state oil firm Petroléos de Venezuela S.A. (PDVSA) to set up a refinery in Guangdong province,[140] a $9 billion refining project between Kuwait Petroleum Corp. and Sinopec,[141] and a 2010 agreement between CNPC and Rosneft to build a $5 billion refinery in Tianjin.[142]

Linking up with foreign NOCs boosts commercial interests, as it provides access to feedstock supply of cheaper heavy and sour oil and refinery-petrochemical investments in China. But it also protects China's oil security through long-term crude oil supply contracts, and it is less likely that foreign NOCs will cut supplies to refineries in which they hold investment stakes.[143] China-Venezuela relations illustrate how heavy and sour oil, crude oil refining, infrastructure projects, and long-term supply contracts increasingly conflate commercial and strategic interests. Former president Hugo Chávez of Venezuela repeatedly stated that he seeks to boost crude exports to China and to diversify the amount shipped to the United States. Since 2007, China has extended loans amounting to more than $28 billion through the CDB to boost China-Venezuela energy ties.[144] Others report that China and Venezuela closed out 2010 by signing more than $40 billion in energy-related deals.[145]

Given the shipping distance, higher costs, and the lack of refining capacity in China to process Venezuela's heavy crude, Venezuela's crude oil exports to China have been limited. In April 2009, Venezuela reported oil sales to China of 380,000 b/d, although Chinese customs data show China received only 130,000 b/d.[146] Not only is the economics of transporting and refining oil challenging, but China's NOCs probably prefer to sell their overseas oil production or acquisitions in Venezuela locally or regionally and on the international petroleum market to boost profits, rather than shipping crude oil back to China. Indeed, this fits the broader picture of China's NOCs' overseas activities, which will be examined in the following chapter.

Nonetheless, Venezuelan crude oil exports to China have picked up in recent years. China consumed about 2 percent of Venezuela's crude oil in 2005 and roughly 7 percent in 2010.[147] China's oil imports from Venezuela stood at 120,000 b/d in 2008, but by 2011 it was reported that Venezuela was exporting about 460,000 b/d to China, about 20 percent of its total oil exports.[148] Although various numbers have been reported, and although Venezuelan export data and Chinese import data have differed in the past, the China-Venezuela energy partnership is growing. Shipping costs have probably been one factor preventing large exports of Venezuelan oil. However, it was recently reported that VLCCs that offload in the Gulf of Mexico can be used to carry Venezuelan crude and that China's state-owned tanker fleet can be put into service for the transport of Venezuelan oil exports.[149]

Chinese governmental loans-for-oil deals have boosted crude oil exports, as Venezuela is servicing its debt by shipping oil to China, but these deals have also facilitated Chinese participation in infrastructure and exploration projects in Venezuela. An important game changer in their energy ties is the joint venture between CNPC and PDVSA to develop a giant 400,000 b/d refinery in Guangdong province that, when it becomes operational in 2013, will be capable of receiving oil produced from the CNPC and PDVSA joint venture in Junin Block 4 in the Orinoco Belt, which will produce up to 400,000 b/d of extra heavy crude.[150] Whether Venezuela will succeed in expanding its crude oil exports to China by 1 mbd, as stated by its leaders, remains uncertain and depends on whether the Chinese NOCs prefer to ship the oil back to China. But building the refinery capacity to process Venezuelan heavy crude oil is an important first step, and it will provide a hedge and more opportunities to safeguard China's energy security.

PRICING

Chinese government control over energy prices is a major disincentive for foreign firms to invest in China's downstream sector, since the Chinese firms will have a competitive edge: the government directly subsidizes state refiners' losses and issues added tax rebates on fuel imports. Under pricing rules introduced in 2009, a "fuel price change is justified if the moving average price of a basket of crude oil rises or falls by 4 percent during a one-month review period." By June 2011, the government was considering shortening the review period for fuel prices to ten days and narrowing the 4 percent change in crude costs that justify a fuel price adjustment.[151] Indeed, the government has gradually reformed its price controls since the late 1990s. Periods of surge in international petroleum prices have challenged the government's control of ex-refinery prices, and rising inflation has undercut its commitment to price reform. In 2007, gasoline and diesel prices were increased by 10 and 7 percent, respectively. In the six months following the price increase in October 2007, crude oil costs rose 33 percent, leaving Chinese gasoline and diesel prices on average 20 and 40 percent below U.S. levels at the end of April 2008.[152]

According to an IEA report from January 2009, "unconfirmed reports suggest that if crude oil trades below \$80/bbl, domestic ex-refinery prices will fluctuate freely; between \$80/bbl and \$130/bbl, refining margins will be curbed to minimize the impact on retail prices; above \$130/bbl, the government will set retail prices."[153] Again, it was reported that this price-control formula was slightly adjusted in June 2011, with China's NOCs now allowed to set oil product prices when crude is between \$40 and \$130 per barrel. Even though new price formulas and mechanisms have been introduced, the government reserves the right to constrain price rises in extreme circumstances and to use price-control mechanisms as an instrument to curb inflation.[154] This view was reiterated in a June 2011 report, stating that the NDRC takes "into account other factors such as inflation and fuel supply and demand when making pricing decisions" and that the government has showed "increasing reluctance to lift fuel prices when crude prices topped \$100 because the country was battling high inflation, forcing refineries to cope with dwindling margin or even incur losses."[155]

With prices still lagging behind international prices, China's NOCs have faced massive financial losses in the domestic downstream sector.[156] It was reported that "Chinese state oil majors have had to increase production of petroleum products [in the first quarter of 2011], probably by directive of the central government, to fill a big gap in supply resulting from drastic cutbacks by private refiners in East China, where operations have been cut to as low as 30% of capacity due to dismal margins." Not only are Chinese NOCs forced to engage in unprofitable refining, but "petroleum imports continued to expand in March [2011] as Chinese oil companies purchased more cargoes from the international markets to maintain ample supply of fuels to local industries."[157] In this case, then, commercial interests and the pursuit of profits do not trump the security of supply and government goals to set fuel prices. While the government subsidizes the NOCs for the reduced retail prices for gasoline and diesel, it is not enough to compensate for their losses in the downstream sector.

It has been suggested that China's NOCs can shift some of their refinery operations overseas to escape price controls and enhance returns, either through joint ventures or wholly owned plants. Such shifts have not yet occurred on a large scale. The government can react to such moves and force NOCs to import product equal to what would have been produced domestically, which would erase most of the economic gains from investing in overseas refining projects.[158] Again, this provides another example of how the NOCs are controlled by the government so that the pursuit of profits does not undermine the security of supply. In contrast to the NOCs' overseas investments in upstream projects, which have been encouraged by the government because they combine both security and profit interests, overseas investments in the downstream sector do not provide the same hedge to China's energy security.

CONCLUSION

China depends on coal for roughly 70 percent of its energy use, and coal will dominate the energy mix for the foreseeable future. Abundant coal reserves are an important foundation for China's energy security, together with the fact that China remains the fifth-largest oil producer in the world. China produces just over 4 mbd of crude oil, which accounts for almost 10 percent of China's energy mix. Although China has become the largest energy consumer in the world, when adding hydro, gas, nuclear, and renewable to the energy mix, China's indigenous production still accounts for almost 90 percent of its total energy consumption. This is a remarkable achievement, and even if domestic energy production should fall to 80 percent of its total energy consumption in the years ahead, and even if China's net oil import gap continues to grow, China's energy security in terms of domestic energy production can be sustained. Maintaining such a favorable energy mix, from a security perspective and with the environmental tradeoff, is considered as a costly but important short-side hedging strategy.

Nonetheless, demand is rising, and China's domestic oil production is unable to meet the increasing demand, which has resulted in a growing net import gap of crude oil, now accounting for about 11 percent of China's energy mix. China has also started to import natural gas and LNG as well as coal. In the early 2000s, after leaders became aware of the increasing reliance on oil imports, the central government developed a number of strategies to hedge against oil supply disruptions and high oil prices. The long-term strategies that have largely been advanced and implemented during the last decade show that the central government has been active and successful in shaping China's energy security policy. As China's net oil import gap increased dramatically, risks were identified, and hedging strategies were developed to manage the situation. The government also sought to diversify the energy mix; promote energy conservation, efficiency, and renewable and alternative energy; and control fuel prices.

Finally, the buildup of SPRs and refinery capacity are two important domestic hedging strategies against peacetime risks, and the conversion of coal to synthetic fuel provides China with extra insurance and can be used to develop oil products during a wartime contingency. Building SPRs is a traditional risk management strategy to deal with price volatility and oil supply disruptions. The strategy of boosting China's refinery capacity is a strategy that links commercial and strategic objectives, enhances China's ability to manage oil supply disruptions, and encourages long-term supply contracts between China and the major oil producers. As the next chapters will demonstrate, by combining the advantages of an energy mix largely based on indigenous sources and domestic hedging measures with a global search for overseas sources of petroleum production—and with an emphasis on diversifying and safeguarding China's petroleum transport routes—China has developed sophisticated hedging strategies that insure against peacetime risks and wartime threats to China's energy security.

4

THE GLOBAL SEARCH FOR PETROLEUM

TODAY WE ARE deluged with stories about China's energy security interests and the emergence of China's national oil companies (NOCs) as a major force in the world energy markets. This chapter focuses on the global search for petroleum—and profit—by Chinese NOCs. It explores how Chinese NOCs have gained access to the world energy market along with how China has diversified its sources of petroleum imports and has acquired overseas assets and production rights (equity oil) that provide a hedge against potential supply disruptions.

Overseas equity production can partly be a hedge against the possibility that a war, conflict, or uprising in an oil-rich region will limit the availability of oil in the international petroleum market and partly be a hedge against exceedingly high oil prices negatively affecting China's economy. Thus, China's global search for petroleum is an important aspect of a broader hedging strategy against oil supply disruptions that includes securing overseas equity upstream production, building up a state-owned tanker fleet, developing cross-border pipelines, maintaining a favorable energy mix, developing an SPR, and increasing China's refinery capacity.

The phrase "going abroad for business" or "going out" captures how the NOCs initiated investments and exploration projects abroad in the 1990s and became the main force behind securing petroleum deals overseas.[1] By the turn of the century, the Chinese government was taking a stronger interest in energy security issues, as a result of the rapidly increasing net oil import gap, the shifting international security environment, and higher oil prices. The government became an important driver behind the "going abroad" strategy through cheap loans, subsidies, long-term plans, and diplomatic support. Thus, though initially advanced by the NOCs for profit reasons, "going abroad" was now advocated by the government for strategic and security concerns. Clearly, both the NOCs and the government have been drivers of China's global search for petroleum, and this chapter argues that both profit and security considerations have been important. The NOCs have been the main actors in China's overseas search for petroleum, but as the Chinese leaders became more concerned about China's growing net oil import gap, the government has become more involved and has established core strategies and policies. These strategies, embedded in China's Tenth Five-Year Plan in 2001, were put forward by the State Economic and Trade Commission (SETC), the State Development and Planning Commission (SDPC), and the National Development and Reform Commission (NDRC) in the early 2000s. These long-term plans show a preference for developing hedging strategies that mix strategic and commercial interests in China's energy security policy.

China cannot insulate itself from market mechanisms, and China's dependence on the international petroleum market will grow substantially in the foreseeable future, despite the overseas upstream investments by Chinese NOCs. In addition, Chinese NOCs are not following a neomercantilistic strategy whereby all overseas upstream production is shipped back to China. Instead, Chinese NOCs are selling most of their overseas production on the international petroleum market to boost their profits. So why should China invest in overseas upstream production if it ends up buying most of its oil in the international petroleum market?

Getting a stake in the lucrative international upstream production market has been a core driver behind the NOCs' "going out" strategy and its search for profit. Simultaneously, overseas upstream investments and equity oil production and loans-for-oil deals provide a hedge against supply disruptions, exceedingly high oil prices, or a tightening oil market. Of course, China will still "suffer" during a crisis and cannot escape fully from the international petroleum market. Nonetheless, an important fact of life in international politics is that states worry about their relative gains.[2] At times, some gain more than others; at other times, some lose more than others. The overseas upstream investments and production by Chinese NOCs provide insurance that China will be reasonably well off compared to other major oil importers were there to be a crisis in the international petroleum market. Developing the capacity to obtain a disproportionate gain within the international system or from the international petroleum market is an important strategy for many states—and especially for great powers.

Not only does China strategically hedge against supply disruptions and high oil prices, as is evident in its overseas upstream investments and the strategies embedded in the government's long-term plans, but the outcome of China's energy security policy-making process on sensitive questions when the NOCs' search for petroleum deals clashes with broader national interests can also be characterized as a hedging policy. As Chinese decision makers take security/strategic and profit/market considerations into account, strive for consensus on how to react and respond to growing insecurity in terms of energy supply, and seek to balance the interests of the main actors in China's policy-making apparatus, the energy security policy outcome, especially in sensitive and strategically important cases such as Iran and Sudan, resembles a hedging policy.

Both China's strategy toward Sudan and Iran are driven by security and profit considerations, and they both reflect how China is hedging its bets in its energy security policy. Sudan illustrates how the NOCs' search for profits has been a main driver in the "going abroad" strategy. Indeed, corporate interests in Sudan have partially undermined broader Chinese national interests, for instance prior to the Beijing Olympics in 2008. The Chinese government has tried, with varying success, to direct NOC business activities in Sudan, to demonstrate that the government seeks to shape its NOCs' overseas activities. More importantly, with about 200,000 bpd of equity production in Sudan, which is one of the NOCs' largest overseas oil projects, the overseas upstream investments can provide a hedge against a crisis in the international petroleum market.

Iran differs from Sudan because Chinese NOCs are not allowed to make equity investments in the upstream market. Instead, Teheran requires that all foreign companies sign buyback agreements if the companies pay for everything up front, and, once the field comes on line, they are repaid with revenue that the National Iranian Oil Company (NIOC) earns from exporting oil produced by the field. Thus, China's NOCs do not have a market share over equity production that during a time of crisis could potentially be shipped back to China. Therefore, this is a different hedging strategy. The central government cannot strategically hedge by instructing its NOCs to ship oil produced in Iran back home because the Chinese NOCs have no equity oil in Iran. Nonetheless, close ties with Iran can be strategically valuable if uprisings, instability, or conflicts spread to key Arab oil-producing states. China can also profit and secure its oil supply by maintaining close relations with Teheran even when Iran is being isolated or sanctioned by the international community. The Chinese government has therefore been hedging its policy toward Iran by refusing to take a strong position either for or against Iran. On the one hand, it carefully protects its petroleum interests in Iran and its broader ties with the Teheran regime; on the other hand, it emphasizes that China's ties with Iran cannot undermine its relations with the United States, Saudi Arabia, or the European Union. Chinese NOCs have probably invested much more in Iran than in Sudan. This reflects the fact that Iran produces much more petroleum and offers better opportunities for exploration and development. Simultaneously, Chinese NOCs have also been hedging their bets, seeking to balance their interests in Iran with their business activities in the United States, Europe, and Arab states.

The following four sections address these issues in depth. The first section surveys the overseas investments in petroleum projects by Chinese NOCs, the government-backed loan-for-oil deals that China has signed with a number of petroleum suppliers worldwide, and the goal of Chinese NOCs to diversify the supply of petroleum through overseas production. The data and estimates provide a backdrop for examining China's global search for petroleum. The next section discusses the main drivers behind China's overseas petroleum activities. The third and fourth sections look at Sudan and Iran respectively as case studies to investigate China's global petroleum strategies and China's energy security policy making.

CHINA'S GLOBAL SEARCH FOR PETROLEUM

At the end of 2010, China's oil consumption stood at close to 10 mbd.[3] The U.S. Energy Information Administration (EIA) anticipates that China's growth in oil demand will continue to represent a major portion of the projected world growth in demand.[4] In its 2009 *World Energy Outlook*, the International Energy Agency (IEA) boosted its projections on China's oil demand, stating that in 2030 China's demand will be 16.3 mbd, which will account for 42 percent of world growth over the next two decades.[5] This was updated in the IEA's 2010 *World Energy Outlook*, projecting that China will import 79 percent of the oil it consumes by 2030, thereby accounting for a larger incremental increase in oil demand than any other country.[6] Chinese authorities have estimated that by 2020 60 percent of China's oil demand will be imported.[7] With the cooling of the Chinese economy in 2012, however, there has been a sharp deceleration in the momentum of Chinese oil demand compared to previous years. Despite the difficulties in predicting future energy demand, the issue of ensuring reliable imports of oil features prominently in China's political and security discourse and influences China's diplomatic and strategic calculations.[8]

OVERSEAS INVESTMENTS

CNPC has dominated overseas investments, both in number and value, with roughly twice as many investments and projects as Sinopec.[9] CNOOC has moved from investing in roughly one-quarter of the total number of overseas investment projects in 2006 to investing in one-third in 2009, but the value of its investments still lags far behind the two other major Chinese NOCs.[10] In December 2010, CNPC, the parent company of the Hong Kong–listed PetroChina, announced that it was operating eighty-one projects abroad in twenty-nine countries.[11]

The considerable advances in overseas oil investments during the last few years have been facilitated by large loans from two of China's most important state banks—the China Development Bank (CDB) and the Export and Import Bank of China (EIBC). By 2009, it was estimated that Chinese NOCs had a stake in roughly two hundred upstream investment projects in fifty countries.[12] While it took fifteen years before Chinese NOCs' total investments reached US$30 billion in 2007, this number doubled between 2007 to 2009. By September 2009, Chinese overseas petroleum investment was more than $60 billion.[13]

Large-scale deals continued in 2010: the three NOCs committed over $9 billion to acquisitions, accounting for nearly 20 percent of the value of global deals in the first quarter of 2010.[14] It was further reported that Chinese NOCs, along with other smaller companies from China, spent at least $47.59 billion to acquire oil and gas assets around the world from January 2009 to December 2010 and spent $18.2 billion on M&As in 2009, which accounted for 13 percent of the total global oil and gas acquisitions ($144 billion) and for 61 percent of all acquisitions by national oil companies ($30 billion). In 2010, Chinese NOCs again spent approximately $29.39 billion, with more than half invested in Latin America ($15.74 billion).[15]

LOAN-FOR-OIL DEALS

Another important aspect of China's global search for petroleum is the more recent loan-for-petroleum deals that the Chinese government, Chinese NOCs, and state-owned banks have signed with key petroleum-producing states around the world. CDB is the key player behind China's loans for long-term oil and gas supplies.[16] China stepped up its loan-for-petroleum deals in the aftermath of the 2008 financial crisis, with $54 billion going to countries such as Russia, Venezuela, Brazil, Kazakhstan, Angola, and Ecuador during the first three quarters of 2009.[17] These deals continued in 2010; the Chinese government and banks had large cash reserves and were looking to diversify their investment portfolios away from treasury bonds and financial products and toward commodity markets. One of the largest deals was the $20 billion loan-for-oil deal (over ten years) with Venezuela announced in April 2010, which was expanded in 2012, with loans reaching $32 billion.[18] From the beginning of 2009 to December 2010, CNPC and Sinopec were involved in twelve loan-for-oil deals with nine countries, worth an estimated $77 billion.[19]

China continues to pay billions for oil deals around the world, and there are no signs that it will halt its investments and loan deals in the years ahead.

One report from 2010 estimates that China's overseas oil deals have the potential to deliver more than 7.8 billion barrels of oil to the country over the next several years.[20] But adding the total amount of reserves in the fields in which Chinese oil companies have invested to the volume that will be supplied to China through the loan-for-petroleum contracts is problematic. Some of the investments are not yet productive and might take decades to become so, and it remains questionable whether some countries will deliver on their contracts.[21] Former president Hugo Chávez, for example, has long stated that Venezuela will shift more of its exports to China rather than to the United States, but the number of statements, energy accords, and agreements that have been signed between Venezuela and China during recent years have had little material impact, and it may be some time before Chinese NOC activity in Venezuela develops. The United States remains the main buyer of Venezuelan crude, and despite Venezuelan officials stating that their country shipped 400,000 b/d to China and that they hope to increase the amount to one million b/d before long, Chinese data for 2009 show imports of Venezuelan crude oil and fuel oil at just under 200,000 b/d.[22]

MARKET PRICE FOR IMPORTS

Price remains another opaque issue in some aspects of China's energy security policy. For example, the price on the oil that China imports from Iran since sanctions were introduced, after the Chinese began to squeeze the Iranians (discussed below), and after contracts were renegotiated in 2011 and 2012 is difficult to access. Similarly, it is hard to get insight on the terms and conditions under which the tankers that ship Iranian oil to China operate and how they are insured. Most writers argue that China is largely paying market price for its crude oil, gas, and petroleum-product imports, on oil produced overseas by Chinese NOCs and shipped back home, on oil swapped on the international market, on petroleum that is imported through pipelines or by railway, and on most of the oil and gas China receives through loan-for-petroleum deals. Iran is currently an exception when it comes to how China's NOCs in general operate in the international petroleum market, and the oil that Rosneft shipped to China by rail between 2005 and 2010, which was facilitated by a $6 billion loan from CDB and China Eximbank, was probably an exception to the contracts that China will be able to negotiate in the continental pipeline "market." This loan was supported by 48.4 million tons of oil export from February 2005 through December 2010.[23] The price was the monthly weighted average price of Brent crude, with a discount of $3 per barrel. Under pressure from Rosneft, CNPC agreed to reduce the discount by $0.675.

In 2009, Rosneft signed a long-term contract with CNPC for the delivery of 180,000 b/d of crude from January 2011 through December 2030 to support its $15 billion loan from CDB. Transneft will deliver 120,000 b/d, which it is purchasing from Rosneft to support its $10 billion loan from CDB.[24] Both supply contracts are based on or linked to market prices, but they differ in the specifics. The earlier contract between CNPC and Rosneft was on a much more fixed basis; the later contract is a more standard-term contract linked to global benchmarking pricing. The 300,000 b/d that China will receive from Russia through the Eastern Siberia–Pacific Ocean pipeline (ESPO) spur will be determined monthly, based on the price of oil quoted in Argus and Platts at Kozmino, the terminus of the second leg of the ESPO pipeline.[25]

Although the first contract from 2005–2010 turned out to be a favorable deal for the Chinese and probably is the exception that confirms the rule, it illustrates one of the ways that China and its NOCs have been able to use long-term future contracts to hedge bets in the international oil market. There are fewer sources on the price paid for the oil delivered by the Kazakhstan-China pipeline and what will be paid for oil arriving through the Myanmar pipeline;[26] however, it is unlikely that China has signed similar future contracts and long-term oil supply deals similar to those CNPC was able to secure from Rosneft in 2005.

OVERSEAS OIL PRODUCTION

In addition to the uncertainty surrounding the amount of Chinese overseas investment and to some extent the price that China is paying for its imported crude oil and petroleum products, data on Chinese overseas oil production or equity oil also vary. Numbers on Chinese oil production overseas during the 2006–2008 period were roughly close to 675,000 b/d in 2006,[27] 780,000 b/d in 2007,[28] and 850,000–900.000 b/d in 2008.[29] In 2010, *China Daily* reported a huge jump in the figures for 2009. Drawing on statements by CNPC, it was reported that CNPC had increased its overseas production by 12 percent, reaching a record 69.62 million metric tons, or close to 1.4 mbd, after gaining access to more fields in Kazakhstan and Canada.[30]

There is a large jump between the 2008 figures of no more than 900,000 b/d and CNPC data reported by *China Daily*, which were closer to 1.2 mbd for 2008, even if the 12 percent annual increase is subtracted from the 69.62 million metric tons.[31] Moreover, the *China Daily* figures only account for CNPC overseas production, thus indicating that the boost in overseas production might be even larger, since the numbers above for the 2006–2008 period refer to the total overseas production of all Chinese companies.

Chinese NOCs have increased their international reserve base and by 2010 probably had overseas output of more than 1 mbd. It is believed that the Chinese government had previously set a target for the NOCs to secure 1 mbd of overseas production by 2010. It is expected that by 2013 China's NOCs net international oil and gas output will reach 1.5 mbd.[32] In December 2010, however, CNPC reiterated that its overseas oil and gas production hit a record high in 2009, at 69.62 million metric tons of crude and 8.2 billion cubic meters of natural gas, up 12 and 22 percent respectively, and it announced that the company expected to produce 85 million tons of oil in 2010, or 1.7 mbd, from its overseas assets.[33]

Some of the variation in the figures may be explained by the difference between equity oil production and overseas oil production. Equity oil deals refer to an oil company's share of the total amount of oil produced by a project and standard industry production sharing agreements that give foreign oil companies the right to sell a specific share of the oil it produces.[34] Nothing about the equity deals that Chinese firms have signed is fundamentally different from those signed by their IOC peers. Equity can be sold freely, and many reports indicate that a large proportion of Chinese NOCs' overseas equity oil production is being sold in the international market to the highest bidder in order to boost profits, rather than being shipped home.[35]

Some studies and reports use the wording "equity oil" and "overseas production" interchangeably.[36] It is complicated to distinguish between equity and overseas oil production each time one writes about China's global international oil policy and some of the roughly two hundred overseas upstream projects that Chinese NOCs were involved in. When this book refers to Chinese NOCs' overseas oil production, it includes equity oil. Accordingly, comparing and contrasting "overseas equity shares"[37] with "overseas production"[38] and "overseas oil and gas production" is inaccurate. It is to be expected that overseas oil production is higher than overseas equity oil production.

Another explanation for the gap in the numbers cited above and the difficulties in estimating the exact overseas production of Chinese oil companies is that the companies may include in their total numbers the production of partners in joint-production projects.[39] Finally, some data and numbers may have been distorted because of different measurements. Volume in gallons and U.S. barrels or weight in tons is normally used. The simple rule of thumb for conversion is that a barrel per day is roughly equal to 50 tons per year, but the relationship varies according to density and thus according to product.[40] In short, the data on overseas oil production by Chinese NOCs are difficult to access, but based on the literature examined and interviews conducted it is reasonable to conclude that the overseas production of Chinese NOCs now stands at roughly 2 mbd, with equity oil production accounting for approximately 1.4 mbd.

Of course, the oil that Chinese NOCs produce overseas or the equity oil is insufficient to meet Chinese demand. However, the equity oil is not insignificant. China has equity oil production in about twenty countries, though Chinese equity shares are primarily located in four—Kazakhstan, Sudan, Venezuela, and Angola—which together account for equity production of 820,000 b/d.[41] It is also important to note that among the remaining twenty countries, which together bring China roughly 1.4 mbd of equity oil production, only equity production in Iraq, Oman, and Yemen would be affected by conflict in the Middle East, and even then Oman and Yemen have a long coastline facing the Arabian Sea. So, in principle, most of the equity oil could be shipped back to China during any crisis short of a war involving China.

DIVERSIFYING OIL IMPORTS

The Chinese government has sought to diversify its oil imports. Winston Churchill succinctly stated almost one hundred years ago that energy security is embedded in "variety and variety alone."[42] For oil-exporting and oil-importing countries alike, diversification in terms of both sources of oil and increased use of energy supplies is an overarching principle for energy security, and China is no exception.[43] The emphasis on diversification is important for both security and profit concerns and is shared by the government and the NOCs.[44] Indeed, China's sources of oil imports have gradually been extended to most regions of the world, its import routes have expanded, its energy mix has grown more diverse, and its refinery capacity has been improved.[45]

China still depends on Middle Eastern oil for more than 50 percent of its oil imports, but it has been successful in diversifying and has managed to find alternative sources of supply. With oil consumption more than tripling in China since the 1990s, China has become the world's largest oil importer, and China's reliance on imported oil has increased from 30 to 50 percent during the 2001–2007 period. But China's oil imports from the Middle East dropped from 50.9 percent in 2003 to 44.6 percent in 2007, whereas imports from Africa increased from 24.3 percent in 2003 to 32.4 percent in 2007.[46] China also increased its oil imports from Europe and the Western Hemisphere, and there was a significant drop in imports from the Asia-Pacific region.[47] More recently, *Global Times* reported in 2010 that China now imports more than 50 percent of its oil supplies from the Middle East, indicating an increase from 2007.[48] Given that the IEA estimates that China's growth in oil demand for 2010 and 2011 will represent about 37 percent of the projected world growth in demand and that the Middle East remains the main region for oil production and reserves, it is remarkable that China has managed to keep its reliance on Middle Eastern oil to about 50 percent during the last decade. Security of supply considerations and the policy of hedging against supply disruptions are key factors accounting for this.

Of course, diversification has not simply been a result of strategic or security considerations; it has also been driven by economic incentives, since it is more difficult for China's refineries to process Middle Eastern crude oil, which tends to be heavy and sour. The light and sweet crude from China's African oil suppliers, such as Angola, Sudan, and Libya, or from Kazakhstan and Russia, is similar to the oil produced in northeastern China and therefore is easier to refine. Nevertheless, even though Indonesia's Minas crude is light and sweet, China's oil imports from Indonesia have dropped significantly because of the rapid depletion of that country's domestic petroleum reserves, while Saudi Aramco has boosted its exports of heavy and sour crude to China by developing and investing in refining capacity in China.[49]

For a long time, it has been estimated that both the United States and China will be heavily dependent on the Middle East to supply a large share of their future oil needs. The IEA predicts that as of 2015, 70 percent of China's oil imports will come from the Middle East. Despite the Department of Energy's forecasting for many years that the United States will depend heavily on Mideast oil, U.S. oil imports from the Middle East are declining. Nonetheless, China and the United States need to manage their differences and promote cooperation and common interests as the world's largest oil consumers.[51] Conversely, recent reports suggest that North America has become the fastest growing oil and gas region in the world, and it is likely to remain so for the rest of the decade and into the 2020s. U.S. net petroleum imports are declining. As North America becomes a major hydrocarbon exporting center, it arguably may become the "new Middle East."[52] If such estimates and projections come through, and if the United States becomes more energy independent, it might change its major role at the center of the global energy order as the guarantor of supply lanes globally and as the protector of main producer countries in the Middle East and elsewhere. As other great powers potentially seek to fill the vacuum, such development might create more tension, friction, rivalry, and economic and security challenges than if the United States remained a leading power in sustaining the international energy order.

Although the diversification of sources, routes, and mix in China's energy supply remains important, this study argues that diversification is only one starting point in examining China's energy security policy and only one aspect of the insurance mechanism embedded in a hedging strategy. Consequently, as is illustrated throughout this volume, a hedging strategy encompasses more than the traditional policy of diversifying energy sources, routes, and mix; it also incorporates broader economic, political, military, and diplomatic tools and strategies. It is acknowledged that many energy experts assert that a traditional diversification strategy is embedded in economic, political, military, and diplomatic considerations, but it is nonetheless argued that these considerations are not sufficiently examined in much of the existing literature that deals with the domestic, global, maritime, and continental aspects of China's energy security strategy. More importantly, this book contends that a hedging strategy is more comprehensive. For example, a hedger will not only diversify by investing in different stocks and currencies, but simultaneously he will short sell a number of stocks and currencies, to limit market risks. This provides the hedger with more tools to limit and manage risk than an investor who only attempts to diversify.

In sum, the exact number of Chinese NOC investments overseas may be difficult to estimate, many loan-for-oil deals still need to materialize, and the terms of production rights or equity contracts signed abroad are sometimes concealed from public scrutiny. In addition, when Chinese NOC overseas oil production is being sold, how and at what prices the NOCs sell their oil products, import their crude oil petroleum products, and obtain petroleum through loan-for-oil deals or swaps on the international oil market remain at times opaque. Nevertheless, the trend is clear, and as the IEA writes, "it is hard to overstate the growing importance of China in global energy markets."[53] China has managed to diversify its oil imports, and Chinese NOCs are becoming more important players in the international petroleum market—and they are producing overseas more than 2 mbd.

DRIVERS BEHIND THE "GOING ABROAD" STRATEGY

We find within the literature that looks at the role of China's NOCs in the international petroleum market a number of alarmists arguing that China and its major oil companies are locking up supplies, taking oil off the market, dominating the international petroleum market, asserting a competitive advantage because they are sponsored by the state, and that China has chosen to secure energy supplies less by relying on commercial interests and more by following a neomercantilist strategic approach that pursues ties with countries shunned by most of the world as pariah states.[54] This characterization of China's energy security policy has "promoted an image of China Energy, Inc. in Tokyo, Seoul, New Delhi, and Washington."[55]

It is argued that bilateral Chinese oil diplomacy could lead to Chinese policy responses that thwart effective multinational initiatives in areas like conflict resolution or could hinder efforts to prevent the proliferation of sensitive military technologies.[56] China's rise as a peer competitor to the United States has long been on the agenda in U.S.-China relations and in Washington, and China's role in world energy markets and global energy geopolitics adds new contentious issues to the relationship.[57]

Advocates of the strategic and alarmist view are challenged by those with more of a market orientation. These writers point out that China's oil companies have shown a willingness to take oil from places others avoid, what the international oil companies (IOCs) see as low-hanging fruit. Chinese NOCs are investing in the few remaining underdeveloped oil regions of the world, and their investments in high-risk countries add new capacity to the world's tight energy market. By extracting oil from off-limits places such as Sudan and by operating with higher risk in "problem states," it is argued that China's NOCs contribute to a better-supplied world oil market.[58] Studies show that some of the largest Chinese natural resource procurement arrangements predominantly help expand, diversify, and make the global energy supply system more competitive.[59]

It is noted that overseas production by Chinese NOCs is largely sold for profit locally or on the international petroleum market. Even if Chinese oil production in Sudan were shipped back home, that would mean less that China needed to buy, for example, from West Africa, thus leaving more available for others to purchase.[60] Accordingly, the U.S. Department of Energy observes that "even if China's equity oil investments 'remove' assets from the global market, in the sense that they are not subsequently available for resale, these actions merely displace what the Chinese would have otherwise bought on the open market . . . the effects of these purchases should be economically neutral."[61] In short, from the viewpoint of consumers in the United States, Europe, and Japan, Chinese investments in the development of new energy supplies around the world are not a threat but rather something to be desired because it means there will generally be more petroleum available.[62]

The Chinese government has emphasized the importance of international cooperation and has promoted investment in new capacities both at home and abroad.[63] Thus, some argue that China's increase in oil imports should be treated as a normal phenomenon and as part of China's growing interdependence with the rest of the world.[64] Through the liberalization of international and domestic energy markets, one can prevent disruptions of oil supplies by improving the flow of information, maintaining strategic oil stockpiles, and promoting investments in new capacities.[65] Advocates of a more market-oriented approach emphasize growing interdependence and cooperation in maintaining stability in the world oil market. Energy-dependent countries are bound closer together economically, which can have positive spillover effects on the overall geopolitical climate and of improving political relations between countries.[66] They also point to closer ties between China and the IEA, in addition to the growing relationship between Chinese NOCs and IOCs and NOCs in other countries, as evidence of increased cooperation.[67]

LINKING STRATEGIC AND MARKET MECHANISMS

Arguably, the picture is complex, and China's energy security policy is shaped by both strategic and commercial incentives.[68] This study contends that a hedging framework best incorporates the mix of these conventional approaches to account for China's energy security policy. Chinese decision makers draw on both security and profit considerations to develop energy strategies, which can be captured by a hedging framework that shows how Chinese leaders strategically aim to manage risk and uncertainty and how the outcome of the interactions between advocates of strategic or market considerations in the decision-making process leads to a hedging policy.

In contrast to the literature that emphasizes the mercantilist and strategic natures of China's overseas petroleum policy, there is evidence to suggest that Chinese NOCs during the early phase of the going-abroad period in the 1990s went abroad out of profit considerations. The important role of the Chinese NOCs in the going-abroad strategy and their search for profits have been highlighted by a number of scholars and experts writing about both the 1990s and the 2000s.[69] Initiated by CNPC, the overseas expansion of Chinese oil companies was largely driven by challenges at home, including domestic petroleum production shortfalls, depleting reserves, and rising production costs.[70] CNPC, the pioneer behind the going-abroad campaign, was compelled to seek foreign capital and technology to develop domestic resources, and so it turned to the international market for resources to meet domestic demand. It is worth remembering that the Chinese NOCs' going-abroad strategy in the early and mid-1990s coincided with a general opening of the Chinese economy and a strong emphasis on trade and market access.[71] This process was certainly facilitated by the Chinese government, but CNPC and other oil companies took advantage of these reforms to develop their businesses overseas.

The Chinese government and its leaders, such as former premier Li Peng, former president Jiang Zemin, and former premier Zhu Rongji, remained influential in promoting the going-abroad strategy as part of the reform and restructuring of state-owned enterprises during the 1990s.[72] The strategy was part of the broader policy of global engagement and was reinforced in early 2001 by Zhu. Accordingly, the overseas oil and gas equity purchased by Chinese oil companies has been an important part of the political, diplomatic, and economic game for the Chinese government.[73]

One researcher who has emphasized the central and often independent role of the NOCs in the going-abroad strategy acknowledges that there was a shift toward more strategic thinking and a stronger involvement by the Chinese government in the going-abroad strategy at the turn of the century.[74] This shift gained momentum for both domestic and external reasons. Domestic consumption rapidly began to outstrip production after 2000, resulting in an increasingly larger net oil import gap. China's oil refining capacity improved, which made it easier for China to import more types of oil. Reform of the pricing system encouraged oil refineries to increase imports, and thanks to a nationwide campaign against oil smuggling between 1998 and 2001, China's customs statistics more accurately reflected the actual volume of oil imports.[75]

Externally, the September 11, 2001, terrorist attacks against the United States, the U.S. war in Afghanistan beginning in 2001, U.S. involvement and increasing presence in Central Asia, and the U.S. invasion of Iraq in 2003 all contributed to higher international oil prices and consumption beginning rapidly to outstrip production. These factors all reinforced the Chinese government's concern about energy security during the 2000s.[76]

Although "going abroad for business" was a major driver behind the overseas expansion of Chinese NOCs in the 1990s, the picture becomes more mixed by the late 1990s and early 2000s. As we have seen, China's energy policy was already on the radar screen of the central leadership beginning in the late 1990s, and by the turn of the century the central government regained initiative over the petroleum policy from the three oil companies.[77] Judging from how China's energy security policy has developed during the last decade, there is evidence to suggest that the stated goals of China's leaders and Chinese government energy institutions have had a major influence on both strategy and policy.

As the next section will show, in Sudan and Iran China's national and corporate interests have at times been in conflict. Although it is difficult to find conclusive evidence, it will be argued that the Chinese government has used its authority to control the NOCs in these two sensitive cases and has hedged its bets on an important issue that inextricably links energy security, corporate interests, and broader national interests. Of all the scholarly and media attention given to China's worldwide search for energy supplies, the cases of Sudan and Iran figure most prominently. Thus, an investigation of China's policy toward these two countries will provide important insights into the evolution of China's global energy security policy.

SUDAN AND SOUTH SUDAN: FUELING THE DRAGON?

Oil is fundamentally intertwined with China's involvement in Sudan and South Sudan.[78] But China's search for oil in Africa also needs to be understood in a historical context and in view of China's emphasis on trade with

African countries during the last decade.[79] Communist China traditionally nurtured relations with African countries and advocated shared interests in anticolonialism, antihegemonism, and South-South collaboration, as former Chinese premier Zhou Enlai proposed in the aftermath of the Bandung conference in 1955. In addition, diplomatic support from countries in Africa has been important for China's aim of isolating Taiwan diplomatically within the international community.

Oil was discovered in southern Sudan in the late 1970s. Western oil companies, however, withdrew from Sudan when the conflict between the southern rebels and the northern Khartoum-based government erupted and intensified in the early 1980s. CNPC entered Sudan in 1995, taking advantage of U.S. economic sanctions and the vacuum left after the Western companies withdrew. It began operating in Block 6 in Sudan's Muglad basin and soon thereafter secured a 40 percent share in the Greater Nile Petroleum Operating Company (GNPOC).[80] At the time, Sudan was still importing oil, and the Khartoum government agreed to give China generous concessions for its investments, because of the ongoing civil war and its desperate financial situation. China's investments facilitated Sudan's shift to becoming an oil exporter in 1999, with China allegedly investing more than $8 billion in Sudan's oil sector, including four upstream oil projects, three downstream projects, an oil terminal near Port Sudan, and a 1,500- kilometer pipeline and another pipeline stretching more than 1,300 kilometers linking CNPC's numerous wells in southern Sudan to northern processing plants.[81] How China's NOCs' investments in Sudan have been funded remains opaque, and it is difficult to determine how much China's NOCs gain from Beijing's backing. The NOCs were created from state assets and continue to receive financial support from the Chinese government. China's high level of foreign exchange reserves has been available to NOCs seeking overseas investments, and the NOCs have capitalized on China's loans-for-oil agreements and loans, donations, and trade deals between China and oil-rich countries in the developing world. It is debated whether the NOCs have had access to cheap bank loans—since the terms of at least some of the loans have not been available[82]—but it is reasonable to conclude that the NOCs, which are among the largest and most profitable firms in China, have benefited from state-owned banks eager to invest their surplus.

Sudan is important to China's overseas energy security policy because it supplies 6 to 7 percent of China's oil imports and was the sixth-largest source of Chinese oil imports in 2010.[83] Several writers point to China's behavior in Sudan as evidence of a "strategic" approach, noting that China's energy security policy has been driven by a fear of U.S. and Western companies' dominant position in the energy market. [84] As explained above, a strategic approach focuses on direct investments in overseas sources of energy and cultivation of special relationships with energy-exporting states through trade, arms exports, aid, investments, and diplomatic support.[85]

China's close relationship with the Khartoum government has been consolidated through investments in infrastructure, trade agreements, weapons sales, and diplomatic support and protection.[86] Sudan's chronic instability is of concern to Beijing, because Sudan allegedly hosts up to ten thousand Chinese workers, some of them decommissioned from the army and charged with protecting China's investments.[87] China's close relationship with the Khartoum government has led it to obstruct, dilute, or abstain from UN resolutions on sanctions and other punitive steps initiated by the United States and favored by the broader international community in order to put pressure on the Sudanese government to resolve the Darfur crisis.

Nonetheless, both market incentives and policy hedging have been evident in China's Sudan policy. Most of China's overseas oil production is sold on the international energy market. In 2011, Chinese oil companies controlled about 40 percent of Sudan's oil assets, and CNPC's operations in Sudan were now the largest part of its overseas oil business.[88] Indeed, Sudan has become one of the most lucrative overseas production sites for CNPC, giving it huge incentives to make the most profits from its production. Although difficult to measure, it has been argued that in 2006 only 24 percent of the Chinese oil produced in Sudan was shipped back to China and that roughly 65 percent of the oil produced in Sudan was sold to Japan.[89] More recent reports suggest that nearly 60 percent of the oil CNPC pumps in Sudan is now exported to China.[90]

As the international community began to highlight the atrocities and humanitarian disasters centered in the Darfur region of Sudan in 2007, China's energy interests there and its close ties with Khartoum became a liability. An international campaign began to brand the planned 2008 summer Olympic games in Beijing "China's genocide Olympics" because China was dragging its feet in the United Nations, shielding the Sudanese leader Omar al-Bashir, supplying Sudan with arms, and maintaining close energy ties with the regime.[91] Accordingly, China hedged its energy security policy in Sudan and initiated subtle adjustments to its policy. The Chinese government showed an interest in not letting its strategic and corporate interests in Sudan undermine its broader national interests and demonstrated a willingness to avoid opportunity losses from its involvement in Sudan. China then balanced its strategic and corporate interests in Sudan with a hedging policy that recognized that "more than energy is at stake."[92]

China, aiming to protect its investments and oil production in Sudan, has been reluctant to support UN initiatives to persuade the Sudanese government to accept a UN peacekeeping force and has abstained from UN Security Council (UNSC) resolutions. However, contrary to popular belief, there is no correlation, either positive or negative, between China's willingness to support UNSC action and its bilateral trade with the country in question.[93] Thus, the notion that China's oil interests in Sudan or Libya override its willingness to support UN action is difficult to sustain.[94]

Throughout early 2007, Beijing started to play a more active role in finding a solution to the Darfur crisis through its ambassador to the United Nations, Wang Guangya. The NDRC removed Sudan from the list of countries with preferred trade status and the list of countries in which Chinese NOCs were encouraged to invest. CNPC ignored these guidelines and purchased more assets in Sudan after 2007. This shows that the NDRC does not fully control the NOCs' overseas investments,[95] but the new guidelines were one additional factor demonstrating how the Chinese government had initiated changes to its Sudan policy in 2007. More important to the argument developed throughout this book, there is a difference between encouraging Chinese NOCs not to invest and the central government instructing the NOCs to ship more of its overseas oil production back home during times of crisis in order to hedge against supply disruptions, limited availability, and large increases in affordability.

Shifting to such a hedging policy is possible because China has developed a strategy that emphasizes overseas equity oil production, diversification, boosting China's state-owned tanker fleet, developing cross-border pipelines, developing SPRs, and enhancing China's refinery capacity. Accordingly, it is maintained that the party-state apparatus is able to control NOCs in times of emergency or crisis that threaten China's supply security. The Sudan case did not threaten China's supply security, but broader national interests were undermined, and so the central government adjusted and hedged its Sudan policy to be more conducive to managing risks to China's diplomatic interests and prestige.

China sought to get firsthand reports from camps of internally displaced persons in Darfur and appointed Ambassador Liu Guijin as special envoy to Africa.[96] This work culminated in Khartoum's June 2007 decision to accept a combined UN and African Union (AU) peacekeeping force in Darfur. Liu reportedly stated that "Beijing had been using 'very direct language' as well as its 'own wisdom' to persuade Khartoum to accept the AU/UN hybrid force."[97] Undoubtedly, pressure from African countries and the West contributed to a shift in Beijing's approach to Sudan.[98] China was probably also becoming increasingly concerned about another civil war erupting between the Southern People's Liberation Movement (SPLM) and the National Congress Party (CNP) in Khartoum and about the conflict in Darfur spreading across the border into Chad and compromising Chinese oil and economic interests there.[99]

Another important factor was that China's status and prestige were at stake, and the Chinese government was not willing to undermine its international reputation by protecting limited energy interests in a discredited regime. In other words, the Chinese government was committed to not letting corporate interests determine national interests.[100] China has broader national interests that trump control over oil and commercial interests. But because most of the oil produced by China's NOCs is sold in the international market, one may ask: why should China take all the criticism and foreign policy pressure that the Sudan issue caused it in its very important relationships (the United States, EU, AU, and Japan), when a large percentage of the CNPC oil production in Sudan is not shipped back to China but is instead made available on the international market?[101]

By becoming too closely associated with the Sudanese government, China was taking on risks of opportunity losses. Being seen by the world community as supporting rogue states has a cost.[102] Indeed, China was facing diplomatic risks and undermining its international status by shielding Sudan from UNSC resolutions. Thus, China's diplomatic steps in 2007 indicate a willingness to develop a hedging policy by "buying" more "longs" (China becomes a responsible player in international affairs and cultivates a market approach to energy security) in order to offset "shorts" (the detrimental effects of a strategic approach to secure energy supplies). The predicaments facing China's Sudan and energy security policy reflected profit and security considerations and showed that relying on oil deals from rogue states or a short strategy is highly risky and may not be economically sustainable over the long term.[103]

At the same time, the strategic steps or the short strategy that compromised China's diplomatic standing and prestige, initiated to hedge its heavy reliance on the international energy market and improve China's global search for petroleum, secured for China a prominent role in finding a diplomatic solution to the Darfur crisis. China certainly has strong interests in the Sudanese oil sector, but China is of critical importance to Sudan, since about 65 percent of its exports go to China.[104] Thus, once it was in China's interest, and once the hedging strategy was adjusted from less reliance on commercial interests to more emphasis on broader national interests, as witnessed by the diplomatic and political steps taken in 2007, China developed a hedging policy whereby it could put pressure on the Sudanese government to accept a UN/AU hybrid peacekeeping force, thereby improving China's diplomatic standing and simultaneously maintaining existing ties with the Sudanese government and safeguarding strategic and corporate interests in Sudan. Through its diplomatic initiatives, which changed from being reactive and defensive to becoming proactive and offensive, China enhanced its diplomatic standing and reinforced the perception that it is a valuable diplomatic player in finding settlements to regional crises on a global level.

More recently, China's special envoy for African affairs, Liu Guijin, told reporters that China had "done a lot of work to persuade" the north to implement the peace agreement from 2005 and allow the referendum that was held in January 2011, in which the south voted by an overwhelming majority to split from the north.[105] South Sudan declared its independence from the north on

July 9, 2011—splitting most of the nation's oil reserves, located in the south, from most of the oil refineries and shipping facilities, located in the north. CNPC oil fields are located in the south and straddle the border between the north and the south—a border that has yet to be clearly defined. Clashes erupted between northern and southern forces in May 2011, and Khartoum seized the main town of Abyei in the border region.

Whether or not China put pressure on Khartoum is uncertain, but China is keen to protect its energy interests and its relationship with the newly established state of South Sudan. When the northern Sudanese leader Omar al-Bashir arrived in China on June 28, 2011, for the first time since 2006, it was only one day after the UN Security Council voted unanimously to send a 4,200-strong Ethiopian peacekeeping force to Abyei to ease tensions.[106] Despite concerns over China's role in Sudan, a senior U.S. official said in early July 2011 that China has "played a helpful role in encouraging Sudan to move forward with reconciliation with the south."[107] China has strong interests in both Sudan and South Sudan. After a conflict over oil revenues, South Sudan closed down its oil production in January 2012, a moved that reportedly deprived China of 260,000 bpd. Some of the worst military clashes since South Sudan's succession took place in May 2012. Because of its unique position, China managed to maintain close ties with both sides throughout the conflict and played a significant role in the mediation between the parties during peace talks in Addis Ababa on May 29, 2012. Undoubtedly, politics and economics—or security and profit—are inseparable in relation to China's aims in Sudan and South Sudan.[108]

The oil reserves in southern Sudan are landlocked and will depend on the infrastructure in the north for years to come, which suggests that cooperation may be facilitated between north and south. However, there are speculations that some Western companies or companies from India, Malaysia, Singapore, and Japan might build an oil pipeline running south to the port of Lamu in Kenya.[109] China will also face the dilemma over whether to participate in building such a pipeline. This may fuel competition between Chinese NOCs and compromise China's relations with northern Sudan, although Bashir has insisted that southern independence will not affect the relationship between Beijing and Khartoum.[110] Hedging is a likely policy outcome, as China continues mixing energy security and profit consider-ations with broader diplomatic interests with both Sudan and the interna-tional community.

A hedging strategy cannot eliminate all risks. The Chinese government can not prevent that China's oil-for-loans deals and financial support to petro-leum-producing states or the NOCs overseas investments are undermined by conflicts or nationalization in oil-rich countries. CNPC has suffered a resources cost from investing in trouble spots like Sudan and Libya. The cost of pursuing a hedging strategy that includes investments in rough regimes has often been conflict with other countries, and that undermines China's diplomacy. Nonetheless, China has gained recognition from not obstructing UN intervention in Sudan and Libya, despite its petroleum interests in these countries, which can be considered bonus profit from the short strategy of investing in overseas oil production.

IRAN: A STRATEGIC PARTNERSHIP?

Similar mechanisms can also be seen in China's approach to the Iranian nuclear issue. Iran is strategically located between the Caspian Sea and the Persian Gulf. It has the world's second-largest proven reserves of both con-ventional crude oil and natural gas and is the world's fourth-largest crude oil producer and the world's third-largest oil exporter.[111] Negotiations have taken place on the construction of a 620-mile pipeline in Iran to transport oil to the Caspian Sea, which can link up with the Sino-Kazakh pipeline to China.[112] Another option being explored is to build a pipeline from Iran to Pakistan and then to transport petroleum products by rail, road, pipeline, or ship to China.[113] If these plans are realized, China will have additional land routes for its energy supplies, thereby avoiding sea routes and the narrow straits of Hormuz and Malacca.[114]

There are good reasons to be skeptical about these "pipedreams."[115] Because of attacks on infrastructure and workers by insurgent and separatist groups in Pakistan's Baluchistan province and the unstable situation in the Kashmir region, it has been argued that the security risks will undermine any pipeline project. It is also estimated that a pipeline will be highly costly compared to supplying seaborne oil, as oil will have to be pumped from sea level up to the 15,400-foot-high Khunjerab Pass and then into China and farther to reach China's main market on the east coast.[116] A Silk Route of pipeline petroleum supplies from Iran to China, which coincides with an Iranian foreign policy that focuses on "developing an eastern orientation," may seem attractive to some decision makers and pundits in both China and Iran, but overcoming the economic, security, diplomatic, and strategic challenges will be difficult.[117]

Leaving aside the ambitious pipeline projects, Chinese NOCs have taken advantage of the window of opportunity presented by U.S. and EU sanctions against Iran.[118] U.S.-based companies are prevented from doing business with the Iranian government, and the United States has put strong pressure on companies in other countries to abide by the U.S. unilateral sanctions. Many European companies are concerned about their "reputational risk and potential loss of business opportunities in the U.S. market."[119] One report notes that over the course of the past decade, Japanese firms, under U.S. pressure, combined with Iran's onerous investment terms (the "payback" system), have divested significantly from Iran's oil and gas industry. Japan, which once had a 70 percent interest in the Azadegan oil field, reduced its interests to 10 percent. As they have pulled out, China has moved in, and according to the numbers examined in this book, Chinese investment in Iran probably stands at roughly $30–40 billion. In comparison to a number of Western oil companies, Chinese NOCs have "lower capital hurdles" and are able to obtain favorable financing terms through their ties with Chinese state-owned banks, which make it less difficult to overcome the Islamic Republic's so-called buyback system for foreign investment and its notorious reputation for renegotiating "final" agreements.[120]

Chinese state-owned companies and NOCs have invested heavily in Iran's petroleum sector, but it is difficult to verify what actual commitments and MOUs have led to investment. A much publicized MOU for a $70 billion oil and gas agreement with Iran was signed in 2004, whereby Sinopec will buy 250 million tons of LNG from Iran over the next thirty years and develop the giant Yadavaran field. After commissioning the Yadavaran field, Iran committed to export 150,000 b/d of crude oil to China for twenty-five years.[121]

A few years later, Sinopec effectively severed the link between the LNG MOU and the development of the Yadavaran field. Instead, Sinopec signed a $2 billion buyback agreement for Yadavaran in December 2007, with work starting in 2008 and with production expected to reach 185,000 bpd. So far, nothing has come out of the LNG MOU.[122]

Multibillion-dollar deals have been signed to develop fields such as North Azedegan through a buyback contract worth $3.5 to $4 billion and an MOU to invest $2.25 billion in a buyback contract to develop the South Azedegan field (one of the world's largest oil fields). There is a preliminary agreement to develop phase 2 of the South Pars field (ranked one of the world's largest gas fields), with CNPC slated to invest $4.7 billion, in addition to a tentative deal to invest $6.5 billion in Iran's refineries.[123]

As pointed out above, it is difficult to reconcile the conflicting figures related to Chinese NOC investments in Iran. One report drawing on Iranian sources claims that over one hundred Chinese state-owned companies operate in Iran and that between 2005 and 2010, Chinese firms signed $120 billion worth of contracts with the Iranian hydrocarbon sector.[124] Another report argues that although Chinese NOCs only have invested a small fraction of its commitments, Chinese state-owned oil companies have over the past few years signed memorandums committing to invest more than $100 billion in Iran's energy sector.[125] In 2010, Iran's deputy oil minister for international affairs, Hossein Noghrehkar Shirazi, reportedly said in an interview with Iran's semiofficial Mehr News Agency that "Chinese companies have invested about $29 billion in Iran's upstream oil and gas sector and another $10 billion in the country's downstream energy sector, including gas, petrochemical and refinery construction projects."[126] There are also large differences in the numbers used by researchers. The International Crisis Group estimates that investments amounted to roughly $19 billion *only* in 2009, whereas Bo Kong estimates that Chinese oil companies invested $10.763 billion in the Middle East and North Africa across the *entire* 1992–2009 period.[127]

Clearly, one ends up with different numbers when comparing Chinese state-owned companies' investment in Iran's hydrocarbon sector with more limited Chinese NOC investments in specific projects or fields during certain periods. The main source of these disparities stems from the fact that some numbers are based on MOU agreements without a binding contract or a finalized agreement.[128] As seen, the much-publicized investment in the Yadavaran oil field in 2004 never really took off. The project is still at the exploration stage, and the Chinese only committed a fraction of the originally agreed-upon amount. A large portion of the estimated value of the deal, allegedly worth $70 billion, was the cost of importing 25 million tons of LNG, but in December 2007 the deal was converted into a buyback agreement at an estimated cost of $2 billion. However, a purchase and sales agreement for the LNG was still not signed.[129]

Similarly, although the Mehr News Agency reported that CNPC had started development of phase 2 of the giant South Pars natural gas field, in accordance with a $4.7 billion contract signed in June 2009, Dow Jones Newswires reported in 2010 that CNPC is still studying the project and has not started drilling.[130] Although the combined estimated cost—$27 billion—of the five largest oil and natural gas development projects that Chinese NOCs are pursuing in Iran is close to the $29 billion that Iran's deputy oil minister reported in July 2010 that China had committed to Iran, many of the projects are either still being negotiated or are in the early stages of development. Accordingly, Iranian officials often exaggerate China's commitments and investments in Iran in an effort to show that sanctions and Western efforts to isolate Iran are not working.[131] Nevertheless, Chinese NOCs have made large investments and important commitments to make further investments in Iran's energy sector.

The limited competition thanks to sanctions against Iran also provides an opportunity for Chinese NOCs to ship gasoline to the isolated Iranians at what is expected to be a premium price.[132] Despite Iran's large petroleum reserves, it lacks a refinery capacity and relies on the international market to secure 40 percent of its domestic gasoline requirements.[133] Although the United States in the spring of 2010 unilaterally imposed sanctions against Iran and used its leverage to win support for another round of UN sanctions, it was reported that CNPC exported 600,000 barrels of gasoline to Iran worth $110 million, and Sinopec's trading company, Unipec, agreed to ship some 250,000 barrels to the country via a third party in Singapore.[134] Chinese NOCs were also shipping gasoline to Iran from terminals in the United Arab Emirates, and the statistics show that the NOCs were making profits, with Iran paying a 25 percent premium for its gasoline imports.[135] Accordingly, Chinese NOCs have filled some of the void left by the departure of the IOCs that were operating in Iran's energy sector and have taken advantage of the loopholes in the UN and unilateral U.S. and EU sanctions.

Finally, China is one of the largest buyers of Iranian oil, and Iran is an important source for Chinese oil supplies.[136] China is Iran's top oil client, and Iran was the third-biggest foreign source of crude oil to China in 2009, supplying 11.4 percent of China's total imports.[137] With sanctions hitting Iran hard in 2011 and 2012, China became an even more important customer, but Chinese crude imports from Iran fell almost 30 percent from a high in June 2011 to July 2012. The forecasts in August 2012 estimate that China's oil imports from Iran will pick up again by the late summer of 2012, with some reports suggesting that China's July imports will be close to 600,000 b/d. The fluctuation in oil imports from Iran suggests that security and profit considerations shape Chinese energy policy, and it is important to analyze these developments more closely.

IRAN AS CHINA'S HEDGE

Despite commitments to large investments in Iran's hydrocarbon sector, and notwithstanding that Iran offers Chinese NOCs significant business opportunities and that Iran has been a major supplier of Chinese crude oil imports, China has supported four rounds of UNSC sanctions against Iran. Resolution 1737 was adopted on December 23, 2006, after Iran's failure to suspend its nuclear program as demanded by Resolution 1696 of July 2006. In March 2007, the UNSC unanimously adopted Resolution 1747, furthering the sanctions on Iran and imposing a ban on arms sales. One year later, China voted for UNSC Resolution 1803, which approved a new round of sanctions against Iran for refusing to comply with resolutions 1696, 1737, and 1747.[138] Finally, after months of negotiations, new sanctions were imposed in June 2010 under UNSC Resolution 1929, which focused on the economic, military, and political activities carried out by the Islamic Revolutionary Guards Corps.[139] Although it has been suggested that such sanctions are an ineffective means of curbing Iran's nuclear ambitions, China joined the other five permanent members of the UNSC in approving the sanctions instead of siding with Brazil and Turkey, the two of the fifteen countries on the UNSC that voted against sanctions.[140]

China has supported UN sanctions under the condition that they exclude sanctions on Iran's energy sector. This shows, on the one hand, that China does not intend to take a strong stance for or against Iran and that China seeks to protect its commercial, diplomatic, and strategic ties to Iran. On the other hand, this also illustrates China's unwillingness to undermine its relations with the United States and other Arab countries by moving too close to Iran.[141] Overall, close ties with Iran will provide a strategic hedge for China if in the future turmoil or conflicts affect the availability and price of crude oil in the international petroleum market. Thus, China's policy toward Iran will seek to benefit from the commercial opportunities but also to preserve broader strategic and foreign policy goals, that is, finding a balance between China's interests in Iran's hydrocarbon sector and maintaining cooperation with the West and the Arab states.[142] Finally, China has benefited strategically and militarily from the U.S. preoccupation with the wars in Afghanistan and Iraq. As the United States increasingly focuses its attention toward the Pacific Ocean and on counterbalancing China in East Asia,[143] maintaining the Iran issue as a thorn in the side of the U.S. strategic outlook will probably be part of China's strategic calculations.

In assessing China's stand on the Iranian nuclear issue and NOC involvement in Iran, it is interesting to note that Iranian oil exports to China dropped significantly in the first half of 2010. Compared with the same period of 2009, China cut its imports of Iranian crude oil by 31.23 percent.[144] Some traders in the international petroleum market said the reductions had little to do with sanctions and attributed the drop in imports from Iran to relatively higher Iranian official selling prices compared with rival Middle East grades.[145]

Other commodity market analysts found this puzzling given that China's import needs and demand increased considerably in the first half of 2010.[146] Indeed, Iran was the only country among China's top ten suppliers that saw its exports to China fall in the first six months of 2010.[147] One important issue is the potential linkage between deliberations on new and tougher sanctions against Iran; the pressure on China from the United States, the EU, and a number of Arab states to support sanctions on Iran; and the significant drop in Iranian crude oil exports to China in the first half of 2010.

As the United States pushed to secure international support for harsh sanctions on Iran in the spring of 2010, China's commercial interests in Iran became subject to broader foreign policy priorities. Maintaining a good relationship with Washington is a top priority in Beijing. In early April 2010, it was reported that President Obama had called former president Hu Jintao regarding the sanctions. Even if the evidence is circumstantial, it has been speculated that Obama allowed the Chinese revaluation issue to slide in order to win greater cooperation on increased sanctions, thereby "deferring a decision on whether to accuse China of manipulating its currency."[148] In addition, the "warming trend" in the relationship was boosted by the announcement that President Hu would attend Obama's nuclear summit in Washington in April 2010.

Other countries supported U.S. pressures and signaled their concerns to Beijing about Iran's nuclear developments.[149] In particular, Saudi Arabia and the United Arab Emirates expressed willingness to leverage new sanctions on China.[150] It has therefore been argued that, in an attempt to decrease reliance on Iranian oil and secure Chinese agreement to sanctions, the United States has been encouraging key Arab states to boost oil exports to China.[151] A former U.S. decision maker and senior adviser with excellent knowledge of the Saudi position, developments in the Middle East, and China's policy-making apparatus has argued that "the Saudis need no encouragement for selling its oil to China." Because Saudi Arabia in December 2002 dropped its priority formula, because of its opposition to the war in Iraq, China had already become a priority market.[152] Therefore, market factors, not U.S. pressure, are seen as the key drivers behind Saudi Arabia's increased oil exports to China.[153]

This view is supported by commodity market analysts who have noted that trade flows suggest that Saudi Arabia might be "buying" market shares in the case of China. Saudi Arabia may have offered three inducements to gain China's vote at the United Nations. First, the Saudis may have allowed Chinese NOCs to invest in Saudi Arabia in order to cement economic and political ties. Second, China may have been offered price discounts on its oil imports.

Third, "there might be something afoot in the way of a long-term joint owner-ship of strategic stocks of Saudi-originated crude oil stored in China and kept 'off-books' by both countries."[154] With Saudi oil exports to China increasing to more than 1 mbd in 2009, a figure representing about 20 percent of China's total oil imports and nearly double the number of barrels imported daily by China in 2008, China's market for Saudi oil became larger than that of the United States.[155] Accordingly, whatever the nature of the deal between Riyadh and Beijing, with China becoming Saudi Arabia's largest costumer, the king-dom had its own incentives to gather China's support in isolating Iran.[156]

Commercial and market incentives have fueled ties between Saudi Arabia and China, but it is unlikely that China will abandon Iran as a source of crude oil supplies and instead choose to rely on Saudi oil. Indeed, the export/import numbers for 2011 show that Iranian crude exports to China have returned to earlier levels or more in the past months. Overall, the figures show that the Iran-China oil trade grew 50 percent from the first half of 2010 to the first half of 2011.[157] China is not going to give up on Iran either as an oil supplier or as a location for investment, but the significant drop in the first quarter of 2010 points to a pattern whereby China is hedging its bets and mixing profit and security considerations. It also suggests that the central government may have directed NOC activities in Iran. After the United States imposed uni-lateral sanctions on Iran, it was reported that the Chinese government infor-mally instructed firms to slow down, and according to one industry official with knowledge of China's overseas oil and gas activities, "the political pressure came directly from the government."[158] China has clearly restrained itself in dealing with Iran. Not only could China's NOCs profit more from picking up cheap oil from Iran as sanctions began to hurt Iranian export, but China could have used the opportunity to increase its SPR significantly. Instead, China has been scouring the world for crude to make up for the lost Iranian oil, and Saudi Arabia, Iraq, Oman, Angola, Venezuela, and Russia have seen significant rises in their exports to China.[159] China has also ensured that they have cut ship-ments sufficiently to abide by the U.S. sanctions and stay eligible for a waiver.

Chinese oil import from Iran was reportedly down almost 30 percent from July 2011 to June 2012, after it had picked up from the second half of 2010 to the first half of 2011. While political and diplomatic pressure and China's strategic and security interests again contribute in explaining this pattern, the primary factor in accounting for the drop in oil shipments in 2011–2012 is likely to be market factors and price disputes. Since the end of 2011, when U.S. and EU sanctions began to sting, Iran's crude export has probably fallen as much as 1 mbd. This has forced Iran to sell at below-market prices and provided China's NOCs with an opportunity to renegotiate contracts and extract new deals. Thus, China's attempt to squeeze the Iranians largely accounts for the drop in Iranian oil export to China in the first half of 2012. Simultaneously, China's NOCs operating in Iran, such as Sinopec, have been hedging their bets, wary about not breaching U.S. sanctions and undermining business opportunities and collaboration with U.S. energy companies. Strategic considerations by the Chinese government are also likely to play an important part as China seeks to strike a balance between maintaining its ties with Iran and the Arab states and finding an appropriate equilibrium between sustaining cordial relations with the United States and more aggressively opposing the U.S. "pivot" and rebalancing toward the Asia-Pacific region.

In the second half of 2012, it looks as if China's oil imports from Iran will reach record-high levels of close to 600,000 b/d, reflecting both the fact that this is a profitable trade for the NOCs and that the Chinese government is less concerned about accommodating Washington, as U.S.-China relations have become more confrontational. Without its hedging strategy, however, China would have struggled to boost its oil imports from Iran after July 1, 2012, when sanctions were tightened as the EU issued an insurance ban on Iranian oil export. This made it almost impossible to finance the deals with Iran through normal market mechanisms, find tankers, and arrange insurance coverage to ship the crude. Western countries had already banned its oil trade with Iran, but the steps taken by the EU forced Iran's remaining key customers in Asia, that is, oil and shipping companies in China, Japan, India, and South Korea, into talks with their governments to secure insurance coverage for tankers shipping Iranian oil. The Japanese government quickly decided to self-insure, the Indian government stalled but eventually agreed to provide insurance coverage for their own vessels, and South Korea more or less stopped importing, as its government was unwilling to arrange insurance coverage for their shipment.

The details are difficult to come by, and experts in the shipping industry in China have said that the insurance issue is "very sensitive," but the Chinese government has probably stepped in to provide some form of coverage and insurance, especially after the P&I Club Hong Kong stated that they would not provide insurance after July 1, 2012. It has been reported that Iran is using its own tankers to ship oil to its Asian clients, making Iran liable for the insurance costs. Iran, however, simply does not have the capacity to ship almost 600,000 bpd to China. The National Iranian Tanker Company (NITC) has forty-two tankers, of which twenty-eight are VLCCs. Because of the sanctions and the significant drop in Iranian oil exports, it is believed that around 60 percent of the NITC fleet is currently engaged in storage. With a sailing time of about sixteen days from Iran to China and some tankers probably still delivering oil to India and Japan (and potentially South Korea, as its refineries are in talks with Iran to resume oil imports by using Iranian tankers to circumvent EU sanctions), it is impossible for China to import all its oil from Iran on Iranian tankers. China has to be using its own tankers to import the amount of oil that traders are reporting. European insurers cover about 95 percent of the world tankers. If it is almost impossible to insure tankers carrying Iranian oil in the market,[160] then the Chinese government would have to step in. Thus, this situation illustrates how the Chinese state-owned tanker fleet can provide an important hedge against emergencies, crisis, conflicts, and embargoes.

The absence of major competitors in Iran presents a growth opportunity for Chinese NOCs, but there are obstacles. Even if the NOCs wanted to move full speed ahead, sanctions would still prevent them from getting the technology they need.[161] Chinese NOCs are most likely willing to continue taking on what one industry analyst describes as "political and commercial risk in Iran mainly as a strategic hedge, with limited expectations."[162] NOC hedging and exit strategies in Iran must also be seen in relation to the more recent growing ties between Chinese NOCs and U.S. energy companies,[163] and reports that China's NOCs have been given greater access to the U.S. energy sector in return for slowing down work on projects in Iran and for supporting UN sanctions on Iran.[164] Chinese NOCs' potential exposure to unilateral U.S. sanctions because of their presence in the U.S. energy sector may have prompted Chinese oil companies to hedge their business activities in Iran. It has been reported that Sinopec has turned down offers of bargaining over Iranian crude, and a senior Chinese oil executive has stated that ties with the United States are more important.[165] U.S. regulators will also be chasing foreign banks involved with Iranian transactions, and investigation is ongoing in the United States about PRC banks handling U.S. dollar remittances. If Chinese companies are "caught," there is the risk that the United States will impose sanctions on China. Thus, more than profit considerations are driving the NOCs, and more than a short strategy shapes the government's. The companies are concerned about their business opportunities if they try to go around the sanctions, and the Chinese government is concerned about spillover effects on U.S.-China relations from how the Chinese companies conduct themselves in Iran.

Commercial and strategic hedging has been part of China's "dual game"[166] or goal of balancing its relationships between the West, the Arab countries, and Iran. China's "tactical hedge," as some label it, gives both sides part but not all of what they want, and the benefits for China come from each side.[167] Moreover, China's "delay and weaken" strategy with regard to U.S. sanctions on the Iran issue also encompasses a hedging policy. By taking into account the preferences of the top members of the CCP, the NOCs, the central government (the State Council and the NDRC), the MFA, the People's Liberation Army (PLA), and other foreign policy actors, it is likely that the outcomes of decision-making processes at the highest levels balance various priorities, including commercial, strategic, and diplomatic interests, into a hedging policy.[168]

Thus, it is improbable that China will choose one side over the other or cut off its energy ties to Iran.[169] Reluctant to follow a policy that might maximize "profits" by putting its eggs all in one basket or by taking higher risks, the Chinese government and the Chinese NOCs both pursue hedging strategies that will allow them to remain reasonably well off.

With China's decision makers mindful that China's core interests very much depend on continued access to Western markets and avoiding long-term damage to China's national and corporate image, China has been reluctant to allow its arrangements to secure oil from rogue states compromise relationships with the United States, the EU, and other Middle Eastern countries, especially Saudi Arabia, all of whom are uneasy about Iranian intentions and nuclear ambitions. The recent U.S. pivot or rebalancing toward the Asia-Pacific region and stronger U.S. commitments to develop its political and military relationship with countries in Southeast Asia, especially Vietnam; its decision to boost its military ties with South Korea; and its intervention in the maritime disputes in South China Sea all may suggest that China will be less forthcoming on sensitive issues such as Iran and Syria. The primary cause for such behavior is not Chinese oil interests but a political and diplomatic response to perceived threats to China's national security interests. It remains uncertain whether China will act as a so-called responsible stakeholder and maintain close relations with the West and Arab states, but China aspires to maintain—and depends on—a stable global energy market.[170] So far, it seems that China has been cautious and determined to hedge its bets and maintain a balance so that the expected costs of a strategic and commercial partnership with Iran do not outweigh broader Chinese national and petroleum interests.

In conclusion, China has pursued a hedging strategy in its global search for petroleum. It has diversified sources, routes, and the mix of its petroleum supplies; has developed ties with "pariah" states; and balanced between strategic, commercial, and national interests to improve its security of supply. This explains China's involvement in Iran and Sudan. Moreover, China's diplomatic position on the Iranian nuclear issue reinforces the argument that China combines its overall hedging strategy to insure that there will be no petroleum supply disruptions with a hedging policy in specific cases to profit from both its commercial interests and from its maneuvering of China into a position whereby it can accommodate the United States, benefit from being viewed as having a more "responsible" role in the United Nations, and consolidate benign relations with the Arab countries and the international community.

5

SAFEGUARDING CHINA'S
SEABORNE PETROLEUM SUPPLIES

C HINA DOES NOT have a navy capable of protecting China's sea lines of communications (SLOCs), and it will not develop such sea power capabilities any time soon. At the same time, any blockade of China's vital SLOCs would likely lead to war. In a wartime situation, everything becomes a security issue, and under such conditions China would likely manage disruptions to its oil supply by relying on its domestic oil and energy production and by curtailing oil consumption that is not strategically important. Hence, the primary reason why China is hedging against oil disruptions through the buildup of a state-owned tanker fleet is in order to manage peacetime risk scenarios and conflicts in oil-rich regions that do not directly involve China.

This chapter has two main objectives. First, it will assess the importance of energy security as a driver behind China's naval ambitions. Second, it will examine how China insures against potential disruptions to maritime oil supplies through a hedging and risk management strategy. It is argued that China successfully hedges against insecurity in the availability, reliability, and affordability of adequate oil supplies not through the modernization of the PLA Navy (PLAN) but through the buildup of a state-owned tanker fleet, which, in a period of crisis short of a war involving China, should leave China relatively better off than other oil-importing countries.

It has often been maintained that energy security is an important factor behind China's naval modernization. Energy security shapes China's strategic outlook, including China's increasing attention to the protection of its vital SLOCs, but energy security cannot be viewed as the primary factor driving China's naval modernization program. Instead, as the first part of this chapter will show, changes in the strategic environment following the end of the Cold War allowed China to shift its orientation toward the sea, as the Taiwan issue preoccupied the PLAN. With the maintenance of stability and deterrence across the Taiwan Strait in recent years, the PLAN has been able to reorient its focus. Foremost have been the goals of protecting Chinese sovereignty and maritime rights, but other objectives include the goal of safeguarding China's oil supplies, trade routes, and SLOCs. Energy security has been largely presented as a "pseudo-national-interest argument" advocated and developed by China's naval nationalists to justify a carrier-based blue-water navy.[1] That is, many experts, military officers, and analysts seek a larger carrier-based navy to boost China's status and prestige, and they justify such ambitions in part under the banner of enhancing China's energy security.

Since a war between the United States and China remains unlikely, and since China will be unable to safeguard its seaborne oil supplies in such a scenario, a more fruitful analysis of China's energy security policy is to focus on contingencies other than a war involving China. Accordingly, instead of emphasizing rather unlikely wartime scenarios, a better explanation of China's insecurity regarding an adequate seaborne supply of oil is offered by looking at peacetime risks. The second part of this chapter divides these risks into two types: first, conflicts or wars in which China is not involved or is not a belligerent. For example, if the turmoil in the Middle East were to spread to Saudi Arabia or Iran, or if a war were to erupt between the United States and Iran, such a situation would most likely trigger rising oil prices affecting affordability, with a potentially negative impact on China's future economic growth. The physical availability and reliability of oil transportation would also be in peril if any other conflict were to erupt in this vital petroleum region. Shortages in supply would then fuel further competition for access to oil. In such a scenario, it is argued that China, based on its hedging and risk management strategy, might begin to ship its overseas oil production home on Chinese tankers, and it would self-insure its own tankers and direct the growing number of state-owned tankers into the conflict area or war-risk exclusion zone, thereby obtaining oil during a period of crisis at a price lower than that on the international market. Such steps will be possible because China is hedging its bets against risks today by gaining access to overseas equity crude oil production and by building a large state-owned tanker fleet to insure against future uncertainties and threats.

Second, nontraditional security challenges, such as piracy, terrorism, and environmental crises including hurricanes, typhoons, or other incidents, may limit supplies or block vital transportation routes. These risks can never be entirely eliminated and will create emergency scenarios that may even require the use of force. But despite these risks, nontraditional security challenges do not pose a severe threat of disruption to China's oil supplies, and it is unlikely that such risks would escalate into a conflict or war. Nevertheless, China has been engaged in cooperation in order to safeguard vital transportation routes at sea and at important choke points.

WAR OVER OIL AT SEA AND CHINA'S NAVAL AMBITIONS

With China becoming increasingly reliant on seaborne oil supplies to maintain its economic growth, China's economic vulnerability is increasing. The scenario most often put forward by alarmist analysts is that the United States and potentially its Asian allies "would likely move to cut off China's overseas 'oil lifeline'" during a conflict over Taiwan.[2] However, Sino-U.S. relations have been hostage to cross-strait developments for decades, and China's leaders and the Chinese military will not allow its wartime contingency planning regarding Taiwan to be affected or dependent on seaborne or overseas oil supplies.

Nevertheless, because anxiety over a perceived vulnerability of seaborne oil imports has become prominent in the literature, it is worth examining the issue more closely. The next three sections will consider this issue. The first section provides a brief historical summary of the drivers behind China's naval developments and ambitions, noting that energy security has not been a dominant factor. The second section discusses the contemporary linkage between China's oil import (in)security and China's search for sea power. It highlights that for the foreseeable future China is unlikely to secure its seaborne oil supply or SLOCs, in particular if the threat comes from the U.S. Navy. It is also pointed out that steps taken to develop a large navy that can challenge the United States for control of vital SLOCs, instead of enhancing China's energy security, are likely to be counterproductive to China's national security. The third section discusses whether China is vulnerable to a blockade of its seaborne oil imports. It is argued that this claim is largely exaggerated and that any blockade, short of a major war, will be ineffective. In short, there is little China can do to withstand a blockade during a war with the United States, but China has adequate domestic energy resources to wage war. In addition, China will not be vulnerable to a blockade during peacetime risk scenarios.

CHINESE NAVAL DEVELOPMENTS

With a strategic culture shaped by land wars and a continental orientation, the focus of the PLAN traditionally has been toward defending the People's Republic of China (PRC) from attack by sea instead of developing sea power to strike, invade, or conquer other countries.[3] Although China's naval modernization intensified in the 1980s, particularly in the aftermath of the Taiwan Strait crisis in 1995 and 1996, China's naval strategy since the establishment of the PRC in October 1949 has largely been defensive.

China's naval strategy has gradually developed from "near-coast defense" prior to the mid-1980s to "near-seas active defense" after the mid-1980s, and thereafter to the advancement of a "far-seas operations strategy."[4] During the "near-coast defense" period, the PLAN focused on supporting land operations in a major war with the Soviet Union and in counteramphibious landing operations from Taiwan-based Guomindang (KMT) forces.[5]

Real changes, both in strategy and in improved naval capabilities, did not occur until the mid-1980s, with the transformation to "near-seas active defense." A grand strategic shift followed after Deng Xiaoping boldly stated in the summer of 1985 that the threat of a major war with the Soviet Union

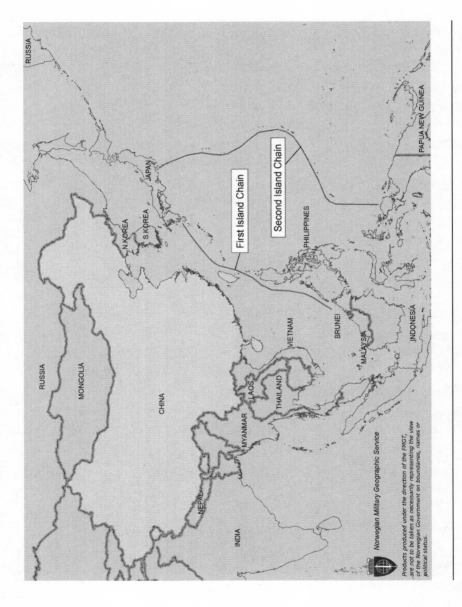

MAP 5.1 Island Chain Defensive Perimeters

had diminished. As a result of this decision, a strategic transition occurred whereby the military moved from a preoccupation with the continental frontier toward putting more emphasis on the seas.[6]

The PLAN benefited from this strategic transition, but it is difficult to estimate the influence of prominent figures in promoting this shift. Admiral Liu Huaqing, the PLAN's commander from 1982 to 1988, had a close relationship with Deng. However, Liu waited until Deng had proposed a new strategic direction before he promoted a new naval strategy.[7] Nonetheless, Liu became influential in implementing the change from the near-coast defense strategy to the near-seas active defense strategy, or what other experts call the new strategic concept of "offshore defense."

By proceeding from a coastal defense force to a more active offshore defense, the PLAN could draw on both its own combat experience and the lessons and strategies pursued during the Cold War by the Soviet Union, another major land power. Inspired by the thinking of the People's Liberation Army (PLA), which focused on the battlefield's interior and exterior lines, the PLAN envisioned operations within and around what has been called "island chains." The first island chain constituted the interior-line defense and offense, "a near and narrow realm where a relatively clear front would develop to define the engagements of the two sides." The space beyond the first island chain "was the far and vast realm for external-line operations."[8] However, it is important to emphasize that the seas are seamless and that operational capability is a more crucial factor than is drawing lines in the ocean.

China's distance-related island chain defensive perimeters paralleled the Soviet Union's "linear ground combat approach to thinking about maritime defense," which is often referred to as a "layered defense." By establishing layers of defense, the Soviet Union aimed to deny the use of the sea to its enemy, the United States.[9] Admiral Liu studied under Admiral Gorshkov at the Voroshilov Naval Academy in St. Petersburg from 1954 to 1958, but it is unclear how much Gorshkov and Soviet naval strategy influenced Liu's thinking.[10]

Liu's thinking suggests that PLA strategists had expanded upon the Soviet's layered defense and had incorporated Chinese "characteristics" to the approach favored by Soviet naval strategists. A more active defense, including exterior-line operations, differs from Gorshkov's thinking.[11] Still, the similarities between Soviet and PLA naval strategy can be attributed to their continental strategic culture, Soviet mentorship, and the operational challenge of "defending the homeland against a force approaching from the sea."[12]

"Antiaccess" or "access denial" is a central goal for the "near-seas active defense strategy."[13] Surveillance, ballistic missiles, naval aviation, submarines, and surface ships are important elements to deny the waters between two hundred and one thousand nautical miles from China's shore to the U.S. Navy and to deter the United States from intervening in a conflict over Taiwan. Failing such missions, an access-denial strategy would delay the arrival or reduce the effectiveness of intervening U.S. naval and air forces.[14] China has put a strong focus on its space-based program, which is critical to open-ocean surveillance, navigation, and targeting. Its land-based ballistic and cruise missiles pose a formidable threat to any enemy operating in China's near seas, and China is working on using short-range ballistic missiles (SRBMs), medium-range ballistic missiles (MRBMs), land-attack cruise missiles (LACMs), and antiship cruise missiles (ASCMs) to attack moving surface ships.[15]

The PLAN and People's Liberation Army Air Force (PLAAF) have traditionally been a weak branch in China's military forces, but progress is now being made. A RAND report on the military balance in the Taiwan Strait concluded in 2009 that China now has the capability to achieve and sustain air superiority over the strait, but this is largely thanks to the role of land-based aircraft and missiles.[16] A force of approximately sixty submarines has been complemented by a modernizing surface force with four new domestically built classes of surface combatants with anti-surface-ship capability and advanced attack weapon (AAW) and antisubmarine weapon (ASW) systems (three destroyers and one frigate).[17] China has also purchased four Sovremenny-class destroyers from Russia designed specifically to target U.S. aircraft carriers. Amphibious warfare capability has expanded, and although experts disagree as to how much attention China has paid to mine warfare, its ability to place and clear minefields has improved.[18]

In short, it can be concluded that "the PLAN in 2010 is developing into a maritime force of twenty-first-century credibility in all warfare areas—even if marginal in AAW, ASW, and force integration."[19] In addition, an emphasis on hardware and capabilities should not overlook the people operating the new ships, submarines, missiles, and airplanes. Open-source reporting and the performance over two years of China's naval task groups operating in the Gulf of Aden (see below) demonstrate the substantial progress the PLAN has made in personnel training. However, China's ambition to develop twenty-first-century sea power will depend on the navy's ability to attract talented, intelligent, and dedicated personnel.

China became a net importer of oil in 1993, suggesting that energy security has not been a major factor in shaping the "near-coast active defense" strategy. The Taiwan issue has been on the top of the agenda and a key driving force behind China's naval modernization, and a core objective of the antiaccess strategy has been to develop the capability to keep U.S. forces from getting close enough to attack the Chinese mainland or from interfering in a PLA attack on Taiwan.[20]

Additional objectives, such as protecting maritime sovereignty claims, patrolling SLOCs in the near seas, conducting strategic nuclear deterrence, and asserting China's status as a major world power, have also been operational goals of the "near-seas active defense" strategy.[21] Taiwan and economic imperatives provide "dual incentives for modernization" of the PLAN.[22] China has become more trade dependent with respect to energy, which increasingly presents the PLAN with a core maritime mission to safeguard the SLOCs. The booming Chinese economy is largely concentrated in the coastal regions. This has been a "spur to naval expansion and modernization."[23]

China has accomplished most of the major objectives set out in its "near-seas active defense" strategy. The PLAN has achieved a "limited deterrent" and maintains sufficient forces for access-denial capability. This means that the United States must act more cautiously close to China's shore or operate at a greater distance from China's coast. In comparison with the Taiwan Strait crisis in 1995 and 1996, China's military modernization has now complicated the operations of the U.S. Navy, especially in a Taiwan contingency.[24] Indeed, Chinese military and political leaders are becoming more confident about China's ability to deter and potentially to succeed in a military conflict over Taiwan, in addition to safeguarding maritime interests in the near seas.

Therefore, Chinese strategy is looking toward "far-seas operations," which Chinese officials believe will not conflict with the goal of forming "a more solid basis for resolving the Taiwan issue" and maritime territorial disputes.[25] This new naval orientation, which will lay the basis for far-seas operations or a blue-water navy by 2020, has been fueled by the argument that China needs to protect its oil supplies and maritime trade routes.[26]

CHINA'S OIL SECURITY AND NAVAL AMBITIONS

Energy security is undoubtedly influencing China's diplomatic and strategic calculations, but the uncertainty of China's future naval ambitions makes it difficult to assess the extent to which energy security is driving China's expanding naval power. Chinese leaders have expressed concern about energy security, and former president Hu Jintao stressed the issue when he addressed the G8 meeting in 2006, declaring that "global energy security has a bearing on the economic lifelines and the people's livelihood of various countries and is of utmost importance to maintaining world peace and stability and promoting common development of various countries."[27]

Former premier Wen, heading the State Energy Leading Group, argued in 2005 that "energy is an important strategic issue concerning China's economic growth, social stability, and national security."[28] The 2006 *China's Defense White Paper* identifies access to raw materials and various other issues upon which economic development depends as major national security concerns. It states that "security issues related to energy, resources, finance, information, and international shipping routes are mounting."[29] The 2008 *Defense White Paper* adds that "struggles for strategic resources, strategic location and strategic dominance have intensified," and the 2010 *Defense White Paper* notes that "non-traditional security concerns, such as existing terrorism threats, energy, resources, finance, information, and natural disasters, are on the rise." However, the papers do not directly address oil (in)security. Indeed, energy and oil security is only briefly mentioned in the 2006, 2008, and 2010 *Defense White Papers*.[30]

The question of reliable and safe imports of oil was allegedly elevated to a national security interest in late 2003, when former president Hu reportedly voiced concern about the security of delivery of energy imports to China, referring to the passage of about 80 percent of China's oil imports through the Strait of Malacca.[31] The state-run media soon dubbed the issue China's "Malacca dilemma," and the *New China News Agency* (*Xinhua*) reported that Hu advocated a revised strategy to deal with the possibility that "some big countries" or "certain major powers" were attempting to control the Malacca Strait.[32]

Speaking to representatives of the navy on December 27, 2006, the *People's Daily* reported that Hu had "called for the building of a strong and modern navy force . . . of vital importance in defending state interests and safeguarding national sovereignty and security."[33] Hu stressed that China should continue moving toward "blue-water" capabilities, but energy security was not mentioned directly in Hu's guidelines, raising the question of whether an oil security mission for the PLAN is specifically sanctioned by China's leadership.[34]

There is clearly a debate occurring in China over China's future oil security strategy. Alarmists and "mercantilists" believe that China must be capable of controlling its foreign oil supply through military means in a zero-sum competition over energy resources. Conversely, "free marketers" and pragmatists favor market mechanisms, consider mutual interdependence to facilitate common and absolute goals, and stress that a large-scale military and naval buildup would undermine China's successful peaceful rise.

The alarmists have taken advantage of Hu's alleged concern about the "Malacca dilemma" to promote the buildup of China's naval power. Since roughly 80 percent of China's seaborne crude oil imports arrive via the Malacca Strait, it is argued that "it is no exaggeration to say that whoever controls the Strait of Malacca will also have a stranglehold on the energy route of China. Excessive reliance on the strait has introduced a serious potential threat to China's energy security."[35] The alarmist writers consider that the narrow Malacca Strait would be "easy to blockade," becoming "the strategic throat of China's energy and economic security." Accordingly, "whoever controls the Strait of Malacca . . . effectively grips China's strategic energy passage, and can threaten China's energy security."[36] Other Chinese voices within the alarmist camp call for China to focus its attention on sea power to protect its "resource security," arguing that China's "fate is connected to naval modernization." "Without an aircraft carrier, an interruption of natural resource imports would plunge China's economy into a crisis, blockading the rise of China."[37]

A number of international observers contend that the alarmist voices shape China naval ambitions and that "commerce and energy . . . compel Beijing to cast anxious eyes on [SLOCs] as China will focus on a Mahanian-driven bid for 'command of the commons.'" As a result, it is maintained that "the security of energy supplies transiting critical sea lanes has become a top foreign-policy priority for China."[38] Anxiety over safeguarding the seaborne oil flow to China has led Beijing to build "powerful naval forces that can secure the nation's sea lanes of communications stretching across the Pacific and Indian Oceans."[39] Hence, with imported oil accounting for more than half of China's current oil consumption, protecting oil supplies and securing the sea lanes for a commerce-driven economy is seen by many as an important task in the "far-sea operational" strategy and a driving force behind China's naval ambitions: "China's emergence on the global stage as an economic power and as a net importer of oil has had a significant impact on China's maritime strategy. In order to protect oil and other trade routes, the PLA(N) is beginning to develop the foundations of a naval capability that can defend sea lines of communications (SLOCs)."[40]

Another author notes that the increasing need to import oil for China's growing economy "highlights the issue of energy transportation safety and threats to its sea lanes of communications (SLOCs), prompting Beijing to invest more in national defense, especially naval building."[41] Some envisage China's future energy supply as "overly dependent on the sea lanes" and argue that the fact that the "U.S. might cut them off as a result of the deterioration of Sino-U.S. relations over the Taiwan issue drives much of Beijing's modernization of its navy and air force."[42]

Pragmatists and free marketers in China challenge the alarmists' propositions and convictions by pointing out that China's seaborne oil supply represents less than 10 percent of its total energy usage; that China's ability to secure and control the SLOCs will remain limited; and that, instead of benefiting Chinese security and energy interests, a carrier-based blue-water navy may undermine China's national interests.[43] China's domestic sources of coal (68 percent), oil (10 percent), gas (4 percent), nuclear energy (1 percent), and hydropower (6 percent), in addition to alternative energy sources such as wind, solar power, and biomass (1 percent), supply China with about 90 percent of its energy requirements. Two researchers at the Energy Research Institute of the National Development and Reform Commission in Beijing, China's leading governmental energy body, note that "China depends, to a remarkably low level, on overseas energy resources with net imports taking only 8.2 percent of the total energy consumption."[44] Rather than arguing that China's energy security fate is connected to naval modernization, these researchers maintain that it is crucial

> for China to invest more capital and human resources in the research and development of energy technologies and improve its independent inno-vation capability, especially to encourage the technologies of clean coal, nuclear power, new energy automobiles, renewable energy resources and important energy conservation technology, strive to make breakthroughs in key technologies and therefore ensure China's energy security.[45]

Energy savings and efficiency have also been highlighted by President Hu, the Politburo of the Central Committee of the Chinese Communist Party (CCP), and other influential governmental bodies. Hu reportedly said "the Party and society must realize the importance and urgency of energy effi-ciency. . . . Saving resources and energy are crucial for China in pursuit of sustainable development and economic and national security. It requires more attention and effort."[46] "Free marketers" such as Zha Daojiong at Peking Uni-versity and Zhao Hongtu at the China Institute of Contemporary Interna-tional Relations (CICIR) are not only strong supporters of market mecha-nisms and interdependence but also share the belief that the domestic aspect remains the crucial element and will be the key issue in determining China's future energy security policy.[47]

Energy security is not the primary factor driving China's naval moderniza-tion, notes Zhao. There are "no signs that expanding the navy force to safe-guard the SLOC has already been promoted as one part of China's national energy strategy or policy."[48] There is also little evidence to suggest that China will be capable of protecting its seaborne energy supplies in the face of U.S. naval dominance during a time of war. Not only have Chinese military and naval officers stressed in interviews with the author that the PLAN will remain inferior to the U.S. Navy, but Western professionals also emphasize this point. As one analyst closely following the PLAN modernization has argued, the "only country that could seriously disrupt merchant traffic destined to or from China is the United States," and according to this observer, it is unclear what the PLAN could do about it. "It will be many years before the PLAN is able to operate surface ships independently at sea in the face of a hostile United States."[49] Another leading U.S. expert on the U.S. and Chinese navies concludes that "the security of such global SLOCs demands international cooperation because the PLAN does not possess the forces to defend other than coastal SLOCs."[50]

Pragmatists and free marketers in China therefore both emphasize that China should seek to cooperate with other naval powers to protect the SLOCs because such a strategy will best secure China's seaborne oil supplies. Zhao Hongtu contends that multilateral cooperation is the most effective way of dealing with a complex maritime environment that faces a number of risks.[51] It is maintained that the forces of globalization and economic interdependence are pushing China toward a focus on common security challenges in the maritime domain. Thus, energy security should be framed within the realms of "geo-economics" and "economic threats and market solutions," writes Zha Dajiong. "Economic interdependence . . . serves as perhaps the single most powerful deterrent against embargo or blockade by China's neighbors." In short, argues Zha, "China has no choice but to learn how to live in a world of (complex) interdependence."[52]

Some of the researchers who see market mechanisms and cooperation as the keys to China's energy security also warn that a more aggressive Chinese naval posture and an ambition to control China's SLOCs may be counterproductive to China's successful peaceful rise strategy.[53] Zha has noted that the "primary factor in maintaining the stability-based security China has enjoyed for the past three decades is China itself."[54] A naval buildup will increase tensions and potentially fuel a regional arms race, creating less security for China. Although most states in East Asia view China's rise positively and are reluctant to see China as a threat, China's more recent assertive diplomatic stand and military posture have led some regional countries to balance against China's power by increasing their military buildup and developing closer ties with the United States.[55]

Instead of embarking on a strategy to build an ocean-going navy that might undermine China's energy security and overall foreign policy objectives, resources can be allocated differently. For example, China's national interests, maritime rights, and energy security can be enhanced by combining an access-denial strategy with a limited power-projection capability, which can be used to safeguard the global commons,[56] and efforts to improve China's domestic energy supply infrastructure, manage demand, utilize energy efficiency, and develop alternative energy sources, which are at least as important as securing the SLOCs and foreign supplies.[57] China's ties with a number of countries around the Indian Ocean, what some refer to as the "string of pearls" strategy, often consolidated through arms sales and political, economic, and diplomatic support, will not "neutralize the SLOC defense challenge" or "insulate China from an energy-related challenge."[58] More importantly, the evidence simply does not support a "string of pearls" strategic ambition by Beijing.[59]

It has been convincingly demonstrated that the linkage between oil security and naval power is largely grounded in a pseudo-national-interest argument. Naval nationalism—"the demand for great-power status and domestic legitimacy"—rather than security drives China's naval ambitions and shapes evaluation of the policy-making process on the need to improve capabilities and on the perception of China's national interests.[60] The alarmists' intermingling of oil security and national interest serve naval nationalism by providing justification for the PLAN budgets, modernization plans, and procurement priority.[61] Growing nationalism—and nationalists claiming that expanded naval power-projection capability is essential to protecting oil imports and maritime sovereignty rights—has pressured the Chinese leadership to construct a power-projection navy centered on an aircraft carrier.[62] But the Chinese nationalists and Western analysts rationalizing China's aircraft carrier ambitions have not demonstrated that Chinese energy security interests require naval power projection or that a carrier-based navy can make China more secure and safeguard its seaborne energy supply.[63]

Chinese "alarmists" are skeptical about U.S. dominance of the sea lanes and the potential U.S. ability to interfere with or blockade China's oil supply in times of hostilities, in addition to voicing concerns about uncertainties in the international oil market. But in challenging U.S. maritime supremacy and attempting to preempt Chinese vulnerability to a seaborne energy supply, China needs to expand its naval power. As one analyst puts it, "China should acquire naval facilities and expand its naval power in the Indian Ocean and into the Persian Gulf, where it would face no resistance."[64] As previously pointed out, it is only the U.S. Navy that can threaten China's energy supply. The advocates of expanded Chinese naval power do show how expanded naval power will neutralize the U.S threat to China's SLOCs or discuss whether China can develop such a capability, especially if the United States continues to develop its maritime capabilities. Thus, China's naval nationalists merely assert that development of a blue-water navy will make China more secure, without critical examination of either the necessity or feasibility of such a project.[65]

In short, China's naval nationalist assertions are characterized by grand expectations of China's ability to develop naval power to protect Chinese interests, extreme assessments of the threats to those interests, and expansive definitions of Chinese global security interests.[66] In addition, although "naval nationalists" and alarmist writers maintain that the possession of an aircraft carrier would help secure sea lanes vital to China's oil imports and China's future rise, they often ignore the fact that the buildup of Chinese naval power-projection capabilities, in particular an aircraft carrier, might negatively affect China's security environment.

WHY FEAR A BLOCKADE?

After the U.S.–led blockade and embargo of China during the 1950–1971 period interfered with oil shipments to China, there have been few attempts to disrupt oil and other goods flowing to China, with the exception of the *Yinhe* cargo ship incident in 1993.[67] The likelihood that U.S.–China relations will return to the level of the Cold War hostilities of the 1950s and 1960s is remote.

Based on estimates of the naval forces required to blockade tankers passing through the Strait of Malacca, only large navies, probably only that of the United States, would have a sufficient capability to enforce a blockade of the nine-hundred-kilometer waterway that borders Malaysia, Singapore, Indonesia, and Thailand.[68] Even for the mighty U.S. Navy, enforcing a blockade of China's oil shipments without triggering a large-scale war may not be as straightforward as many commentators, analysts, and observers seem to think. A limited blockade, a distant blockade, a close blockade, or precision strikes on China's oil infrastructure and installations all would be difficult to implement without escalating into a major conflict with China.

Some of the worries of Chinese leaders and military pundits can be mitigated by looking at the map. Clearly, there are several ways to circumnavigate the Strait of Malacca if it were closed.[69] Even if a power managed to blockade the alternative routes through the Lombok, Sunda, Makassar, and Mindoro Straits, tankers could circumnavigate Australia and approach East Asia via the Pacific. It is estimated that, depending on whether tankers are rerouted through the Lombok Strait or all the way around Australia, sidestepping a blockade of the Malacca Strait would result in four to sixteen days of disrupted oil shipments to East Asian consumers. This would drive up shipping costs by about $2 a barrel but would also make it more complicated to achieve the goals of a distant blockade and would necessitate additional forces.[70] The key point, then, "is that the oceans of the world are seamless, and stopping traffic once . . . operating on the high seas is very difficult."[71]

More importantly, short of a major war, implementation of blockades of vital choke points such as the Strait of Malacca would face immense challenges. In today's oil market, it is very difficult to know precisely where oil shipped from Africa or from the Middle East bound for East Asia will end up. Price and speculation often determine the destination of oil cargo. Normally, the oil on tankers shifts hands dozens of times while at sea. The flexibility embedded in today's oil trade complicates matters and makes it more difficult to execute a distant blockade. A tanker's destination can shift after inspection, the bill of lading can be manipulated, shipping documents can be forged, and the oil cargo can be "parceled out," with one tanker carrying oil bound for several consumers.[72]

Since China has largely been relying on third-party operators, it would be difficult to isolate tankers bound for China from those bound for other Asian and East Asian states. Consequently, China may successfully be able to "hide" in the market, as any blockading power would have to block, for instance, all fifty or so of the oil tankers that pass through the Strait of Malacca daily.[73]

In addition, since the United States might be unable to identify tankers bound for China, it might well end up interrupting oil supplies to some of its closest allies in the region, thereby potentially undermining diplomatic support for the crisis that precipitated the blockade in the first place. China has become the largest trading partner with most states in Asia as well as with all the major powers, with the exception of the United States. Because China maintains a vital role in the international trading system, any power intending to impose a blockade will immediately face pressure to end the blockade rapidly. Even were the blockade conducted with little harm to third-party countries, which is inconceivable, pressure to end the blockade would rise steadily as economic damage mounted. Historically, blockades have taken months or even years to register their full effect. In the meantime, the blockader risks alienating its allies and undermining its international status and prestige.[74]

China, of course, would not be a passive observer of any attempts to blockade or interrupt its energy supplies. Despite the lack of military power-projection capabilities to counter a distant blockade by the U.S. Navy, China possesses growing diplomatic, economic, and military means to hamper or deter any attempts to blockade its energy imports. Its diplomatic clout is significant, and unless China commits wrongdoings on such a scale as to unite the international community against it (with a conflict over Taiwan hardly being sufficient), alliances with states sharing its energy and economic interests cannot be ruled out. Such steps will undermine and raise the costs for the blockader.[75]

The role of the littoral states will also complicate any control or blockade of choke points. How would Indonesia, Malaysia, and Singapore react to an outside power that breaches its sovereignty and territorial integrity to enforce a blockade of the vital straits? Will the littoral states act in concert with the United States against China, or will the blockading power need to dominate them to control navigation in the strait? Indeed, one of the best ways for

China to protect its SLOCs and maritime trade "is to maintain good diplomatic relations with trading partners and littoral states adjacent to them."[76]

Tankers and crews that refuse inspections will have to be boarded by marines and be escorted, which operationally will be very challenging given the number of ships that pass through the vital choke points on any given day. Few marines have the necessary knowledge to operate and take oil tankers to a marshaling area. Then, what should the blockading power do about seized tankers, the cargo, and their crews? Selecting an area would be problematic if the littoral states refuse to cooperate. Even if the littoral states openly assist the blockading power, there are few harbors deep enough to allow the entry of VLCCs. In addition, both the crews and cargo are often multinational. Ship owners and owners of the cargo can be expected to protest to their respective governments, which presumably would put pressure on the blockading state(s) to release the ships. This, in turn, might undermine diplomatic support for the blockade, effectively leading to a situation whereby blockading states will have to seize the same ship more than once.[77]

Finally, short of a major great-power war, it is unlikely that the blockading states could sink uncooperative tankers that refused to be stopped and inspected. The environmental destruction, the diplomatic repercussions resulting from endangering the lives of the civilian crew, and the high value of the cargo all represent risks that may deter any blockading power if the stakes are not sufficiently high.[78]

Other alternatives such as a close blockade, a blockade by convoy, a supply-side blockade, or precision strikes against key Chinese energy installations and infrastructure would be dangerously escalatory or would exhaust resources and thus be ineffective or unfeasible.[79] Since most experts estimate that the PLAN is capable of an offshore defense to protect the littoral SLOCs and coastal areas, it would be an act of recklessness on the part of the blockading power to place warships close to the Chinese coast without defending these forces by striking military installations and targets on the Chinese mainland, at sea, and in the air. Such an engagement with Chinese forces may trigger the outbreak of a major war, undermining the initial limited objective of the blockade.

To ensure compliance with an energy embargo against China, a blockade by convoy might be initiated in order to escort tankers bound for neutral and friendly Asian states. Excluding Taiwan and the Philippines, roughly 10 mbd of oil, or five VLCCs (carrying roughly two million barrels of oil each), per day is required for an operation to supply Japan and Korea. A round-trip sailing time of twenty days between Singapore and Japan and Korea for each group of five VLCCs can be expected. With a two-day turnaround, twenty-two separate convoy groups with escorts, replenishments, and maintenance ships will be needed, while simultaneously also enforcing and upholding the blockade against China. In short, the logistical requirements to supply U.S. allies and East Asia's largest non-Chinese consumers "would overwhelm most or even all navies" and would require "the active cooperation of neighbouring states."[80]

A potential supply-side blockade would be an even more daunting task, mainly since China receives most of its oil from a flexible international oil market and because contemplating using force against noncooperative oil suppliers would broaden the conflict considerably. Historically, embargoed states will receive oil even from embargoing states if they are able to pay, as demonstrated by the 1973 Arab oil embargo. Hence, it can be concluded that a supply-side blockade would be neither effective nor feasible.[81]

A blockade is often initiated because it relies on a limited use of force that can be withdrawn quickly, with little permanent damage done. Striking energy infrastructure on the Chinese mainland to make an energy blockade more effective is essentially "antithetical to the purpose of a naval blockade."[82] Precision strikes on Chinese energy facilities and territory would force Chinese leaders to consider any necessary means, including the use of nuclear weapons, to protect China. Much more than oil and energy security must be at stake before a blockading power would contemplate putting Chinese leaders under such extreme escalatory pressures. Judging from China's recent buildup of its strategic petroleum reserves, located near China's coastline and mainly constructed as aboveground tank farms (see the map in chapter 3), it seems that China does not regard the destruction of its energy infrastructure as a key vulnerability.

Last but not least, many alarmist writers seem to forget that China currently relies on seaborne oil imports for about 10 percent of its total energy needs. Petroleum can be transported to China overland through pipelines or by rail or truck. In 2011, China also produced just under 50 percent of its oil needs domestically, with domestic production standing at roughly 4.1 mbd. Accordingly, with a hedging strategy that emphasizes alternative supply routes, SPRs, a favorable energy mix from a security-of-supply perspective, and domestic oil production that is more than ten times greater than what the United States needs to fight wars and operate militarily on a global scale, it is clear that it is not necessary for the PLA to conduct military operations on account of oil.

In short, a distant blockade would not prevent supplies of oil from reaching China; a close blockade would be dangerously escalatory and potentially would undermine the limited objectives of the blockade; a blockade by convoy would logistically and operationally overwhelm the blockading state, including the powerful U.S. Navy; and precision strikes on Chinese oil installations and infrastructure would be unthinkable short of a total war. In addition, seaborne oil imports constitute only about 10 percent of China's total energy mix. This suggests that China would be able to withstand a blockade that normally would take years to be effective anyway. It is therefore difficult to imagine a limited-war scenario or a crisis that would justify such actions by any blockading nation, and "contrary to what appears to pass for conventional wisdom among naval analysts and observers in the PRC, China is not fundamentally vulnerable to a maritime energy blockade in circumstances other than global war."[83]

An indication that blockades and wartime contingencies are not a dominant focus of Chinese energy security policy can be observed by examining China's ambition to develop a state-owned tanker fleet that can transport more than 50 percent of China's oil imports. To many experts, this seems to undermine Chinese interests: a large state-owned tanker fleet would make it "easier for the United States to determine which ships are carrying oil to Chinese ports."[84] Then why does the Chinese government support the development of such a fleet? Below, I will argue that China's ambition to develop a large state-owned tanker fleet is an important hedging strategy to manage the risks of disruptions to its oil supplies during a potential conflict or crisis in which China is not directly involved, for instance, if another conflict were to erupt in the Middle East.

PEACETIME RISKS

In 2007, only about 10 to 20 percent of China's oil shipments were imported on tankers flying a Chinese flag.[85] Thus, China could hide in the international market, thereby complicating any peacetime attempts to blockade oil shipments. However, in its five-year plans China has a stated ambition to develop its own large state-owned tanker fleet. Expanding China's shipbuilding capacity and developing its own tanker fleet makes little sense from a purely economic perspective, since there is an overcapacity in the tanker market, and the shipbuilding costs in China are estimated to be higher than in Korea and Japan because of lagging technology. But a state-owned tanker fleet and commercial shipbuilding, which can support warship construction, are considered strategically important and create military, diplomatic, and economic spillover effects. The Chinese government has used subsidies to offset the costs of maintaining a less-competitive shipbuilding industry and in order to prevent bankruptcy among state-owned tankers operating in a bearish tanker market, similar to how the Chinese government has promoted less-cost-effective cross-border pipeline projects over relying on seaborne energy supplies.

The percentage of seaborne oil imports carried by Chinese-flagged tankers was set to rise to 50 percent in 2010 and 75 percent by 2020—and these are percentages of ever-increasing absolute quantities of oil.[86] The strategy thus makes China's seaborne oil supply potentially more vulnerable, since the PLAN lacks the sea power and military capabilities to convoy or escort Chinese tankers during a blockade safely. Nonetheless, wartime contingencies and blockades may not be the main concern in China's energy security strategy and policy making.

It is difficult to imagine any state interrupting oil supplies to China unless a conflict or war was imminent. China possesses enough comprehensive military and economic power to deter most states from waging war against it, thereby giving leaders in Beijing the confidence to focus primarily on peacetime risks. As emphasized above, China is not as vulnerable to a blockade, short of a major war, as many analysts would assume. Instead, from the perspective of strategic, commercial, legal, and military factors, hedging and risk management considerations drive this seemingly contradictory objective of developing a state-owned tanker fleet.

HEDGING WITH A STATE-OWNED TANKER FLEET

A Chinese flagged and state-owned tanker fleet can be a strategic advantage in a potential conflict in which China is *not* involved, for example, if there were a conflict between the United States and Iran. A hedging strategy is developed primarily to deal with peacetime contingencies and risks. Conversely, states seldom hedge when confronted with a direct threat to their core national interests.

Any future conflict in the Middle East will most likely result in rising oil prices. Under the CCP's autocratic leadership, it can be assumed that Chinese leaders could pressure Chinese state-owned shipping companies and even Chinese ship owners to cooperate with the Chinese national oil companies (NOCs) to ship Chinese-produced overseas oil back home and to operate in a war exclusion zone in order to access terminals during a time of crisis. To date, such emergency measures have not been taken, not even during the Iraq war in 2002 and 2003.

However, China did not posses a large state-owned tanker fleet in 2003, and China is now much more dependent on imported oil, and will be even more so in the future, than it was a decade ago. Concern about energy security is currently a top priority among Chinese leaders. Awareness of the issue of energy security was largely the result of events since 9/11, the U.S. military presence in Central Asia, and the wars in Afghanistan and Iraq.

Thus, if China's oil supply were to be endangered because of a war or a crisis, it is possible that the government would instruct its tanker fleet and NOCs to provide oil to the homeland. As noted throughout this book, there are many examples whereby the government has instructed the state-owned companies to operate strategically instead of pursuing their own commercial interests, whether concerning the development of an SPR, domestic ex-refinery prices, or the construction of cross-border pipelines, and not to allow corporate interests to undermine national interests in sensitive diplomatic cases such as Iran and Sudan.

During the electricity shortage and rising coal prices in January 2008, the Chinese government used its authority to ensure adequate energy supplies. Described as China's worst-ever power shortage, the state media reported that the government "put in place a two-month ban on coal exports." The Ministry of Communications declared that "all thermal coal exports will be suspended. Where there is a need, *all international shipping capacity will be diverted for domestic transportation requirements.*"[87]

Consequently, instead of selling its oil on the international market for a profit during a period of war in the oil-rich regions, and instead of ship owners favoring foreign clients for profit in a tight tanker market, Chinese NOCs might be forced to ship their oil produced overseas back home on Chinese tankers. As a result, despite disruptions to the physical availability of oil in the conflict area, the Chinese government can "secure" its oil supplies during a time of crisis at a lower price and with lower shipping costs than if China were to import all its oil through the international market during any crisis period. Competitive NOCs, which do not undermine broader foreign policy goals by their daily activities, and a large tanker fleet are part of a hedging strategy to provide oil security insurance both during peacetime and times of crisis.

If necessary, the Chinese government will be able to "call on" its NOCs, its state-owned tankers, and even other major Chinese shipping companies to serve Chinese national interests during a period of crisis despite the fact that China's NOCs and major shipping companies, such as China Ocean Shipping (COSCO), China Shipping (CSCO), and China Merchant Steam Navigation Cooperation (CMSNC), are large corporations and commercial actors that have become increasingly profit oriented.[88] Depending on the tankers' location, it might take weeks to put the vessels into service, but that would only delay the implementation of the operations somewhat. Based on the experience in the lead-up to the Iraq War in 2002 and other conflicts, such as in Libya, there will probably be some form of advanced warning prior to another war in the Middle East.[89] This gives the Chinese government a window of opportunity to inform and prepare tanker operators for oil deliveries during periods of crisis. Tankers may be chartered out in the international market, but contracts are usually flexible, especially in times of crisis or war, which means that the Chinese tankers can be diverted from their charters and immediately put into service.[90]

Pressuring Chinese tanker companies into service and forcing cooperation between tanker operators and Chinese NOCs requires that China's NOCs maintain and increase their overseas oil production and that China build a large state-owned tanker fleet. The government has explicitly supported the goal of securing overseas production and long-term supply deals. Chinese leaders are aware that overseas production and seaborne supplies might be cut off during any war with the United States. Accordingly, the most plausible way that China's overseas production can provide oil-supply security is if the oil, which normally is sold on the international market for profit, is shipped back to China as part of a hedging strategy during any major disruption in the international oil market.

As we saw in chapter 3, the Chinese government previously set a target for the NOCs to secure 1 million b/d of overseas production by 2010 and that over the "next three years . . . the net international oil and gas output [will] . . . approach 1.5 million boe/d."[91] The U.S. Energy Information Agency (EIA) estimated that China's NOCs' overseas production of 850,000 b/d in 2008 represented approximately 29 percent of China's total oil production in that year.[92] An 2011 IEA report notes that China's NOCs expanded their overseas equity shares from 1.1 mbd in 2009 to 1.36 mbd in the first quarter of 2010.[93] Based on the previous years' increases in Chinese overseas equity oil production, if we estimate that China's equity production has reached 1.4 mbd, then overseas production will account for nearly 30 percent of China's oil import supply.[94] Remarkably, China overseas equity oil production has almost kept pace with China's extraordinary demand for oil imports, even as China has moved, in less than twenty years, from a net exporter of oil to becoming the world's largest importer. Despite such an extraordinary growth in investments and production, it is unlikely that the NOCs' overseas production will be able to meet future Chinese demand. Nonetheless, the equity oil is not insignificant since most of the overseas oil production is located in countries that can handle crude carriers, so in principle, short of any war involving China, most of the equity oil could be shipped back to China in a time of crisis.

Moving from selling and buying locally to shipping equity oil back to China might be difficult if Chinese NOCs are operating with long-term contracts on its oil trade. However, Chinese companies are largely trading on the spot market.[95] Similar to the Chinese tankers carrying crude oil for foreign clients during peacetime, contracts in the oil market are flexible and changeable during times of crisis. If China's NOCs were to ship their overseas production back home, it would be a challenge for China's refineries to process all of the oil produced overseas efficiently. Heavy crude oil, for instance from the Middle East and South America, is more difficult for some Chinese refineries to handle than the low sulfur oil coming from areas such as Sudan, West Africa, and Russia.

To meet this challenge, as pointed out in chapter 3, Chinese oil companies have expanded and upgraded their refining capacity. The refining capacity may still impede the strategic benefits of shipping all Chinese equity oil home or processing some of the oil that Chinese tankers pick up during a crisis.[96] One way to work around this problem is for the NOCs to swap more of their oil production on the international oil market. There is no escaping the fact that future supplies will be tied to market access, as China's NOCs' overseas oil production will not be able to keep up with Chinese demand.[97] But the equity overseas production, shipped back to China on Chinese tankers, will make a difference, although it may be small, and China is likely to be relatively better off than other major oil importers during a period of crisis in the international petroleum market, and its economy might even be reasonably well off, thereby contributing to minimizing a potential economic downturn and the risk of social instability, thus realizing the main objectives of China's energy security policy.

More importantly, a large state-owned tanker fleet will provide an important insurance or hedge in a crisis situation even if China's NOCs' overseas production remains inadequate and cannot be shipped back to China. For example, if a conflict were to erupt between the United States and Iran, major shipping insurance companies would declare the Persian Gulf a war-risk exclusion zone. In such a situation, most shipping companies intending to operate tankers in the exclusion zone would have to insure the cargo with their governments.[98] Chinese state-owned tankers "could conceivably self-insure and forego paying insurance premiums in order to maintain continued oil delivery service to the home country."[99] As discussed in chapter 4, China could also self-insure and use its tankers to work around sanctions or embargoes against one of its major suppliers. This means that China could obtain oil in the market at a lower price than if it were to use third-party tankers during a wartime contingency or emergency. Not only can Chinese leaders force state-owned tankers and Chinese ship owners to operate in a conflict area with lower insurance premiums, but, more importantly, China will gain access to oil supplies by having the capacity to operate where other countries and ship owners refuse to operate or are prevented from operating. This is reinforced by the buildup of a state-owned tanker fleet that increasingly gives China direct control over an increasing percentage of the international tanker fleet. A more dominating role within the world's tanker fleet also provides the Chinese state-owned companies with an opportunity to profit from a crisis in the international petroleum market, as it can be expected that not all of the Chinese tankers will be diverted for domestic transportation requirements. Spare capacity can also be put into service for states considered as "partners" or to improve Chinese diplomatic ties, thereby providing the Chinese government with more leverage in its foreign relations.

In its emphasis on building a large state-owned tanker fleet, it is still a challenge to speed up technology, investment, and infrastructure to build a sufficient number of VLCCs and ULCCs. Most vessels in the Chinese tanker fleet are small and old, and they are better suited for coastal and local trade than for international shipping. On average, they are twenty years old; there are few large tankers in the fleet. But this situation is rapidly changing: the government has launched an aggressive tanker-building campaign.[100] Asian nations still dominate the global shipbuilding industry. South Korea, Japan, and China are the leading shipbuilders and build almost all of the world's new oil tankers, with orders in the Asian region accounting for an impressive 97 percent of the world's new orders in 2005, and since then the number of orders has continued to grow.[101]

Asia's global shipbuilding lead is increasing, as South Korea, Japan, and China continue to improve their capacity. China's shipbuilding industry reported a 113 percent rise in the first six months of 2006 over the same period of 2005. This was approximately 17 percent of global shipbuilding in 2006 and reflected China's goal of acquiring more than 25 percent of the world shipbuilding market.[102] Writing in late 2008, two Chinese researchers from Shanghai Jiaotong University estimated that to reach its target of transporting half of China's seaborne oil imports by its own fleet, China will need at least seventy additional VLCCs.

Domestic shipbuilding is growing because the Chinese government is encouraging domestic shipping companies to order VLCCs from domestic shipbuilding companies and has promised financial support to shipyards building VLCCs.[103] The Chinese government has taken a number of measures to promote its long-term ambitions for the shipyards and the state-owned tanker fleet, which have included favorable tax conditions, securing a steady flow of steel, facilitated joint ventures with foreign partners, and investment funding reforms that allowed shipbuilding companies to raise capital for shipping industry development from public issues or corporate bond sales.[104]

In 2002, CMSNC, holding half of China's tanker capacity and its only VLCCs, delivered only 7.3 percent of its cargo to China. Only 3.78 percent of the crude oil imported from the Middle East was carried by domestic fleets. But by October 2008, Chinese shipping companies operated thirty-six VLCCs and eleven Panamax tankers, and forty-nine additional VLCCs were under construction, which suggests that the government has been successful in achieving its goal of building a large tanker fleet.[105]

To support shipbuilding, encourage Chinese NOCs to import oil using Chinese tankers, and persuade Chinese shipping firms to prioritize Chinese clients, domestic shipping companies and oil companies have been encouraged to enter into long-term transportation contracts with one another. The four major oil transporters, CMSNC, COSCO, CSCO, and Yangtze Transportation Co. Ltd., have all contracted with Sinopec for crude oil transportation, and Petroritan has signed a long-term transportation contract with Zhuhan Zhenrong Co. Ltd., a crude oil importer in Guangdong province.[106] In accordance with the preferences of the twelfth five-year plan, the Fourth International Tanker Shipping Summit in Hong Kong in June 2012 pointed out that China imported 38 percent of its oil with Chinese tankers, a significant rise from January 2010, when *China Daily* reported that foreign tankers were shipping more than 80 percent of China's crude oil imports.[107] Even when taking into account estimates that suggest a much lower percentage, this still indicates a significant rise in oil carried by domestic tankers.

While the cooling of the Chinese economy in 2012 and an ongoing bearish tanker market suggest that Chinese shipyards eventually are finding it difficult to take on new orders and that the Chinese shipbuilding industry may face a downturn, it is likely that Chinese state-controlled companies will place further significant orders with the yards.[108] A leading Chinese shipping expert with excellent sources in the industry confirmed in September 2012 that COSCO just placed orders for five plus five VLCCs at subsidized prices and that China Shipping Group and China Merchants Group will place orders of a similar size to keep state shipyards going.

For years, the government's strong incentive to build domestic tankers strained the overall capacity at Chinese shipyards. One option, then, was to pick up secondhand tankers in the international market. Capitalization in the shipping industry is low, and many tankers are owned by small family syndicates. The aftermath of the financial crisis in 2008 therefore provided China with a window of opportunity to purchase, merge, or take over other shipping companies and tankers.[109] By acquiring secondhand tankers, China may forestall the danger that China's aggressive tanker buildup program could outstrip demand, driving tanker rates down.[110] Shipping rates in 2011 dropped as the number of oil tankers in service rose 2.8 percent from March 2010 to March 2011, suggesting that the overcapacity in the tanker market will last.[111] More importantly, China has not pursued a strategy for picking up tankers secondhand, indicating a preference for the spillover effects and profits that a domestic shipbuilding industry offers. Most tankers that are sold secondhand are old, which makes them less attractive for Chinese state-owned tanker companies with a long-term strategic objective and the goal of becoming leading companies in the tanker market. Finally, as steel prices remain relatively high, some tanker companies prefer to scrap their old tankers.

Chinese tanker firms, not unlike the Chinese NOCs, pursue the more profitable approach of being chartered out in the market and are not strictly controlled by the central government in their daily business activities. Just as to be cost effective China's NOCs prefer to sell their upstream oil production on the local or international markets and prefer to purchase oil on the market for downstream refinery production in order to obtain the best prices, economic incentives also motivate Chinese tanker owners to operate in the international shipping market and largely to serve foreign clients.

In 2007 and 2008, an estimated 90 percent of China's oil shipping capacity served foreign clients, which points to the fact that China's ship owners operate where the highest profits are to be made.[112] As with the Chinese NOCs, the shipping companies have benefited from the government's concern with energy security and have used the government's interest to promote profit and commercial goals.[113] Additionally, shipbuilding is a strategic industry that boosts the entire industrial chain, stimulates investment, and, most importantly, creates a significant number of jobs, similar to investments in large pipeline projects. Job creation is an important stabilizing factor for the central government, but China also has a window of opportunity to invest in shipbuilding when wages are still low and credit is available. Spillover effects on other maritime and oil industries and China's naval buildup are also expected as China expands its deep-water port facilities and onshore oil-handling infrastructure to accommodate seaborne oil imports by VLCCs.[114] The Chinese state-owned and private companies seek to benefit from the government's aim to boost the shipping industry, and the government "likely believes that it is hedging its bets against future threats to oil shipments by supporting a large tanker buildup."[115]

It is clear that a mix of security and profit considerations are driving the tanker buildup, and it is interesting to note that China is not following the most cost-effective strategy. For several years since 2009, there has been an overcapacity in the international tanker market, and rates have been very low. China's ambition to develop a large tanker fleet is likely to sustain the crisis in the tanker market, and Chinese tankers and shipyards are probably losing money and are expected to continue to do so in the future. Indeed, it would have been cheaper for China's NOCs to charter tankers in the international market and for the Chinese government to let its shipping companies pick up tankers in an oversupplied tanker market rather than build new tankers at cost-ineffective Chinese yards. However, just as China is willing to pay a premium on oil delivered through cross-border pipelines, China is also willing to buy security through the development of a state-owned tanker fleet that can provide a hedge against oil-supply disruptions during conflicts in oil-rich regions that do not directly involve China and when sanctions are imposed on China's major oil suppliers.

LEGAL CONSIDERATIONS

Strategic and profit factors are not the only drivers behind building a large state-owned tanker fleet. International legal norms also support the buildup of such a fleet. PRC-flagged tankers enjoy legal protection, since the Chinese government could claim that its sovereignty has been breached if these tankers are interdicted, which would justify and legitimize an armed response.[116] This legal protection may deter some countries from interdicting China-bound tankers. More Chinese state-owned tankers might also reduce the risk that other major powers would put diplomatic and financial pressures on operators from smaller countries, such as the Bahamas, Greece, Denmark, and Norway, in order to force them to end oil deliveries to China.[117]

Finally, as noted above, a state-owned tanker fleet will most likely undermine Chinese strategic interests during a war or a blockade because a blockading power would more easily be able to identify tankers going to China. Thus, until China is able to develop and deploy successfully its military capability overseas in a wartime contingency, a hedging strategy that focuses on peacetime risks and building a state-owned tanker fleet will be an important part of China's strategy to improve energy security and to insure against supply disruptions in the international petroleum market.

MANAGING RISKS AND NONTRADITIONAL SECURITY CHALLENGES

These last sections explore peacetime risks to China's energy security, not potential wars (without Chinese involvement) or crises that might affect the physical availability and affordability of petroleum supplies in the international energy market. China has become more involved in managing such risks that may disrupt its oil supplies, such as preventing accidents at sea, antipiracy missions, humanitarian relief, and antiterrorism operations.

ACCIDENTS AT SEA

According to two Chinese energy experts at the ERI and the National Development and Reform Commission (NDRC), the main worry is not primarily a blockade of the Malacca Strait but the fact that this "crucial transportation channel . . . will be even more crowded in the future." This will increase the "threats of oil tanker accidents, pirates and terrorist attacks (computer system attacks included) as oil and LNG trade escalates day by day."[118]

At the Phillips Channel near Singapore, the Strait of Malacca narrows to 2.8 kilometers (1.5 nautical miles), creating one of the world's most significant traffic bottlenecks. About 33 percent of global seaborne crude oil moves through the 965-kilometer (600-mile)-long strait. With approximately 15 mbd of oil flowing through the Strait of Malacca and a daily oil flow of 17 mbd going through the Strait of Hormuz, these vital oil choke points are vulnerable to collisions, grounding, or oil spills. Any accidents run the risk of forcing closure of the strait, thereby interrupting oil supplies, forcing the use of alternative routes, and driving up shipping costs, which affect affordability.

Accidents are rare, considering that at least 94,000 ships pass through the Strait of Malacca each year, but a Taiwanese oil tanker carrying 58,000 tons of naphtha fuel collided with a Greek-managed bulk carrier in the strait in August 2009. Malaysian police told reporters that the burning ship and oil spill were not a threat to other ships in the area and were not disrupting traffic.[119] Yohei Sasakawa, the chairman of Japan's Nippon Foundation, an organization involved in navigational safety in the strait, was still concerned that "there is a potential for accidents to happen. If it involves crude oil tankers, there will be a major spill. This could disrupt international trade."[120]

The prospect of safer navigation through the Strait of Malacca has grown since regional governments implemented a dual-traffic system in 1997. International shipping associations, the International Maritime Organization (IMO), and the International Association of Independent Tanker Owners

(IAITO) have been discussing measures to increase safety and avoid accidents. The debate over getting users and states to agree on who pays and how much to use the strait and to contribute toward the costs of maintaining buoys and other navigational aids in the area is not new. During a London meeting in December 2007, an agreement was reached between the Nippon Foundation in Japan and the roundtable of international shipping associations to set up a voluntary fund to pay for navigational aids. The Nippon Foundation agreed to pay one-third of the estimated costs of $28.2 million for the first five years, and the Republic of Korea, the United Arab Emirates, and Greece pledged their support.[121]

Although China did not contribute directly in this process, in September 2005 it agreed to sponsor a $2.6 million project to replace several lighthouses and beacons along the eastern coast of Sumatra that had been damaged by the 2004 Indian Ocean tsunami.[122] Bilateral and multilateral diplomacy have been used by the Chinese government to promote cooperation with the littoral states to secure the Strait of Malacca. In July 2007, during a state visit to Indonesia, former president Hu signed an agreement to conduct "various joint activities in such areas as navigation security, maritime security, ship building, naval cooperation, and maintenance of security across the Strait of Malacca." A similar cooperative program was initiated with Malaysia in December 2005, and the Chinese government has also worked with the Association of Southeast Asian Nations (ASEAN), the Association of Southeast Asian Nations Regional Forum (ARF), and the Council on Security Cooperation in the Asian Pacific Region (CASCAP) in order to improve the safety of passage through the vital strait.[123]

ANTIPIRACY MISSIONS

Piracy, hijackings, and robberies constitute a large risk to shipping, driving up insurance rates on vessels passing through shipping lanes that are known to be vulnerable to such incidents, thereby affecting the affordability of shipped goods. It is estimated that piracy costs the global economy $7–12 billion per year. According to the International Maritime Bureau (IMB), the total number of reported actual and attempted attacks decreased from 445 in

2003 to 293 in 2008, but the number then increased to 406 in 2009 and 489 in 2010, with most of the attacks taking place in the Gulf of Aden and off the coast of Somalia.[124]

Although there was a sharp increase in piracy in Southeast Asia after the Cold War, from six to seven attacks annually before 1989 to fifty in 1991 and 469 in 2000, recent figures show that the number of incidents has dropped dramatically in the last few years. Indonesia reported 121 attacks in 2003 and twenty-eight in 2008, and, according to the IMB, there were twenty-eight actual and attempted attacks in the Malacca Strait in 2003 and only two in 2008. It is argued that national, bilateral, and multilateral measures largely account for the drop in piracy incidents.[125] The littoral states of Indonesia, Malaysia, Singapore, and Thailand have all taken steps to address piracy. These measures include more naval and coast guard patrols, improved surveillance, increased contacts between command centers and the shipping community, and a focus on improving the living standards and welfare of those living in areas that border the key sea lanes.[126]

The littoral states have coordinated their sea patrols, launched "eye in the sky" aerial surveillance, formed an intelligence exchange group, and set up a joint coordination committee to organize bilateral cooperation. Collaboration with other regional actors, including China, has been enhanced through capacity building, training, and technical assistance, but patrols in the Strait of Malacca have been confined to the littoral states.[127] A multilateral framework has also been established through the Regional Cooperation Agreement on Combating Piracy and Armed Robbery Against Ships in Asia (ReCAAP). Seventeen countries are contracting parties to the ReCAPP; Norway was the first nonregional country to participate, with Denmark and the Netherlands later joining to enhance collaboration. Composed of staff from all the participating countries, the ReCAAP's main functions are to improve information sharing, facilitate operational cooperation, analyze patterns and trends in piracy and armed robberies, and support the capacity-building efforts of the contracting parties.[128] Finally, the IMO has initiated a cooperative mechanism between users of the strait and the littoral states, implementing Article 43 of the United Nations Convention on the Law of the Sea (UNCLOS).[129]

Steps taken primarily by the littoral states and secondarily by the international community, including China, which has been one of several actors engaged in cooperative measures to increase safety along vital SLOCs in Southeast Asia, indicate that it is possible to manage and limit the risk of piracy. This stands in stark contrast to the situation in the Gulf of Aden and along the eastern coast of Somalia. In 2012, however, there was a resurgence in piracy in Southeast Asia, which suggests that piracy's underlying social and economic causes have not been adequately addressed, but attacks in the Gulf of Aden decreased significantly, suggesting that escort and antipiracy missions have had an effect.

There are a number of key differences between the attacks in Southeast Asia and those along the coast of Somalia. National and bilateral measures have been ineffective in curbing piracy and hijackings.[130] Most of the attacks in the Gulf of Aden and along Somalia occur on the high seas, with small, high-speed, open boats that operate from a "mother ship" carrying supplies, personnel, and equipment to facilitate attacks at a great range from the shore.[131] In contrast, Southeast Asian pirates are less organized and conduct opportunistic raids closer to shore. Indeed, most of the attacks in Southeast Asia take place from anchored vessels or vessels in port.[132] Southeast Asian waters are relatively confined, in comparison to the wide-open ocean along the Somali coast, thereby making Southeast Asia more prone to hit-and-run attacks.[133]

Another difference between Southeast Asia and East Africa and the Gulf of Aden is that many of the attacks that occur along the East African coast are attacks on large ships, in which the vessels are hijacked and the crews taken as hostages for ransom. According to the IMB, in 2008, out of forty-nine vessels that were hijacked and 889 crew taken hostage worldwide, forty-two were hijacked by Somali pirates, who took 815 crew hostage.[134] Somali pirates are well armed, to intimidate the target vessels and their crew, and they often operate in daylight. By comparison, hijackings for ransom are rare in Southeast Asia; pirate activities generally involve robberies to steal valuables, and they are conducted secretly under the cover of darkness. Firearms are seldom used.[135]

The most important difference is the lawlessness on land in Somalia and the inability of the Somali government to respond to increased piracy. Such disorder is generally not found in Southeast Asia. For instance, there are very few places, if any, in Southeast Asia where pirates can take a large vessel, hold it and its crew for ransom, and block its recovery. Thus, order at sea depends on order at land.[136] An important lesson from Southeast Asia is that national and regional capacities to increase law enforcement onshore must be improved to reduce inland enclaves of pirates. Such measures so far have proved difficult to implement in the Horn of Africa. Instead, robust measures by international governments and navies have been necessary to enable the restoration of the safety and security of this major trading route.[137]

China supported the October 7, 2008, unanimously adopted UN Security Council Resolution 1836 and the December 16, 2008, Resolution 1851 under Chapter VII of the United Nations Charter, which "called upon States with naval vessels and military aircraft operating in the area to use, on the high seas and the airspace of the coast of Somalia, the necessary means to repress acts of piracy in a manner consistent with the 1982 United Nations Convention of the Law of the Sea."[138] With Somalia's transitional federal government ready to consent to other nations' navies to enter Somalia's territorial waters and use "all necessary means" to repress acts of piracy and armed robbery, soon thereafter China decided to deploy two navy destroyers and a supply vessel to the Gulf of Aden.

The PLAN's first operational combat deployment far beyond China's territorial waters marked a milestone for the Chinese military. The two destroyers—the *Haikou* 171 and the *Wuhan* 169—arrived in the Gulf of Aden in January 2009, operating along with supply vessel 887, the *Weishanhu*, and carrying about eight hundred crew, including seventy special-operations troops and two helicopters. According to the *New China News Agency (Xinhua)*, during the three-month mission the first Chinese fleet escorted 206 ships, including twenty-nine foreign merchant vessels, and successfully rescued three foreign merchant ships.[139] China replaced its first flotilla of ships with a second flotilla in April 2009 and a third in July 2009, illustrating the success of the deployment.[140] China's twelfth escort group replaced the eleventh in the summer of 2012, and so far the PLAN has escorted more than 4,700 ships from all over the world. The destroyers and frigates involved in this deployment are some of China's most sophisticated and modern surface warships, indicating that

China regards its first potential combat mission beyond its territorial waters an important and prestigious task.

The attacks off the coast of Somalia have shocked the international maritime community; even VLCCs have been hijacked.[141] Rear Admiral Du Jingcheng, the task force commander of the first flotilla, told reporters that "the primary missions of the destroyers . . . [was] to protect Chinese merchant ships, especially tankers with crude oil, that traverse the gulf."[142] Both the Ministry of Defense spokesperson Hu Changming and the Ministry of Foreign Affairs spokesperson Liu Jianchao stated that the "main mission [was] . . . to protect the safety of Chinese vessels and their crew," noting that 20 percent of the 1,265 Chinese ships passing through the area had come under attack and that seven hijackings had involved Chinese ships and crew.[143]

The PLAN's protection of Chinese and foreign merchant vessels will contribute to energy security both for China and for the international petroleum market, and it will improve the security of the SLOCs in the global commons. The deployment shows that China has developed a military capability that can protect its growing global economic interests and overseas presence.[144] This highlights a connection between military power and safeguarding Chinese energy security interests. At least two other factors should also be considered. First, by patrolling international waters and contributing to international security and stability, China is behaving as a responsible permanent member of the UN Security Council and meeting its international obligations as a great power.[145] This gives China status and prestige and boosts its image as a peacefully rising and developing country.

Second, the escort mission in the Gulf of Aden provided the PLAN with an opportunity to gain experience in long-range military deployments. Although the mission can be considered to be an escort task somewhere between policing and conducting a military operation, the mission provided an important test of combat readiness, long-range logistic supply, real-time C4ISR connection with PLA headquarters in Beijing and Haikou, surveillance, navigation skills, and lengthy deployment at sea. Operating alongside and collaborating with advanced foreign navies allows the PLAN to obtain intelligence information.[146] Additionally, by protecting its overseas interests, the PLAN deployment demonstrates to the Chinese public that there are returns for China's high military spending and promotes the view that China's growing capabilities will ensure world peace and mutual development.[147]

HUMANITARIAN RELIEF OPERATIONS

Naval vessels from the United States, India, and China participated in cooperative humanitarian relief activities in the aftermath of the 2008 tropical cyclone in Myanmar. This was the first time that the PLAN had carried out a maritime humanitarian relief operation. Again, such missions will improve both China's national interests and its energy security. First, China's participation in maritime natural disaster relief operations contributes to enhancing China's status and prestige in Asia and promotes the image of China as a responsible stakeholder in the international community. Future relief missions will allow for more cooperation between the PLAN and regional navies and between China and other naval powers, and they can expand into multilateral maritime cooperation to safeguard the global commons.[148]

Second, it is acknowledged that maritime natural disasters may have security implications and affect the international petroleum market. Hurricane Katrina, which hit the Gulf of Mexico and the southern portion of the United States in August 2005, had a damaging effect on oil production, refinery capacity, and the international petroleum market. It is estimated that along China's 18,000-kilometer coastline, more than one hundred natural disasters occur at sea each year.[149] Accordingly, the experiences gained from China's participation in humanitarian relief operations will benefit missions to deal with natural disasters within China and will improve China's energy security and risk management capability.

MARITIME TERRORISM

Narrow straits, big ships, large ports, and unprotected coastlines have facilitated or have been the target of terrorist attacks in the past. In contrast to pirates seeking money, terrorists are often pursuing political goals. Any potential collusion between pirates and terrorist groups may endanger the transportation of nuclear, biological, or chemical weapons of mass destruction (WMD) at sea. A number of measures have thus been taken in order to prevent nuclear terrorism and the proliferation of nuclear material. In July 2003, in the aftermath of the September 11, 2001, terrorist attacks, China signed the Container Security Initiative (CSI) with the United States, whereby cargo from Chinese ports must be inspected in advance before being shipped to the United States. On July 1, 2007, China, together with about 150 other states, also joined the International Ship and Port Facility Security Code (ISPS) of the IMO.[150] To cooperate to prevent nuclear proliferation, China also joined the Global Initiative to Combat Nuclear Terrorism (GICNT).[151]

CONCLUSION

With China becoming increasingly reliant on seaborne petroleum supplies, it has been argued that China is vulnerable to SLOC security. But insecurity associated with the transport of China's petroleum imports has been inflated. Although seaborne oil accounts for roughly 85 percent of China's oil imports, it represents only about 50 percent of China's total oil consumption—or about 10 percent of China's energy mix.

To evaluate the claim that China is vulnerable to a maritime blockade, it is necessary to differentiate between wartime and peacetime scenarios. Too much of the literature disregards this important distinction. Bluntly put, if there is a war between the United States and China, all bets are off, and China's coast will most likely face a blockade by the U.S. Navy. Consequently, China cannot expect to rely on seaborne oil supplies, since China will not have the naval power capability to withstand a blockade by the United States during a major war.

But China may be vulnerable to a limited blockade because of its dependence on the Strait of Malacca. It was therefore emphasized that a blockade of the Strait of Malacca, together with a number of other alternative blockades, short of a war, would not specifically prevent supplies of oil from reaching China.

Accordingly, instead of focusing on the threat of war, limited war scenarios, and a blockade, it is more useful to focus on peacetime risks that, even though they cannot be eliminated, can be managed through hedging strategies. Under different risk scenarios, the buildup of a large state-owned tanker fleet may be of strategic value if a conflict in which China is not involved were to erupt.

Prior to or during such a crisis, the Chinese government can call on their oil and shipping companies to work in tandem with the government. If oil supplies to China can be maintained through a hedging strategy, the country might be able to manage the risk of interrupted supplies and rising prices and could "secure" oil supplies during a conflict in an oil-rich region. It could do so at a lower price (by shipping China's overseas equity oil production back home) and at lower shipping costs (by using its state-owned tanker fleet) than if it were to import all its oil supplies through the international market during a crisis.

The oil that Chinese NOCs produce overseas is insufficient to meet Chinese demand. But having a state-owned tanker fleet will make it possible to operate in a high-risk area of conflict and to access oil terminals that other shipping companies may not reach. Finally, by having its own tanker fleet, China can self-insure and avoid the high insurance premiums that it would necessarily incur after a war-risk exclusion zone is declared by the major insurance companies.

A large tanker fleet is important because it provides profits and benefits by generating revenue, boosts the competiveness of the Chinese commercial fleet, and has important spillover effects from shipbuilding, which can be beneficial to the entire industrial chain, by stimulating investments, creating a significant number of jobs, and improving China's infrastructure and the shipbuilding capacity necessary for a naval buildup. In addition, a large state-owned tanker fleet is an important hedge that will provide security during a potential crisis or conflict in which China is not directly involved.

The shipping industry and the PLAN seek to profit from peacetime risks to China's energy security. For the shipping industry, this boils down to benefitting from the government's goal of having more Chinese-flagged tankers to boost the transportation security of Chinese seaborne oil supplies; emphasizing the importance of the shipbuilding industry for jobs, competence, infrastructure, and technology as an important factor to the government's goal of sustaining economic growth and domestic stability; and aiding China's leaders' ambitions of developing China into a maritime power. The PLAN uses energy security in a pseudo-national-interest argument to legitimize and justify a larger navy under the banner of protecting China's national interests. The PLAN can also channel some dimensions of its naval buildup into risk management tasks that will enhance China's energy security and simultaneously promote both cooperation and a benign view of the PLAN.

Risks other than war include accidents, piracy, environmental catastrophes, and terrorism. The Chinese government has taken a number of steps to enhance collective maritime security and safety, and the PLAN, through its antipiracy missions and by conducting maritime humanitarian relief operations, has carried out tasks to police and protect the SLOCs. These steps have contributed to energy security, economic prosperity, increased cooperation, and a benign security environment.

It is important to remember that energy security cannot be completely "secure." China cannot safeguard its petroleum supply through military means. Instead, through its maritime hedging and risk management strategies, by collaborating with other navies to maintain maritime security, by combining elements of security and profit considerations, and by promoting the buildup of a large state-owned tanker fleet, China has insured against the risk of supply disruptions and has developed the capacity to manage risks to its energy security.

6

CHINA'S CONTINENTAL PETROLEUM STRATEGY

A CHINESE CONTINENTAL petroleum strategy based on pipeline supplies, upstream assets, and loan-for-energy deals is presently taking shape. The development of cross-border pipelines has not been a cost-effective and profit-maximizing strategy but instead an insurance strategy that seeks to hedge against a potential disruption in China's seaborne petroleum supplies. In the West, the Kazakhstan-China crude pipeline is already operational, and the Central Asian–China gas pipeline runs south of the oil pipeline, pumping gas into China's East-West domestic gas pipeline. China's national oil companies (NOCs) also have large equity investments upstream in Kazakhstan. An oil pipeline has been built from Russia to China in the northeast, and although there are challenges to the proposal for a gas pipeline to China, a deal may eventually be realized. China-Myanmar oil and gas pipelines in the southwest are currently being built and are expected to be operational by 2013. A pipeline from Pakistan to China and an Iran-Pakistan-China pipeline have been considered, and some analysts have suggested that China also intends to extend the Central Asian–China pipeline to Iran to tap into its large reserves.[1] The Pakistani and Iranian pipeline proposals are rather far-fetched projects and face a number of economic, security, and technology obstacles. They will therefore not be addressed in any depth in this book.

More than diversification is driving China's continental petroleum strategy. By distinguishing between the pipeline market and the international market, this chapter will show that China's continental petroleum strategy is a hedging strategy that goes beyond diversification. The continental petroleum strategy acts as a hedge, or a "short" selling strategy, to provide protection if China's reliance on "longs," or the international petroleum market and seaborne petroleum supplies, are disrupted during a wartime contingency or blockade. It is also argued that the long-term contracts and the partly fixed prices that characterize China's continental petroleum deals allow China to hedge against some of the price volatility in the international market and to safeguard China's petroleum supply during possible peacetime risks that would affect availability in the international petroleum market.

Concurrently, it is noted that China often pays a premium for pipeline supplies and that its NOCs forgo some profit opportunities. In most cases, the NOCs would be better off economically relying on seaborne supplies and the international petroleum market rather than developing costly cross-border pipeline projects. Indeed, the NOCs have built oil pipelines even before contracts to fill them have been signed.[2] But this is to be expected in a hedging strategy where China seeks to insure some of its petroleum supplies. Similar to a hedging strategy that combines "longs" and "shorts," China buys security of supplies through its continental petroleum strategy even though it forgoes some profits in the global and maritime international petroleum market thereby. The continental petroleum strategy can be thought of as insurance: although a more costly option, it is one more likely to minimize risks and threats to China's energy security.

While China is buying more security of supply through its continental petroleum strategy in accordance with a short strategy, there are variations in the security of the different pipelines. Just as different short-selling strategies provide different levels of risk and security, so do the pipelines. For example, because of the risk of escalation in a conflict between China and the United States, the Russia-China pipeline is safer, the Kazakhstan pipeline is less safe, and the Myanmar pipeline is least safe during a wartime contingency.[3] The Russia-China pipeline is also less prone to peacetime risks, such as terrorist or separatist attacks, but it is vulnerable to supply-side disruptions during peacetime as a result of pricing conflicts, commercial competition, or political intervention because of Russia's great-power status and its numerous alternative supply routes for its petroleum products. Kazakhstan also has alternative supply routes and can play Russia, China, and the West against one another. But Kazakhstan is not a great power, and it is less likely that its energy supplies would be linked to a broader agenda of great-power competition. Central Asian states have become dependent on China as a market to diversify their export away from Russia and for investments and revenue. They have also sought to use China as a counterweight to Russian diplomatic and security influence. Myanmar has few alternative routes to the pipeline to China and thus depends more on good relations with China than the Central Asian states or Russia. The Myanmar-China pipelines are also less secure because the petroleum being piped is produced offshore or overseas; thus the pipeline is vulnerable to a offshore blockade. Additionally, and of importance to China's hedging strategy against oil supply disruptions, in the case of China's petroleum security, pipelines are more vulnerable during peacetime risks and safer during wartime threats, whereas seaborne energy supplies are safer during peacetime risks and more vulnerable during wartime threats.[4]

The first section of this chapter examines the various pipelines that China has developed and the loan-for-energy deals that it has signed with some of its continental neighbors. It shows that Chinese NOCs first established a foothold in Central Asia. Despite being active in Russia since the 1990s, a real breakthrough on a Russia-China pipeline did not materialize until the late 2000s, when the financial crisis of 2008 became a game changer. China has recently been able to capitalize on its close ties with Myanmar, and China National Petroleum Company (CNPC) is now building a dual oil and gas pipeline from the Bay of Bengal to the Chinese southwestern province of Yunnan. The acquisition and production rights, the pipeline construction, the investments, and the loan-for-energy deals are part of a strategy to diversify Chinese sources and transportation routes for imported petroleum supplies.

Nonetheless, diversification is only one part of China's continental petroleum strategy and only one dimension of its multidimensional hedging strategy. The second section discusses the goals of China's continental hedging strategy. It also emphasizes the important distinction between peacetime risks and wartime threat scenarios and their implications for the role that pipelines play in China's hedging strategy against a growing reliance on imported oil.

The third section examines how a continental petroleum strategy has important spillover effects, ones that are beneficial to China foreign policy objectives on the Asian mainland, that strengthen China's relations with neighboring countries, and that enhance China's security interests and domestic stability. Thus, China's continental pipeline strategy has broader objectives than simply to diversify routes for the import of petroleum. The concluding section summarizes our main arguments.

CONTINENTAL PIPELINE AND PETROLEUM DEALS

KAZAKHSTAN

Acquisitions in Kazakhstan by CNPC began in 1997. On June 4, 1997, CNPC acquired a 60.28 percent interest in Aktobemunaigaz for $325 million. The deal included production licenses for the Zhanazhol, Kenkiyak Oversalt, and Subsalt oil fields; a contract for an exploration block; and a pledge by the CNPC to invest $4 billion in the development of this enterprise over the next twenty years.[5] Two months later, CNPC won a 51 percent interest in the Uzen oil field, promising to invest $2.4 billion, in addition to building a 3,277-kilometer pipeline to link the oil field to Alashankou in the Xinjiang Autonomous Region of China. But the Uzen oil field was never turned over to CNPC, and a pipeline was never built to link the Uzen oil field to China.[6]

Several factors explain these early setbacks in Chinese NOC involvement in Kazakhstan. First, the 1997–1998 Asian financial crisis and the resultant collapse of oil prices made the pipeline economically unviable. In addition, China's net petroleum imports in the late 1990s had not yet increased to the alarming levels that we have seen since the turn of the century. Second, China had pushed the Kazakh government to support its fight against "separatists, extremists, and fundamentalists." With the Shanghai Five cooperation in place in 1996 and settlement of the 1,930-kilometer border with Kazakhstan in 1999, the promotion of energy deals was no longer a crucial factor in broader foreign policy goals.[7] Third, CNPC complained about investment conditions in Kazakhstan, which included heavy local taxes and complicated work permits for CNPC personnel. CNPC was also suspicious about how much oil was available, as Kazakhstan was also considering other long-distance pipelines. These factors resulted in CNPC shelving the proposed pipeline in late August 1999.[8]

CNPC gained important experience from bidding for this rather large international acquisition, and it prepared China's NOCs for a more active role in overseas acquisitions in the years ahead. As noted in chapter 3, the broader external circumstances shaping China's energy security policy shifted within several years, resulting in Chinese NOCs returning to Kazakhstan after the turn of the century. The terrorist attacks on 9/11 raised fears of collateral damage from potential new attacks on the United States or oil installations around the world. More terrorist attacks could disrupt supplies and result in higher oil prices that would affect China's economic growth. The 9/11 attacks and the following U.S. military campaign in Afghanistan, which led to a stronger U.S. military presence in Central Asia, raised concerns in Beijing about U.S. long-term intentions in the region.[9]

The Iraqi war in 2003 fueled more Chinese suspicion about U.S. intentions and reinforced Chinese perceptions about its vulnerability because of its high dependence on Middle Eastern petroleum supplies. The continued use of military force by the United States in a region that supplied China with roughly half of its oil imports in 2002 worried Beijing about the affordability, reliability, and availability of resources as the net import gap increased significantly. This partly stimulated a call for a new strategy to enhance energy security.[10] China had also initiated negotiations with Russia over potential pipelines in Siberia and the Far East, but the negotiations became fruitless when Japan stepped in to compete over the Russian oil pipeline, making China more interested in hedging its bets. Pursuing a pipeline deal with Kazakhstan was thus now an attractive option to insure against a negative outcome in the talks with Russia and to diversify China's petroleum supplies.[11] All these factors shaped China's renewed interest in Kazakhstan.

The initial stage of a Kazakhstan-China oil pipeline, jointly developed by Kazakhstan's state-owned KazMunaiGaz (KMG) and CNPC, commenced

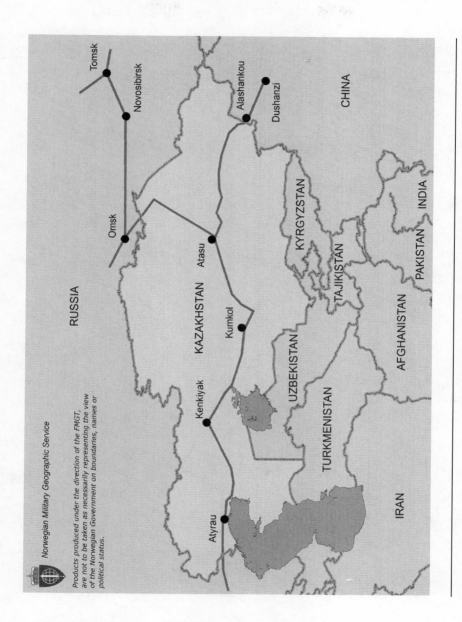

MAP 6.1 Kazakhstan–China Oil Pipeline

in 2002 with the building of two legs of the pipeline, from Kenkiyak to Atayrau in 2002–2004 and from Atasu to the Chinese border at Alashankou in 2004–2005. In May 2004, CNPC agreed to fund the $806 million cost of the 1,000-km-long Atasu-Alashankou crude pipeline. The first section was completed in December 2005 and soon thereafter became operational. CNPC also built a 252-km extension from Alashankou to the refinery at Dushanzi in Xinjiang. This last step was finalized in 2009, when KazStroyService completed the "missing link" between Kenkiyak and Kumkol, allowing Kazakhstan to increase its exports to China.

Supported by state-owned loans and the Chinese government, CNPC paid for 85 percent of the multiple stages of the pipeline construction but participated in a 50/50 joint venture with the KMG.[12] It is expected that the Atyrau-Alashankou pipeline will reach a maximum throughput capacity of 400,000 b/d, whereby CNPC will be able to draw on its production in the Aktobe, Zhanazhol, and Kenkiyak oil fields acquired in 1997 and 2003 and its upstream assets from MangistauMunaiGaz (MMG) in 2009; CNPC will also gain access to production in the Tengiz and Kashagan fields in the Caspian Sea.[13]

Chinese oil firms have achieved a dominant position in Kazakhstan's upstream petroleum sector through a number of acquisitions, including CNPC's additional 25.12 percent stake in Aktobemunaigaz in 2003 (bringing its shareholdings to 85.42 percent), CNPC's purchase of PetroKazakhstan in 2005 for $4.18 billion,[14] CNPC's takeover of MMG in April 2009 (with a price tag of $3.3 billion), and several other acquisitions by CNPC and other Chinese oil companies, such as China International Trust and Investment Corporation (CITIC) Resources, Sinopec, and Kunlun (part of CNPC). CNPC was the second-largest oil producer in Kazakhstan in 2007, producing just over 200,000 b/d, the largest foreign production of oil by Chinese NOCs.[15] CNPC boosted its production in Kazakhstan when it acquired PetroKazakhstan in October 2005, which was Kazakhstan's "second-largest foreign petroleum producer, and the largest manufacturer and supplier of refined products."[16]

Chinese NOCs' aggressive overseas expansion during the last decade to acquire companies, such as PetroKazakhstan in 2005, has been facilitated by favorable loans by the Chinese government through the state-owned banks. The impact of the 2008 financial crisis and the accumulation of cash reserves have strengthened the position of China's leading banks. Chinese NOCs with close ties to the government and to the state-owned banks are therefore well positioned for further overseas M&As that can provide even more access to Kazakhstan's reserves. This is confirmed by China's extensive "loans-for-oil deals" around the world, documented in chapter 3, which included the $5 billion deal by China's Export Import Bank when MMG was purchased in 2009 and the $10 billion deal to Kazakhstan in 2009 by the China Development Bank (CDB).[17]

CNPC had the third-largest liquid production in Kazakhstan in 2009, behind only KazMunaigaz and Chevron.[18] In 2009, Chinese production in Kazakhstan stood at 18.57 mt, or close to 400,000 b/d, indicating that production had almost doubled in two years.[19] Reuters reported in 2011 that CNPC announced that it had produced a record 30 million tons of oil equivalent in Kazakhstan in 2010, or 600,000 bpd.[20] CNPC has become a leading oil producer in Kazakhstan and one of the largest foreign producers, but since it has production-sharing contracts and partnerships, it is hard to say how much CNPC is entitled to of that oil production. Chinese companies operating in Kazakhstan also hold a much smaller percentage of reserves.[21] Combined with the technological advantages of the larger international oil companies (IOCs) and Kazakhstan's desire to diversify its markets and prevent one company from controlling its upstream production, Chinese control of Kazakhstan's oil production seems unlikely.[22] Chinese NOCs are latecomers to the international oil market, and they lack the large project management skills possessed by the major IOCs and are often unable to compete technologically. CNPC has gained valuable experience from the big integrated projects that it operates in Kazakhstan and Sudan and is just now beginning to catch up.[23]

Approximately half of Chinese oil production in Kazakhstan is sent back to China. In 2008, China imported an average of only 115,000 b/d of crude oil from Kazakhstan, both by pipeline and rail, and in December 2007 the Kazakhstan-China crude pipeline was running at only about half its total capacity. This was because of pricing disputes, supply availability, or higher profits from selling it locally or sending it west through other pipeline systems. In June 2010, it was reported that 200,000 b/d was being pumped from Kazakhstan to China every day, representing about 3–4 percent of China's crude oil imports, but with the Kenkiyak to Kumkol leg finished, it is estimated that the Kazakhstan-China pipeline will run at full capacity by 2013–2015, bringing 400,000 b/d to China and accounting for about 6 or 7 percent of China's oil imports. Most of Kazakhstan's oil export of roughly 1.5 mbd is exported via the Caspian Pipeline Consortium (CPC) pipeline, which runs through Russia, but China has now become Kazakhstan's top export destination mainly thanks to the growing capacity of the Kazakhstan-China crude oil pipeline.[24]

The longest leg of the 1,833-km Central Asian gas pipeline from Turkmenistan through Uzbekistan and Kazakhstan to China runs 1,404 km through Kazakhstan. The Kazakh section cost $6.7 billion, and the 30-bcm dual line capacity gas pipeline, with an expected additional supply of 10 bcm, will offer Kazakhstan an opportunity to supply China with gas from its Karachaganak, Tengiz, and Kashagan gas fields and will provide Kazakhstan with a gas route that completely bypasses Russia. Currently, Kazakhstan provides only a small share of the gas heading to China through the Central Asian gas pipeline, but the Kazakh Oil and Gas Ministry has stated that Kazakhstan will boost its gas supplies to China by 25 bcm, which will be facilitated by the construction of a third spur on the gas pipeline.[25] The 10-bcm pipeline, Beyneu-Bozoy-Samsonovka, which is yet to be built, aims to fulfill the domestic Kazakhstani goal of having an interconnected gas market. At present, Kazakhstan cannot supply itself with the gas it produces, and it has to export to Russia. But this pipeline will change this situation. KMG and CNPC reported that they would start building in the second quarter of 2011. This is indicative of the relevance of domestic factors in the host country and of the high priority that Chinese NOCs attach to securing a foothold in Kazakhstan, even at the risk of overpaying.[26]

TURKMENISTAN

Turkmenistan holds the world's fourth-largest gas reserves. CNPC received a license for exploration and development of the right bank of the Amu Darya

River in August 2007. This contract gave CNPC a prominent role in Turkmenistan as the only company awarded a production-sharing agreement (PSA) by the Turkmen government for the development of the country's onshore natural gas resources.[27] CNPC also secured a service contract worth $3.28 billion at Turkmenistan's South Yolotan field. Construction of the 188-km Turkmen section of the gas pipeline to China began on August 30, 2007. After overcoming the obstacles to reach a consensus with the transit countries, construction of the Uzbek and Kazakh section began in June and July 2008. From the groundbreaking to the commission of line A, the 1,833-km pipeline was completed in twenty-eight months, in mid-December 2009. Line B of the pipeline became operational in 2010, and a delivery capacity of 30 bcm per annum will be reached by the end of 2011. The Central Asian–China gas pipeline starts at the Turkmen-Uzbek border city Gedaim and runs through central Uzbekistan and southern Kazakhstan before reaching Horgos in China's Xinjiang Uygur Autonomous Region. The total cost has been estimated at $7.3 billion. By connecting with the second West-East gas pipeline in China, the Central Asian-China gas pipeline is the longest in the world.[28]

A gas sales and purchase agreement, envisaging annual deliveries of 30 bcm per year for thirty years, has been signed between Turkmenistan and China. It is expected that 12 bcm will come from CNPC's Amu Daray River production, and the Turkmen government has promised to develop and produce the remaining 17 bcm and possibly an additional 10 bcm. In the first five months of 2011, Turkmenistan provided China with 50 percent of its total gas imports, shipping 5.7 bcm of gas to China through the Central Asian–China gas pipeline. The pipeline is scheduled to operate at full capacity by June 2012, transporting 30 bcm to China annually.[29] To boost deliveries, the Turkmenistan government signed a loan-for-gas agreement on June 24, 2009, for the CDB to loan $4 billion to Turkmengas to help develop the giant South Yolotan gas field.[30] Because Turkmenistan needs both financial and technical assistance to develop its large gas reserves, the CDB provided another $4.1 billion loan for South Yolotan in 2011. China's financial strength may also facilitate the building of a third Turkmenistan-China pipeline.[31] The Central Asian gas route is the first to bypass Russia. Some of the broader implications of these new energy routes will be examined below.

One of the main drivers behind China's push for a Central Asian gas pipeline was its dissatisfaction with Russia's handling of a potential gas pipeline from Russia to China, similar to China's frustration at the lack of progress on a crude pipeline from Russia to China. In addition, an important factor motivating China to cooperate with Central Asia in terms of gas matters is that Gazprom will not allow participation or ownership of Russian fields, and China prefers equity gas to be able to integrate production vertically.[32] As plans in 2003 for a gas pipeline from Russian reserves in eastern Siberia stalled and became entangled in a struggle over who would control the gas exports, China looked elsewhere for pipeline gas.[33]

The Central Asian gas pipeline is being progressively built up, and the amount of gas going to China through the Central Asian gas pipeline will be roughly half of China's 2008 total gas consumption of 78 bcm. In 2010, domestic production had risen to 95 bcm, and annual consumption reached 107 bcm. The small net import gap of roughly 10 percent in 2010 is projected to grow. Projections indicate that China will import as much as 40 percent of demand by 2020 and over 55 percent, or 132 bcm, by 2030.[34] It is estimated that China's natural gas demand will increase to 176 bcm in 2020 and 242 bcm in 2030, with domestic production growing to 127 bcm in 2020 and 125 bcm in 2030.[35] CNPC expects that China will import 12 million tons of liquefied natural gas (LNG) in 2011 and that pipeline imports will increase to 15 bcm.[36] Long-term estimates by CNPC suggest a sharp increase in demand to 300 bcm by 2020 and imports reaching 80 bcm.[37]

By 2015, it is estimated that China's total gas imports from both Central Asia and elsewhere will be close to 50 bcm in pipelines and 30 billion bcm in LNG contracts, roughly meeting the expected rise. Still, there are a number of uncertainties. First, estimates of China's future oil and gas demand, especially from the International Energy Agency (IEA), tend to be conservative. Second, it is difficult to account for the contribution from unconventional or shale gas. Third, China's gas imports may be complemented by plans to develop two pipelines, with a capacity to export 70 bcm of gas from Russia to China. There are many obstacles to building a gas pipeline from Russia, but an agreement on a crude pipeline was recently reached. Therefore, it is useful to explore recent developments in the Russia-China pipeline.

RUSSIA

China has been working for years to develop the Russia-China oil pipeline. Eventually, China's diplomatic and economic steps and the NOCs' hard work paid off in the aftermath of the 2008 financial crisis. Little came out of the negotiations in the 1990s, even though an energy cooperation agreement was signed during former President Boris Yeltsin's visit to China in 1996.[38] Plans by the Russian oil company Yukos in 2001–2003 to construct a pipeline from Angarsk to Daqing were shelved after the arrest of Mikhail Khodorkovsky in October 2003.[39] With Yukos marginalized, CNPC signed a long-term cooperation agreement with Rosneft in July 2005 and with Lukoil in September 2006. The ties with Rosneft were consolidated through the CDB and the Export Import Bank of China, providing Rosneft with a $6 billion loan, which Rosneft used to fund its $9.4 billion purchase of Yuganskneftgaz, the main production unit of Yukos.[40] Progress was then made on a Sino-Russian energy partnership, which prompted CNPC's vice president Zhou Jiping to claim in March 2006 that the relationship had entered "a new stage of development."[41]

In October 2006, the first Rosneft-CNPC joint enterprise, Vostok Energy Ltd., was established, in which CNPC held a 49 percent stake. CNPC's ties with Rosneft advanced a Russian East Asian energy strategy and promoted support for the building of the East Siberian Pacific Ocean (ESPO) oil pipeline, despite reluctance from Transneft, Gazprom, and Russian Railways.[42] At the same time, CNPC struggled to gather enough Russian support from a partner that would be powerful enough to build a crude pipeline to China. Crude imports by rail, reaching more than 300,000 b/d in 2007 and 2008, were the only shipments of oil from Russia to China, even though this means of transportation is often twice as expensive as pipelines. Price disputes; Russian Railway's opposition to a pipeline; Rosneft, Gazprom, and Transneft's preferences for developing other routes because of lower profitability in the east relative to other options; and the higher average costs required to increase reserves in eastern Siberia compared to western Siberia are some of the reasons for the stalled plans on a Russia-China pipeline.[43] Gazprom's fierce campaign to undercut and prevent TNK-BP from developing the Kovykta field

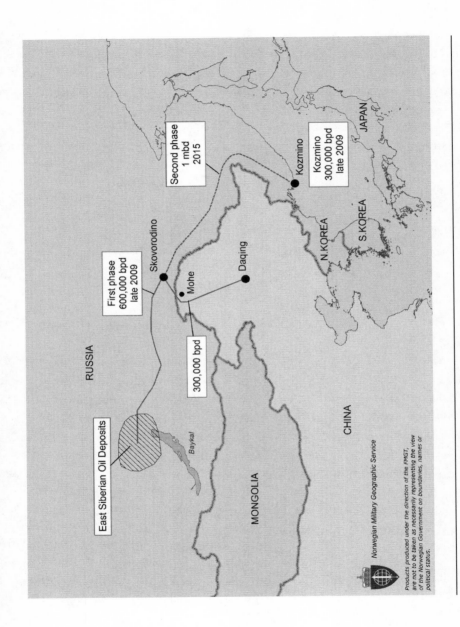

MAP 6.2 East Siberian–Pacific Ocean (ESPO) Oil Pipeline

East Siberian Oil Deposits

First phase
600,000 bpd
late 2009

300,000 bpd

Second phase
1 mbd
2015

Kozmino
300,000 bpd
late 2009

RUSSIA

MONGOLIA

CHINA

Baykal

Skovorodino

Mohe

Daqing

Kozmino

N.KOREA

S.KOREA

JAPAN

Norwegian Military Geographic Service

Products produced under the direction of the FMGT,
are not to be taken as necessarily representing the view
of the Norwegian Government on boundaries, names or
political status.

in order to export to China also is indicative of CNPC's difficulties in building a Russia-China pipeline.[44]

The Russian government has been keen to diversify its energy markets, expanding its production in eastern Siberia and bringing economic development opportunities to the Far East region, although the means to this goal remain unclear and often linked to preserving or enhancing central control.[45] On December 31, 2004, the Russian government approved building the ESPO, and in April 2005 the Russian Ministry of Industry and Energy decided that the pipeline would be built in two phases, but the necessary investments to fulfill these plans were lacking.[46] A real breakthrough on the Russian-Chinese oil pipeline came in October 2008, when former premier Wen Jiabao and his Russian counterpart, Vladimir Putin, witnessed the signing of an agreement between CNPC and Transnet on principles for the construction and operation of a crude pipeline. This deal opened up for Chinese investment in the ESPO project and secured the construction of a spur pipeline from ESPO to the Daqing oil field in China, putting to rest the decade-long Sino-Japanese competition over Russian crude oil and pipeline routes in eastern Siberia and the Far East. Transneft began working on the first 2,757-km stage from Taishet to Skovorodino in April 2006. Completed in December 2009, Transneft estimated that the first stage of the pipeline had a price tag of $14.4 billion. The October 2008 agreement meant that Transneft would construct a 70 km spur from Skovorodino to the Sino-Russian border city of Mohe, and CNPC would invest about $436 million to build a 965-km pipeline from the Russian border to Daqing.[47] The link to China was completed on schedule in October 2010.[48] By January 2011, the pipeline was operating with an initial capacity of 300,000 b/d and with the possibility of increasing the spur supply to 600,000 b/d.[49]

The 2008 financial crisis was a game changer in Russian-Chinese energy relations. Seeking cash and investments to see its oil industry through the crisis, First Deputy Prime Minister Igor Sechin went to Beijing in April 2009 to sign an energy accord with Vice Premier Wang Qishan. The agreement stipulated that the CDB would lend $15 billion to state- run Rosneft and $10 billion to Russia's oil pipeline monopoly Transneft for a reliable supply of 300,000 b/d for the next twenty years, or deliveries of 300 million mt (close to 2.26

billion barrels) over twenty years. It is estimated that the contract between Rosneft and CNPC is worth $100 billion.[50] The interest rate on the CDB's loans to Rosneft and Transneft is the Libor plus a coefficient that fluctuates inversely to the Libor. Rosneft estimates that it will average 5.69 percent over the term of the loan.[51]

The pipeline had an initial capacity of 600,000 b/d, which is slated to grow to 1 mbd by 2012 during the second stage of the project and potentially to as much as 1.6 mbd at a later date.[52] The Russian plan that the second stage of the ESPO pipeline project is to construct a 2,100-km pipeline from Skovorodino to Perevoznaya, later to be moved to the Kozmino Bay on the Pacific coast.[53] This phase is scheduled to be operational by 2012 and will cost around $12 billion.[54] In the meantime, half of the initial exports of 600,000 b/d is transported by rail from Skovorodino to the new $2 billion port of Kozmino, and the other half is piped to China.[55]

After the Russia-China oil pipeline was successfully developed, Russia can deliver oil to China by six different routes: railway, the Russia-China crude pipeline, truck, through the Kazakhstan-China crude pipeline, from Sakhalin, and from oil going to Kozmino that is then sold on the international petroleum market and picked up by Chinese NOCs. China currently imports about 7 or 8 percent of its crude oil, about 600,000 b/d, from Russia, with about 300,000 b/d shipped by pipeline and the remaining imports delivered by rail or imported from Russian export out of Kozmino on the Pacific coast. Accordingly, the Russia-China pipeline provides an opportunity to boost oil imports from Russia, if there is political and commercial will in Russia and China and if Russia has sufficient oil reserves and adequate infrastructure to increase its shipments to China in the future. Having invested in developing capacity and a pipeline so that Russia can increase its oil supply to China provides a hedge against disruptions to China's seaborne oil supplies.[56]

With an energy outlook estimating that "all growth in world energy demand will come from non-OECD countries" and with projections estimating that China will account for 42 percent of the growth in world oil demand over the next two decades, market incentives are forcing Russia to reevaluate its energy strategy and to put more focus on its energy relationship with China and East Asia.[57] Hence, Russia is not ruling out sending additional volumes of western Siberian crude to the ESPO pipeline if there is not enough crude produced in eastern Siberia to fill it.[58]

A RUSSIA–CHINA GAS PIPELINE?

Several factors have prevented a strong momentum to develop the Russian-Chinese gas pipeline. China is not a major consumer of natural gas, which only made up 2.4 percent of its energy mix in 2005.[59] Even though primary use of gas has been growing in recent years and now accounts for roughly 4 percent of China's energy mix, this is far below the world average of about 23 percent in the energy mix.[60] Relying more on gas is important from an environmental perspective, but the Chinese government also understands that diversifying the energy mix by promoting indigenous gas production and developing cross-border pipelines makes China more capable of managing and minimizing risk to its energy security.

Russia's gas monopoly, Gazprom, has emphasized maintaining its monopoly over the lucrative European market rather than diverting much of its resources and interests to the east. The European market generates more revenue, the infrastructure is already in place, and Russia has already signed long-term contracts with European customers.[61] Russia's position in the European market is not only commercially lucrative; it also provides Russia with political leverage within the European Union and has been used as an "energy weapon" against NATO expansion in Eastern Europe and the Commonwealth of Independent States (CIS). Thus, it is likely that Russia will continue to place priority on the European market in the years ahead, even though Russia's energy outlook and strategic priorities may focus more toward the east in the not-too-distant future.

Demand for gas in China has been rising following the government's intention to diversify China's energy mix and to promote gas as a "green" and "clean" energy, hoping to substitute gas for some coal consumption to minimize the environmental impacts of the latter. According to projections by the National Energy Administration Bureau (NEA), China's top energy governmental body, in 2011 natural gas demand will rise by 18 percent to 130 bcm.[62] Only about 10 percent of China's gas demand in 2010 was met by imports.

It is estimated that the import ratio will rise significantly over the next two decades, whereas European demand is expected to grow much more slowly, making China an attractive market for Russia. Promoting natural gas is a top priority in China's Twelfth Five-Year Plan (2011–2015), expecting that natural gas will account for 8 percent of China energy mix by 2015. If this goal is met, then gas imports are expected to rise to 90 bcm by 2015. This is likely to make China a natural buyer of large quantities of Russian gas, thus competing with Europe for the Russian market.[63]

Several steps have been taken in Russia's energy strategy in recent years that show a stronger interest in shifting to a more eastbound flow and potentially to develop gas pipelines to China. "The Energy Strategy of Russia for the Period up to 2030" (ESR-2030), approved by the Russian government in November 2009, shows that plans are developing to accelerate exploration and development of resources in eastern Siberia and the Far East. Several factors account for this shift. First, production in western Siberia has peaked, and the Russian government expects the eastern reserves to make up for the former's gradual decline.[64] Second, Russia is awakening to the shift in gravity toward the east in terms of global energy demand. This will provide commercial opportunities and new markets for Russian petroleum exports. Third, even though Europe will remain Russia's primary concern, strategic considerations are driving Russia toward the east because of geopolitical shifts, growing concern about China's dominance, and great-power politics in Asia. Developing energy reserves in eastern Siberia and the Far East will provide a geopolitical advantage and improve Russian influence in Asia.[65] Fourth, resource development in the east and in Sakhalin will accelerate cooperation with foreign companies and drive domestic development in the Far East. Fifth, by developing reserves in eastern Siberia and infrastructure in the Far East, Russia can improve its bargaining power over Europe by diverting energy supplies to the East.[66]

One of the most contentious issues and a key factor in delaying the construction of a gas pipeline from Russia to China has been disagreement over the pricing formula. Russia seeks to link the gas price to the energy mix or a market spot price. China refuses to pay market price for gas that cannot compete with China's low price of coal.[67] But China has been willing to pay market price for LNG imports, including Russian LNG from Sakhalin. This situation gives Gazprom few incentives to shift its bargaining position. Indeed, recent events suggest that China might be willing to accept a pricing formula whereby the price of gas from Russia would be linked to the price of oil. Conversely, if the Arctic ice cap continues to melt and the Northeast Passage becomes commercially viable as a route for Russian LNG export to East Asia, it might be more cost effective for both Russia and China if Russia ships LNG from the Barents Sea and the Kara Sea instead of developing expensive gas production and pipeline infrastructure in Eastern Siberia and a Chinese pipeline network that will bring gas from the Russian border to China's gas consumption centers in Shanghai and Guangzhou province. It can be expected that China, concerned about both profit and security, will pursue and push for both alternatives.

There have been ongoing negotiations for a pricing formula. No agreement has been reached, but this could be a game changer for a Russia-China gas pipeline. Reports suggest that the two countries may reach an agreement to link natural gas prices to international crude oil prices, which will facilitate further discussions on volume and on the initial contract signed in 2006 calling for two pipelines from Russia to China to transport a total of 70 bcm annually from Russia to China.[68] A recent dispute regarding payment for oil in 2011 under the twenty-year/$25 billion loan-for-oil deal casts a shadow over the cooperation between the world's largest petroleum producer and world's largest consumer. However, oil analysts in Moscow argue that the gas deal will not be sacrificed despite price differences,[69] and close energy ties were emphasized when former president Hu Jintao visited Moscow in mid-June 2011 and when President Xi Jinping met President Putin in March 2013.[70]

CNPC relations with Gazprom were strengthened after President Putin's visit to China in March 2006 and the signing of a preliminary deal whereby Russia would consider building two pipelines to deliver a total of 60 to 80 bcm annually and the Chinese would pay the European price for gas. The two suggested routes were the eastern (Kovykta-Blagoveshensk) and western (Altai-Xinjiang) lines.[71] In October 2009, CNPC signed a memorandum of understanding (MOU) with Gazprom to supply 70 bcm of natural gas to China starting in 2014–2015, with 30 bcm arriving from western Siberia (the Altai gas pipeline route) and the remaining potentially coming from Blagovesh-ensk or via the Sakhalin-Khabarovsk-Vladivostok route to Harbin in China.[72] Despite the difficulties in reaching an agreement on price and disagreements in 2011 regarding the price of crude supplied via ESPO to Daqing, Russian Deputy Prime Minister Igor Sechin reiterated on June 16, 2011, the pledge to supply 30 bcm per year to China via the western route and 38 bcm per year via the eastern route, and he reported that Rosneft had entered into talks with China on increased oil exports to boost the Rosneft and CNPC joint refinery project in Tianjin.[73]

CNPC has also signed an agreement with Rosneft on the basic prin-ciples for establishing joint ventures in China and Russia and an agree-ment on strengthening oil cooperation with Transneft.[74] Still, there is much uncertainty surrounding these deals and whether they will be implemented. In 2006, Chinese frustration was summed up by Zhang Guobao, the head of China's NEA, who stated that Sino-Russian energy relations were "one step forward, two steps back."[75] Five years later, with ESPO being built and a spur to China piping 300,000 b/d, it appears that there has been a change in momentum with respect to Chinese-Russian energy relations.

Chinese NOC participation in the Sakhalin projects is another opportu-nity to develop a gas pipeline to China. Although Gazprom has prevented and interfered with plans to export petroleum to China, when President Alexander Medvedev was deputy CEO of Gazprom, he implied that Gaz-prom might create a joint venture that would distribute gas to China's domes-tic market.[76] In October 2006, Exxon Mobil and CNPC signed an agreement whereby at least 8 bcm of gas from the Sakhalin 1 production site, operating under a PSA, where the natural gas policy does not depend on Gazprom, would be shipped to China's northeastern region.[77] Even if a pipeline from Sakhalin to China does not become operational, oil and gas will still be trans-ported from Sakhalin to China by sea, and this will remain an important part of Russia's Far East energy strategy.

Even if Gazprom were to build a multibillion-dollar pipeline from eastern Siberia to China, one expert argues that Russian threats to the EU about opening alternative routes toward Asia for its gas exports do not appear to be serious, since Gazprom would not be able to increase its annual production by 30 bcm of gas to fill the pipeline.[78] Accordingly, it is contended that Russia is simply bluffing the Europeans by "flashing the unplayable 'China card.'" Gazprom had no intention of investing in the oil and gas field and the infrastructure in eastern Siberia that would be required in order to boost petroleum export to China or in allowing the Chinese NOCs to develop the petroleum reserves themselves.[79]

This argument partly misses the point. It might be correct that to date, or even in the future, Russia, Gazprom, and other major players such as Transneft and Rosneft will not have the capacity, financial strength, or will to deliver on their agreements to develop eastern Siberian reserves and infrastructure. This does not mean that Russia does not have a "China card" to play in its dealings with the Europeans. If Russia has insufficient supplies to deliver on its agreements with the Chinese *and* on its European contracts, or even if the situation is perceived in Europe and Beijing that there are insufficient petroleum reserves to meet demand in *both* the West and the East, then Russia might have stronger leverage in its negotiations with both the EU and China. As one observer notes in commenting on the development of a loan-for-oil deal in February 2009, "there is no way Russia can deliver that amount of oil right now without taking it away from existing export routes to the west."[80] Neither can Russia use shipments of petroleum to the east as a trump against the EU if the infrastructure and reserves are not developed in eastern Siberia and the Far East. In short, if Russia remains passive in the east, it is likely to undermine its leverage in both Brussels and Beijing.

The 2009 loan-for-oil deal shows that the Chinese government and the NOCs are willing to invest to develop eastern Siberian oil and gas fields and infrastructure. Chinese NOCs have struggled, and most likely will continue to face strong opposition, to secure a foothold in the Russian petroleum market, but Gazprom's unwillingness to accept or invite Chinese investments in Russia[81] may become an unsustainable position, as Russia seeks to recover from the 2008 financial crisis and struggles through the 2011 debt crisis and as Russian energy companies are scrambling for cash infusions to weather the credit crisis. With the pathbreaking developments in the North American petroleum market and Russia's already strong position in the European gas market, East Asia and China in particular will become increasingly important markets for Russia. This look toward the East may also be facilitated by the gradual opening of the Northeast Passage. For a long time, there was much skepticism, and rightly so, about a potential oil pipeline from Russia to China; however, the situation has now changed, and it would be unwise to rule out similar developments regarding a gas pipeline.

UZBEKISTAN

Construction of the 530-km Uzbek section of the Central Asian–China gas pipeline began in June 2008. It was built by Asia Trans Gas, a joint venture between CNPC and Uzbekneftegas. Uzbekistan was originally not part of the 30 bcm per year deal between Kazakhstan, Turkmenistan, and China, but on June 9, 2010, former president Hu and Uzbekistan's President Islam Karimov signed a framework agreement whereby Uzbekistan will supply 10 bcm of natural gas to China annually. The two countries will make joint efforts to link Uzbekistan's gas pipeline network to the Central Asian-China gas pipeline.[82] CNPC has also been involved in oil and gas exploration in Uzbekistan since 2006.[83]

MYANMAR

Burma, or today's Myanmar, has longstanding ties with the PRC. Myanmar was the first non-Communist country to recognize the PRC on June 8, 1950. China is Myanmar's largest export market, and in 2010 two-way trade totaled $4.4 billion.[84] Myanmar accounted for 14 percent of China's arms exports between 1990 and 2008.[85] The pipelines and investments in Myanmar's hydrocarbon sector reinforce China's position as Myanmar's largest FDI contributor and strengthen China's ties with Myanmar, having provided 87 percent of the investment inflows into Myanmar in the 2008–2009 fiscal year.[86]

CNPC purchased the Bagan project in central Myanmar, including blocks IOR-3, TSF-2, and RSF-3, from TG World in November 2001. A month later, a contract was signed with Myanmar's Ministry of Energy for exploration of Block IOR-4 in southern Myanmar, obtaining 100 percent holdings in both projects. In 2005, a high-yield gas flow was obtained from well PSC-1 during a production test in Block IOR-4. CNPC signed another contract with Myanmar's Ministry of Energy in January 2007, entering into a PSC with Myanmar Oil and Gas Enterprise (MOGE) and acquiring oil and gas exploration licenses for three deepwater blocks—AD-1, AD-6, and AD-8. The three blocks are located in offshore Rakhine, covering an area of ten thousand square kilometres.[87]

A number of factors explain the construction of a China-Myanmar crude and gas pipeline, including academic interests and ambitions, the parochial interests of local governments, geopolitics, energy security concerns, and competition between China's NOCs. [88] The plans for a pipeline apparently originated among scholars at Yunnan University, writing against the back-drop of President Hu's 2003 speech about China's insecurity of oil supply and dependence on the Malacca Strait. A 2004 article by Professor Yang Xiaohui stresses that a Burma-China pipeline will secure China's petroleum imports by reducing its reliance on the Malacca Strait.[89]

This study, supported by three other scholars from Yunnan University— Li Chenyang, Qu Jianwen, and Wu Lei—emphasized five points in promot-ing the pipeline: it would avoid the Malacca Strait, it would be a shorter route to China's southwest, it would enhance China's influence in Myanmar, the traditionally close China-Myanmar ties would win support for the pipe-line in Myanmar, and the pipeline would be a cheaper option than other land-based alternatives.[90]

This line of reasoning will be explored in the following sections. Suffice it to say here that their proposal won support of the Yunnan provincial govern-ment. With the local government on board, plans were elevated to the highest level of the central government, and approval was eventually secured from the top leadership, including the CCP Politburo. This was a major factor contrib-uting to its success. The NDRC endorsed the pipeline project, naming it one of China's four key import channels. With final approval in December 2007, the NDRC included the Myanmar-China pipeline in the Eleventh Five-Year Plan (2006–2010).[91]

With the consent of the central and provincial governments and China lending its diplomatic support to Myanmar by vetoing UNSC sanctions against Myanmar on January 13, 2007, on June 20, 2008, CNPC signed an MOU with the Myanmar government and Daewoo Combo for the sale and

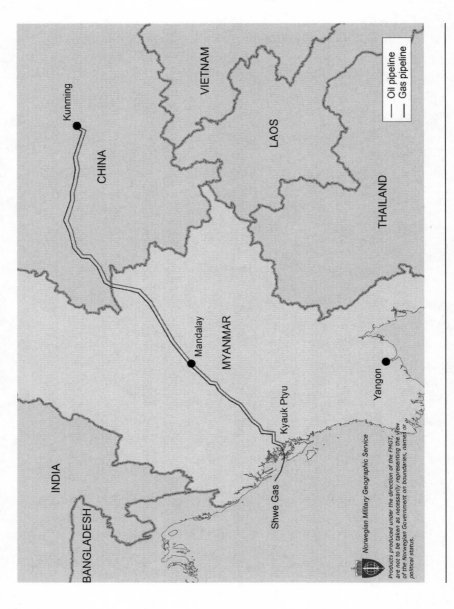

MAP 6.3 Myanmar–China Pipelines

transportation of natural gas to China from Myanmar's offshore blocks A-1 and A-3. An agreement for a gas pipeline to China was signed in March 2009, and an MOU was signed on the oil pipeline to China on June 16, 2009.[92] The Myanmar government's objectives are not explored here, but China's financial package, close diplomatic ties, and CNPC's more viable proposal were probably additional factors in the success.[93]

The oil and natural gas pipeline will run in parallel, with an expected cost of $2.5 billion—an estimated $1.5 billion for the crude pipeline and $1 billion for the gas pipeline. Others argue that the estimated costs for the pipelines and the related refinery and ethylene plant in Yunnan province will cost approximately $5 billion, not including at least $150 to $200 million for the costs CNPC will incur to build a deepwater crude oil servicing dock and an oil storage facility.[94]

CNPC will finance the bulk of the pipeline and hold a 50.9 percent stake, managing the project with the Myanmar Oil and Gas Enterprise. Both pipelines will start at the Kyaukphyu port on the west coast of Myanmar and will enter China at the border city of Ruili in Yunnan province. The 1,100-km oil pipeline will end in Kunming, the capital of Yunnan. It is expected by 2013 to transfer to China as much as 440,000 b/d of crude oil, which will mostly be shipped to Myanmar's coast from the Middle East and Africa. There are also plans to build an oil terminal with 600,000 cubic meters of storage capacity and a port capable of receiving tankers up to 300,000 dwt. CNPC and Sinopec are both looking into building refineries near Kunming and Chongqing municipality and are competing for access to the petroleum market in southwestern China. The natural gas pipeline will extend farther, from Kunming to Guizhou and Sichuan provinces and the Guangxi Zhuang Autonomous Region, running a total of 2,806 km. Beginning in 2012, the offshore site in Myanmar and the pipeline are expected to transport 12 bcm of gas to China annually.

FORGOING PROFITS
AND TAKING OUT INSURANCE

The development of continental pipeline projects has been an important part of China's diversification strategy, but even though diversification is inextricably linked to the continental oil and pipeline deals examined above, the continental diversification strategy is different from how Chinese NOCs have diversified their sources of overseas oil production and imports to places such as Africa and Latin America. For example, differences between the various transportation routes carry security implications. Overseas oil production going to China will need to be shipped back on tankers, whereas continental petroleum imports and production and loan-for-oil deals with China's neighbors will mainly be transported by pipelines, even though a significant volume is currently transported by railway. This has two important implications. First, there is a distinction between the petroleum that China imports through the international petroleum market and what it imports through pipelines, and this distinction allows China to develop a "short" strategy and hedge against some price volatility and supply disruptions. Second, pipelines and seaborne petroleum supplies provide different supply securities during peacetime risks and wartime threat contingencies, again allowing China to hedge its bets. Oil shipped through the Myanmar pipeline should be considered a seaborne route, since Myanmar's domestic crude oil production is limited; seaborne oil supplies will be used to fill that pipeline.

THE PIPELINE "MARKET" AS A "SHORT" STRATEGY

China obtains most of its oil from the international petroleum spot market. This remains a volatile market, with prices in mid-2008 plunging from a high of around $150/b to a low just above $30/b six months later before gradually recovering and stabilizing in the $70–85/b range and moving up to a stable above-$100/b range for most of 2011 and 2012.[95] Despite being linked to and affected by developments in the international petroleum market, the continental pipeline petroleum "market" that is emerging in Asia is largely based on long-term contracts and partially fixed prices through buyer-seller negotiations, even though the pricing contracts are often rewritten.[96]

It has been reported that the oil that Rosneft shipped to China by rail between 2005 and 2010 was facilitated by a $6 billion loan from China that helped Rosneft take over Yukos's assets. With rising oil prices, Rosneft negotiated a revision to the price formula. CNPC agreed to reduce the discount by

$0.675 per barrel in 2007,[97] but it secured a lucrative long-term supply contract at a discounted pricing formula. This illustrates one of the ways in which Chinese NOCs have been able to use long-term future contracts to hedge their bets in the international oil market.

In January 2011, Rosneft decided to extended the contract, but the pricing formula was revised. The 300,000 b/d that China will receive from Russia through the ESPO spur will be determined monthly based on the price of oil quoted in Argus and Platts at Kozmino, the terminus of the second leg of the ESPO pipeline. China has argued for price cuts or concessions based on the fact that it is significantly shorter to pipe the crude oil from Skovorodino to the Chinese border than to deliver crude oil to Kozmino, but there have been no discounts, and CNPC is most likely to pay a premium for its supplies.[98] The "real prize for Beijing was the cross-border pipeline," which has enhanced its supply security.[99]

Given the distinctions between the means of supply, the contracts being signed, and the pricing system, it is hardly surprising that the pipeline "market" is different from the seaborne international petroleum market. In chapter 4, we saw how oil on tankers going to China from the Middle East or Africa may change hands more than a dozen times on its way to East Asia, making it almost impossible to blockade oil shipments to China short of a major war. This differs when it comes to the continental gas or oil pipeline "markets," where the long-term contractual relationship insulates the buyer and seller from outside competitive pressures and makes the "market" highly inflexible.

Therefore, the pipeline "market" in Asia is more political and less affected by commercial opportunities.[100] Although it has been argued that the NOCs' search for profits has been a driving factor behind the "going abroad" strategy and that shipping interests have been influential in shaping the buildup of a large state-owned tanker fleet, political and security concerns seem to be the primary consideration behind the continental pipeline energy strategy. That is not to say that China's NOCs have not been keen to invest in Central Asia, Myanmar, and Russia for profits; that pipelines have not been important in the fierce competition among Chinese NOCs for shares in the domestic oil product market; that local governments have not been looking to use pipelines to boost their provincial standing; or that foreign policy objectives have not played any role. It should be remembered that investors that have developed a hedging strategy can often profit from the "short" side in an investment portfolio.

Simultaneously, the limited continental petroleum pipeline infrastructure in Asia and the inflexibility and distinct characteristics of the pipeline market make it more difficult to profit from overseas production and more likely that petroleum is transported back to China rather than to the international market. Pipelines are expensive to build, and once in place they need to be filled so as to be economically viable. In contrast to Chinese energy companies going abroad to make corporate profits and revenue, investing in continental pipeline projects and loan-for-oil and gas deals have mainly been a strategy to safeguard national interests, with corporate interests a secondary consideration.

BUYING SECURITY

To understand the tradeoff between security and profits in the global and maritime spheres versus in the continental sphere, it is again analytically rewarding to distinguish between wartime and peacetime scenarios. With respect to China's petroleum security, pipelines are more vulnerable during peacetime risks and safer during wartime threats, whereas seaborne energy supplies are safer during peacetime risks and more vulnerable during wartime threats.

Pipelines are less flexible than maritime transportation and therefore more vulnerable to peacetime risks, because of the potential for attacks from terrorists, separatists, and insurgent groups, in addition to environmental catastrophes or accidents. If there were an attack on a tanker or a major accident in the Malacca Strait, for example, other tankers could reroute, whereas pipeline supplies have few or no alternative routes in the case of an attack or an accident, thus possibly leading to more sustained supply disruptions. As an example, the collision in the Malacca Strait between the *Formosa Product Brick* and the *Ostende Max* in August 2009 did not lead to closure of the strait or any disruption in the shipment of oil to China. Tanker flexibility also means that they can continue to operate during environmental catastrophes, such as Hurricane Katrina, which interrupted the supply and transportation of oil in the Gulf of Mexico area in 2005.

As pointed out in chapter 5, there have only been limited terrorist attacks against tankers in Southeast Asian waters. Only two pirate attacks were recorded in the Malacca Strait in 2009. Pirate attacks in Southeast Asia, even in the piracy-prone Arabian Sea or the Gulf of Aden, have not disrupted the flow of petroleum supplies to China and do not represent a significant threat to the supply of oil in the international petroleum market. Conversely, pipelines have been subject to attack by a number of nonstate actors in Africa, Latin America, the Middle East, and Asia.[101] As for China, it appears more likely that insurgency groups, separatists, or terrorists might attack the Burma-China or Central Asian–China pipelines than it is that pirates or terrorists might attack oil tankers in the Malacca Strait.

In April 2010, a bomb blast in northern Myanmar injured several Chinese construction workers who were building a controversial dam. This issue continued to cause difficulties to China-Myanmar relations in 2011.[102] This shows that Chinese interests can be targeted in the conflict between the Myanmar military junta and the various ethnic and separatist groups within that country. Instability can erupt in Central Asia, as witnessed in Kyrgyzstan in April 2010. Insurgency and separatist movements in the Baluchistan region of Pakistan are a major risk to the proposed pipeline from Pakistan to China. There is less instability in Russia's Siberian region and the Far East. Thus, the Russia-China pipeline is likely to be least vulnerable to attacks from terrorists, separatists, or insurgency groups. Finally, pipelines in remote areas are more difficult to protect than tankers on the high seas against peacetime risks. For example, CNPC reports that from 2002 to 2006, thieves drilled into pipelines on 18,382 occasions.[103] This shows that terrorists, separatists, and insurgency groups can more easily target pipelines than tankers on the high seas or even at vital choke points.

Thanks to its monopoly position, Russia has used its gas supplies to leverage against the EU, and it was involved in a gas dispute with Ukraine in 2005–2006 and 2009 and with Belarus in 2006–2007 and 2010.[104] It is unlikely that Russia will blackmail China in a similar fashion. China is a great power that dominates the Asian mainland, and it could easily threaten the Russian Far East, not necessary by military means but simply by opening its border to more Chinese migration. A Sino-Russian conflict or war would not prevent

China from importing oil from the international petroleum market. Oil has always been available in the international petroleum market and is most likely to be available in the future to customers able to pay, and China currently has the money to pay for oil. This does not imply that China is not concerned about relative gains and about pursuing a hedging strategy whereby it seeks to be reasonably well off during a crisis in the international petroleum market. Being rich is not a disincentive for hedging. While the Sino-Russian pipeline is relatively safe and can be easily replaced if cut off by Sino-Russian conflict, when it comes to peacetime risks, pipelines are less secure than oil supplied by tankers and the international market.

In a wartime scenario, pipelines are safer than ships at sea. Though unlikely, were war between the United States and China to erupt, the U.S. Navy could blockade oil shipments to China. China does not possess the military capability to prevent the United States from interrupting China's maritime SLOCs during a war. Yet the United States may be reluctant to escalate such a conflict by attacking China's neighbors in order to destroy pipelines going to China. Accordingly, because of the danger of escalation, it is argued that the Russia-China pipeline would be "safest," the Kazakhstan-China pipeline less so, and the Myanmar-China pipeline the least safe, both because of Myanmar's isolation in the international community and because the oil filling the Myanmar pipeline arrives by sea and the gas production is largely offshore.

Of course, the several thousand kilometers of pipelines in China are vulnerable to precision strikes, but the escalatory barrier is higher when it comes to engaging targets on the Chinese mainland, for example, if China and the United States were to find themselves at war over Taiwan. Moreover, if the United States decides to attack China's pipeline infrastructure, it can be assumed that the PLA's land-air defense would be more capable in defending its territory than the PLAN's ability to defend the sea lanes in the Indian Ocean or the South China Sea. Accordingly, pipelines are more "secure" than sea lanes during periods of war.

Additionally, seaborne petroleum supplies are more vulnerable during wartime scenarios not involving China. A blockade or an attempt to close the Strait of Hormuz would most likely be considered an act of war. It can be expected that the United States will take action to prevent any state from

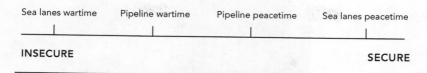

FIG. 6.1

closing the strait, as indicated by previous and current disputes between the United States and Iran. A closure of the Strait of Hormuz during a potential war in the Middle East will cause more vulnerability to China's energy security and reliance on oil imports than the threats facing China's petroleum pipeline supplies during a wartime scenario. Roughly 40 percent of the world's seaborne oil shipments—and more than 40 percent of China's oil supplies—transit the Strait of Hormuz, the world's most vital choke point, with no opportunities for circumnavigation. A closure of the strait thus will cause havoc in the oil market and strike a much greater blow to China's energy security than the bombing of a pipeline supplying China with petroleum or a potential supply disruption resulting from a war with neighboring countries that export oil to China through pipelines. The volume of petroleum going to China through pipelines is significantly lower than the volume of seaborne oil going to China from the Middle East. Thus, during a wartime scenario, China's seaborne petroleum supply is more vulnerable to disruption than is its pipeline supply. In addition, China cannot enforce controls or safeguard shipping in the Strait of Hormuz. As pointed out in chapter 4, China may used its state-owned tanker fleet as a hedge in such a situation, but the supply of petroleum through continental pipelines not affected by the closure of the vital strait also serves as an important insurance or a hedge under such circumstances.

Figure 6.1 demonstrates the security of various petroleum transportation routes to China. Not only is it important to make a distinction between wartime and peacetime scenarios, but, as we have seen, it is also important to examine the degree of security obtained from the various pipelines that will transport petroleum to China. There is evidence to suggest that China can expect reliable supplies through its long-term contracts. Russia has been a reliable supplier of gas to the European market and has delivered on its contract obligations to Western European states willing to pay "market" prices, despite the so-called gas wars with Ukraine and Belarus.[105]

During the Cold War, Russian gas continued to be exported into Central and Western Europe.[106] Pipelines are expensive to build and often not economical to duplicate, thereby creating a long-term mutual interdependency that mitigates against cutoffs. Russia's goal of selling to the Europeans as a reliable, trusted, and safe alternative to the volatile Middle East has been undermined by its disputes with many of the CIS countries and Eastern European states, but Russia is not likely to bully China in similar ways.[107] Ukraine, Belarus, and Moldova are key transit states for Russian gas sold to the EU. These states have some leverage, since both the EU and Russia depend on transit states and cannot always directly work out their differences as a monopoly supplier or monopsony buyer.[108] Russia has been subsidizing former "allies," including the transit states, with petroleum supplies. As some of these states are considering membership in NATO, it is hardly surprising that Russia is unwilling to continue subsidizing these states with cheap gas supplies.

Issues related to transit states and subsidies are not relevant to most of China's continental pipeline petroleum supply contracts. The Russia-China and Kazakhstan-China crude pipelines do not have to deal with transit countries. Even though the Turkmenistan-Uzbekistan-Kazakhstan-China gas pipeline involves transit, there is still an important distinction with Russia's gas pipelines to Europe, since both Uzbekistan and Kazakhstan will be selling their gas to China through the pipeline and therefore will be interested in maintaining the flow of gas and preventing disruptions. The same argument applies to Russian crude oil sold to China through the Kazakhstan-China crude pipeline. If great-power competition intensifies between Russia and China in the future, as a great power Russia may be more capable and willing than either Kazakhstan or Myanmar to breach its contract obligations to China. In such a conflict scenario, during either a peacetime crisis or during a war with one of its continental petroleum suppliers, China will most likely have alternative supply routes and access to oil on the international petroleum market.

The Myanmar-China pipelines can be considered transit pipelines; the oil that will be pumped to China through this pipeline will be shipped to Myanmar by sea. Gas produced offshore and crude oil arriving on tankers at the port of Kyaukryu can be easily identified and be subject to blockades. The production site, terminals, and storage facilities can be targeted with precision strikes from naval warships and submarines, giving Myanmar, and potentially Chinese forces, limited options to defend petroleum production, storage, and transportation sites. Finally, as discussed above, the Myanmar pipelines face security risks from separatist groups in regions through which the pipelines must pass.[109]

A potential pipeline from Pakistan to China faces all of the security risks of the Myanmar-China pipelines, including being supplied by seaborne petroleum at the port of Gwadar, which can be a target for a blockade or precision strike, and internal instability in Pakistan, such as in Baluchistan province and in the Kashmir region, where petroleum installations have been attacked in the past. A Pakistan-China pipeline might overcome the risks at the port of Gwadar if an Iran-Pakistan-China pipeline were to be built, but more transit nodes make the pipeline more vulnerable to supply disruptions.[110]

It has been suggested that China could have built more than fourteen VLCCs, at an estimated cost of more than two billion dollars, for the same price as the Myanmar-China pipeline. Such tankers could have delivered an average of 666,000 b/d of crude versus 400,000 b/d for the planned pipeline.[111] Of course, it takes years to build fourteen VLCCs, but it has also taken years to construct the Myanmar-China oil and gas pipeline. The capacity of the Myanmar-China crude pipeline is estimated to carry 440,000 b/d within a few years, in addition to 12 bcm of gas, but this again illustrates some of the costs of hedging with more expensive pipelines.

There are reasons to believe that there is a cost disparity between maritime shipping and pipelines. Reliance on seaborne oil imports would most likely represent cost savings over pipelines, such as the Myanmar-China or the Pakistan-China pipelines, and probably some costs would be reduced if China favored buying oil shipped by tankers from the international petroleum market rather than building expensive pipelines and inking loan-for-oil deals with Central Asia and Russia.[112] The fact remains, however, that instead of maximizing its search for profits, the Chinese government has been keen to buy security and to manage risk through a continental pipeline hedging strategy. Two energy researchers have emphasized that the building of the crude pipeline from Kazakhstan to China was hardly economically attractive because the capacity of the pipeline was low. Yet they note that "China was willing to invest in this pipeline in order to establish the route."[113] As other analysts observe, China has taken out "an insurance policy on land" and is "willing to pay a higher price in exchange for energy security."[114] Accordingly, the Chinese government might settle for lower returns because of other objectives than profit maximization, such as securing access to energy resources abroad.[115]

Simultaneously investing in a global maritime and continental energy strategy may prevent China from pursuing the most cost-effective and profitable strategy. It may not be a economically sound strategy if the goal is to maximize profits, but the Chinese government has hedged its bets and bought supply security by pursuing more than one strategy so as to minimize risks, diversify, and insure against peacetime risks and wartime threat scenarios, and it has decided to rely on both the international spot market and long-term pipeline contracts, thereby operating with "longs" in the international petroleum market and "shorts" in the pipeline "market."

DOMESTIC INCENTIVES AND FOREIGN POLICY OBJECTIVES

There are more ways to profit than by only relying on the NOCs' overseas production and acquisitions, promoting a state-owned tanker fleet, and pursuing the most cost-effective transportation routes for oil imports. As pointed out in previous chapters, in some cases China's broader foreign policy interests may trump corporate or national energy interests. By supporting UN sanctions on Iran or intervention by UN peacekeeping operations (UNPKO) in Sudan, China may undermine its corporate energy interests, but at the same time it may profit from increased status, prestige, and its role as a "responsible stakeholder" in the international community. Pipelines have clearly been pursued for broader goals than simply diversification of China's petroleum import routes. In the case of the continental petroleum strategy, this is even more salient, as China profits both internally and externally from pursuing "uneconomical" pipeline and oil-for-loan deals with its neighbors. A continental petroleum strategy that is less economically or corporately profitable may still boost national security, domestic interests, and balance the pursuit of security and profits in China's foreign and energy security policies.

INTERNAL PROFITING FROM CONTINENTAL PETROLEUM DEALS

When the Central Asian pipeline plans were being considered in the 1990s and 2000s, some researchers suggested that it would make more sense economically if the oil and gas from the landlocked Central Asian countries were to be "transported to the Mediterranean and then shipped through the normal sea-lanes" to China.[116] At the time, proposals for large-scale pipelines between Central Asia and China were dismissed as nonviable from a cost-benefit viewpoint.[117] As it turned out, the landlocked supplies and infrastructure in Central Asia have been developed in order to enhance Beijing's ability to manage maritime risks and to improve supply security. As we have seen, domestic commercial interests have also been a driving factor behind the continental pipeline plans,[118] suggesting that both profit and security considerations shaped China's continental petroleum strategy. In establishing a hierarchy of causes and arguing that security concerns were the key to developing cross-border and domestic pipelines, it is also important to stress domestic political factors. Indeed, pipeline projects have been used strategically by the central government to develop China's remote and backward regions and to stimulate economic growth through increased public spending.[119]

China's primary focus in its relations with Central Asia and Russia in the 1990s was to secure China's borders in order to prevent any instability from the breakup of the Soviet Union to spill over into China. The next focus was to solve border disputes and enhance cooperation with the Central Asian states and Russia against the three evils of separatism, religious extremism, and terrorism.[120] Initially, energy diplomacy was a means to reach these objectives.[121]

By the turn of the century, when progress had been made on these objectives, China became more concerned about its security of supply. Thus Chinese NOCs more forcefully attempted to gain a foothold in Central Asia, but domestic issues still remained important.

In an analysis of China's search for petroleum deals and pipelines in Central Asia, it is important to note that the development of pipelines offers the government more tools to exert internal control over unstable regions and also to develop these regions.[122] Instability and riots have erupted in China's western Xinjiang Autonomous Region and the Tibetan Autonomous Region. Thus, China's energy security policy in the west and in Central Asia is partially driven by internal security concerns and threats of separatism and unrest.[123] In the "develop the West" strategy, the goal of the pipelines and infrastructure is to bring economic development and thus pacify the Xinjiang region.[124] The twelfth Five-Year plan (2011–2015) seeks to boost investments in development of China's western regions. The NDRC has announced that $100 billion will be invested in twenty-three new infrastructure projects across China's remote and underdeveloped western regions. Taking advantage of the cross-border pipelines from Central Asia to China, Xinjiang is expected to develop refining and chemical manufacturing, oil storage, and engineering and technology services and advance its oil and gas production. Seeking to achieve the goal of increasing Xinjiang's GDP per capita level, the central government's investment in fixed assets is expected almost to double compared with the previous five-year plan (from $160 billion in 2006–2010 to $307 billion in 2011–2015).[125]

China's northeastern region, despite the Korean and Mongolian ethnic groups in the area, is not populated by ethnic groups seeking independence or self-determination. Moreover, this is a relatively well-developed region in China. Thus, concerns about domestic development may not have been a major driver. Still, the belief that pipelines and investments in the hydrocarbon sector will bring economic development and stability, which has been stressed in several five-year plans, is likely to have been an important secondary consideration behind promoting China's energy ties with Russia. As one scholar notes, China seeks to profit internally from a periphery secured through energy deals, building external relationships into a "system of controls," to be strong in the center and defend internal unity by externalizing constitutional imperatives.[126]

Similar internal dynamics were driving the objective of developing pipelines from Myanmar. The mutual interests between the local government in Yunnan province and CNPC probably secured the approval by the central government for the Myanmar pipelines. But Beijing's endorsement of the pipeline project must be understood against two important backdrops. First, as emphasized throughout this book, the central government's concern about China's security of petroleum supplies, which became a driving force behind China's energy security policy after the turn of the century, was a major factor. Second, the central government recognized that the pipeline project could bring development opportunities to China's economically underdeveloped southwestern region and enhance stability.[127] Therefore, the project would advance the broader goal of developing China's interior regions and address the development gap between China's eastern and western regions.

The Myanmar-China pipeline proposal originated among scholars at Yunnan University and was submitted for approval to the central government by the local Yunnan provincial government. Economic and commercial opportunities were a major incentive behind the local government's support, and the pipeline offered CNPC an opportunity to compete with Sinopec for market shares in both Yunnan province and China's southwestern region. A crude pipeline from Myanmar would be a remedy for Yunnan province's frequent oil shortages, would facilitate the buildup of local refineries, and mitigate the 30 percent higher price that Yunnan province pays for its oil products, which it must acquire from distant refineries in Guangdong and Gansu provinces.[128]

Not only could the local government in Yunnan obtain a better price for its oil imports, diversify its oil import routes, push for the development of local refineries, and better manage supply shortages and disruptions, but a pipeline from Myanmar could provide an opportunity for Yunnan province further to open up its economy and become an economical "hub" in China's southwest, thereby boosting the province's strategic importance and potentially "enhancing its leaders' chances of climbing up the country's political ladder."[129] Yunnan province could also become a "bridgehead" to connect China's southwest with Southeast and South Asia and to open up the Indian Ocean shipping route for energy imports and cargo/services trade, thus advancing China's "two ocean strategy."[130] The proposed plans for a new railway from Kunming in Yunnan to the Myanmar coast would help realize such ambitions.

The Yunnan provincial government also won support from CNPC, which was eager to use the Myanmar pipeline option to compete with Sinopec. Originally, Sinopec controlled oil imports to Yunnan through the Maoming-Kunming oil pipeline. Developing a pipeline in Myanmar will allow CNPC to transport its own oil to Yunnan province and to develop refineries to compete with Sinopec for market shares in China's southwestern region. By linking up with the Yunnan provincial government, CNPC agreed to invest a total of $11.7 billion in Yunnan province to "build the Myanmar oil and gas pipelines, refinery, and petrochemical and other related facilities, which will be the largest single investment the province has ever received."[131] All this suggests that the central government believed that the Myanmar-China pipeline could boost internal development in China's southwestern region.

EXTERNAL STRATEGIC AND SECURITY GOALS

Russia's strong position in the European gas market has created tensions with the EU and Eastern European states, but China will not be as dependent on Russian or Central Asian petroleum supplies as the European states are on Russian supplies. More importantly, China's superiority over the Far East border between Russia and China provides Beijing with some leverage.

Russia's aggressiveness toward the Europeans has also been the result of Moscow seeking to prevent competition from other suppliers, which might undermine Gazprom's monopolistic position in the European market. Again, this works differently in China, where Russia is not a monopolist—it is actually seeking to diversify its supply away from relying solely on the Chinese market by building the ESPO and developing reserves in Sakhalin, which means that Russian petroleum can be sold to customers in East Asia and the international petroleum market.

Neither are the Central Asian states or Myanmar solely relying on the China market, and they are both inclined to use aggressive means to keep China as a customer. In short, the petroleum pipeline "market" that supplies

China differs from the European market from both a supplier and a consumer perspective. Additionally, China's energy mix differs from that of a number of leading European states, meaning that China is less dependent on its pipeline petroleum supplies. This fact facilitates cooperation between China as a major petroleum consumer in search of security of supply and China's petroleum-exporting neighbors seeking security of demand.

SINO-RUSSIAN RELATIONS

The Russian and Chinese current "axis of convenience" is largely founded on apprehension about a U.S.-dominated unipolar system.[132] Indeed, their cooperation is driven by the fact that U.S. military capability poses a threat and challenge to their security.[133] However, if China continues its current economic growth and military modernization and expansion, threat perceptions in Moscow may change from a preoccupation with the United States and NATO expansion to China's dominance in Asia. The U.S. military presence in Europe has decreased since the Cold War. As the United States becomes more preoccupied with the rise of China and nontraditional security threats such as the ongoing war against terrorism,[134] U.S. military capabilities in Europe, in addition to those of NATO, may become a secondary threat to Moscow, while competition and rivalry with China in Asia will become a primary concern. As a result, contemporary Sino-Russian cooperation may be unsustainable given the asymmetry in their relationship and potentially shifting threat perceptions in Moscow. The lack of depth in Sino-Russian relations is also illustrated by the fact that Russia and China put more emphasis on their bilateral relationship with the United States than on the so-called strategic partnership.[135] Although it is questionable whether Russia will remain sufficiently powerful, Moscow could gain more from playing China and the United States off against each other politically and strategically in a bipolar world based on U.S.-China relations.[136]

If the United States is confronted with a two-front scenario, facing a rising and more dominant China in Asia and a reemerging Russia that is putting pressure on its neighbors in Eastern Europe, given its continuing efforts to fight terrorism, it is likely that U.S. capabilities and resources will be overstretched.[137] Although Obama's goal of pushing the "reset button" in U.S.-Russian relations has probably been driven by the effort to advance benign diplomatic relations with Moscow to counter Russian objections to a missile defense system, seek Russian support to deal with Iran and Afghanistan, and conclude a new START II agreement, long-term strategic calculations cannot be dismissed. One cannot rule out that the United States might play a "Russia card" to balance against China, just as the United States played a "China card" in the early 1970s to balance against the Soviet Union.[138] Increased Russian leverage may constrain Sino-Russian relations.

With China's looming economic and demographic presence, there is a fear in Russia that China will have de facto control over the Far East and push Russia out of Central Asia. Although the Central Asian states have shown an ability to play the major powers against one another, and although it is unlikely that China will seek to occupy Russia's Far East territorially, one observer has argued that China's rise to dominance on the Asian continent "is the single most difficult issue on Russia's foreign and security policy agenda."[139] Others may point to NATO's expansion in Eastern Europe, instability in the Caucasus, and separatism and terrorism as the primary Russian security concerns, but this does not distract from the point that tensions could still erupt in Russian-Chinese relations.

Energy cooperation will not offset conflict if Sino-Russian security interests diverge. Instead, it would seem that oil cooperation would be more likely a casualty of security conflict. Nonetheless, large oil-for-loan deals and the building of pipelines may be one way of reducing competition and advancing long-term cooperation between China, which seeks security of supply, and Russia, which seeks to diversify its market and enhance security of demand. Energy deals therefore might consolidate Sino-Russian ties, contribute to preventing Russia from moving too close to the United States, and insure that competition between China and Russia in Central Asia does not undermine their overall relationship. Although Russia will not be in a position to use petroleum exports as a "weapon" against China in a way similar to how it has been used against the CIS and EU, energy ties between China and Russia provide Moscow with more leverage in Beijing. It is likely that Russia would have had less influence in Beijing if reserves in eastern Siberia and pipeline projects and infrastructure in the Far East had not been developed. Ensuring "closer energy ties to China" can help "cement Russia's position as a Eurasian (or European and Asian) 'Energy Superpower.'"[140] Russia can also use its position as an important energy supplier to China to bargain against China's growing and favorable military balance vis-à-vis Russia in Asia.

Loan-for-oil deals will also facilitate investments and advance Russia's objective of developing the Russian Far East, thereby giving Russia a stronger presence in a region dominated by China, and at the same time, doing so does not threaten China's interests. The ESPO and the loan-for oil deal agreement illustrates that China and Russia can cooperate on geopolitically sensitive issues even as they compete for influence in Central Asia.[141] Another researcher emphasizes that the ESPO has in many ways locked Russia and China into a producer-consumer long-term relationship.[142] In short, China's energy deals with Russia encompass both security and profiting and resemble a hedging strategy that mixes cooperation and moderate balancing or competition.

Other analysts have stressed that control over the future of Central Asian petroleum flows and production rights have already emerged as a contentious issue between Russia and China.[143] If Russian gas output remains largely the same, around 560 bcm per year during the last several years, it will face a potential supply shortage if it fulfills supply contracts to Europe and meets growing domestic energy demands.[144] Russian supplies are enhanced by Russia's ability contractually to lock up petroleum supplies from Central Asia and to control a substantial part of the petroleum exports from Central Asia, which are shipped through the Russian pipeline system.[145] As one observer notes, it is "urgent that Russia restrict Central Asian production to mainly, or even exclusively, Russian channels."[146] However, Russia's control of petroleum exports from Central Asia has already eroded. As Kazakhstan and Turkmenistan are developing alternative pipeline routes for their exports, Russia faces stronger competition in securing and picking up cheap Central Asian petroleum. Thus, Gazprom can no longer buy large quantities of Turkmen gas at a low price for domestic consumption and then sell its own gas at market prices to European consumers.[147] At worst, Russia may find it more difficult to deliver on its contractual gas commitments to Europe.

How Beijing's influence in Central Asia may undermine Russian interests was underscored during the building of the final stages of the Turkmenistan-Uzbekistan-Kazakhstan-China gas pipeline. Knowing that a new market would open, and after securing a $4 billion loan-for-gas deal with China in addition to seeking pipeline deals with Iran and Europe, Turkmenistan was able to withstand pressures from Gazprom during a dispute over a pipeline explosion in April 2009. Once the Turkmenistan-China pipeline was opened on December 14, 2009, Turkmenistan announced that Russia would resume Turkmen gas imports in 2010 at only 30 bcm, less than half the 70 or 80 bcm agreed to in deals prior to the dispute with Gazprom in 2009 and below the 50 bcm Russia was buying prior to the dispute. Turkmenistan and other Central Asian states have been able to renegotiate substantial price increases on their petroleum exports to Russia during the last several years by using China as an important "vector" against Moscow.[148]

Alternative transportation routes for petroleum exports from the Central Asian states reduce Russia's leverage and increase the bargaining power of the Central Asian states and China.[149] The role of China in Central Asia has therefore undermined two important Russian objectives, namely to maintain and exploit its monopsony position vis-à-vis Central Asian suppliers and to obstruct efforts by Central Asian suppliers to bypass the Gazprom export pipeline system.[150] As some Russian experts note, Moscow seeks to check "China's penetration into Central Asia," and a "'close friendship' between Russia and China would only be a nice cover for the fact that 'our countries will remain competitors and this competition will be very tough.'"[151] Indeed, it is interesting to observe how Central Asian states originally were keen on pulling China into Central Asia as a counterweight to Russian dominance in the 1990s and early 2000s, but now as China has become too dominating they are seeking to pull Russia back in or at least preserve close ties with Russia.

At least two issues should be noted when discussing whether Russia is loosing its grip over the oil and gas in Central Asia and whether Russia dislikes seeing petroleum supplies moving eastward from Central Asia to China. First, in many cases China and Russia are not competing for the same reserves and supplies. For example, fields in the Mary province of Turkmenistan, which have traditionally been used to supply gas to Russia, will now be used to feed the pipeline to China. Gazprom's main resource base for these supplies is the Dalulatabad-Sovetabad fields, not the Samandepe and Altyn Asyr fields, which will be available for the pipeline to China. In addition, the Kazakhstan-China crude pipeline has, at least temporarily, opened up an alternative route for exports of Russian crude oil to China.

Second, although Russia and its major petroleum companies may dislike the competition from Chinese NOCs in terms of exploration, production, and market share in the Central Asian oil and gas market, it might be preferable to Russia that the Central Asian gas moves to the east rather than by alternative routes to the west that might challenge Russia's monopolistic position in the European gas market. Indeed, the China petroleum market that is opening up in the east for Central Asian suppliers may actually support Russian ambitions to undermine any Central Asian–European gas pipeline that could bypass Russia. Faced with the development of reserves in Turkmenistan, "it is better for the gas to go east than west, where it would compete against Russian gas in its primary European market."[152] The China pipeline accommodates Gazprom's interests, as it avoids additional investments in less profitable markets and cements Gazprom's position in profitable markets.

As long as Russia is able to access additional gas in Central Asia so as to deliver on its contractual agreements to Europe and to make sure that Europe-bound Central Asian gas will continue to be transported via Russia, Sino-Russian energy relations may continue to contribute in maintaining a "strategic partnership" able to manage the competition, rivalry, and tensions that frequently erupt in great-power relations.

CHINA AND CENTRAL ASIA

China has been cultivating ties with the Central Asian states since the 1991 fall of the Soviet Union, which opened up China's western frontier both to challenges and opportunities. China's strategy toward the region has predominantly been focused on tackling the challenges: the "three evils" of separatism, religious extremism, and terrorism.[153] Trade relations have steadily increased since the 1990s, and in 1997 CNPC had its first major breakthrough in Kazakhstan. With the signing of border agreements with the Central

Asian states and the establishment of the Shanghai Cooperation Organization (SCO), economic and energy opportunities became a more important element in the broader goal of creating a benign environment with China's Central Asian neighbors.

During the inaugural ceremony of the Central Asian–China gas pipeline on December 14, 2009, the leaders of Turkmenistan, Uzbekistan, Kazakhstan, and China not only spoke about how the gas pipeline would help China meet its energy demands, diversify its transportation routes, and improve its energy consumption mix but also emphasized how the project promoted broader socioeconomic development in the region.[154] It was claimed that "the project has truly realized the balance of interests among energy exporters, transporters and consumers." The leaders even referred to the pipeline project as the "modern silk road," stating that "the most significant point of the modern 'Silk Road' lies in that neighbours in the region find more common interests and cooperative opportunities."[155]

According to the *New Chinese News Agency* (*Xinhua*), President Hu stated that the pipeline is a model of sincere solidarity and mutually beneficial cooperation. It was pointed out that the project employed twenty thousand local workers, or 90 percent of the workforce, and 60 percent of the middle-level managers were local employees. There was also additional funding for student programs and infrastructure development.[156]

Clearly, China's huge loan-for-petroleum deals, its cooperative approach to pipeline construction, and its soft power have led to stronger Chinese influence in the region and have helped to "improve Beijing's security and border relations."[157] Consequently, China is not only profiting in terms of energy security by diversifying its sources, routes, and mix, but it is also profiting in terms of advancing broader national interests through its continental petroleum strategy.

As pointed out in previous chapters, there are risks and costs inherent in both strategic and the market approaches to energy security, or in the pursuit of security and profits, and China will continuously have to reassess its policy to find the appropriate balance. For the foreseeable future, however, it is likely that security and political concerns will remain dominant in China's continental petroleum strategy.

CHINA AND MYANMAR

While the aim of diversifying its supply routes and bypassing the congested Malacca Strait has been put forward by many as an argument for the China-Myanmar pipeline, there is little evidence to suggest that the alternative route enhances the security of China's petroleum supply. Some have argued that the Myanmar pipeline will provide an alternative if there were an accident or another disruption to the transportation of petroleum through the Malacca Strait,[158] but the capacity of the pipeline is limited and it cannot substitute for taking on additional seaborne supplies heading to China. If the pipeline is running at its maximum capacity of 440,000 b/d within several few years and the Malacca Strait were to be closed, how would the Myanmar pipeline be able to accommodate additional barrels of crude oil? The Myanmar-China pipeline provides no alternative for oil shipments to China. Tankers would much rather circumnavigate the Malacca Strait. It is estimated that circumnavigating the Malacca Strait would result in four to sixteen days of disrupted oil shipments, depending on whether tankers are rerouted through the Lombok strait or all the way around Australia, and the additional cost might be as little as two dollars per barrel.[159] However, the 440,000 b/d would probably arrive more rapidly to landlocked southwestern China through the Myanmar pipeline than via a seaborne route.

If any state were to blockade the Malacca Strait successfully, it could easily also blockade ports on Myanmar's west coast. The pipeline probably faces more security risks from insurgencies and separatist groups in Myanmar than tankers do from piracy in Southeast Asia. Indeed, the bomb blasts in northern Myanmar on April 17, 2010, which injured several Chinese workers building a controversial dam project, illustrate the risks of pipelines running through Myanmar.[160] In short, the inflation in the political discourse in China and among international observers suggesting that a Myanmar-China pipeline allows China to bypass the Strait of Malacca and therefore enhances China's energy security rests on flawed assumptions.[161]

In terms of corporate profits for Chinese NOCs, it can be expected that the pipeline will bring additional crude oil supplies to China's southwest region. Competition to develop refineries in the southwestern region, primarily between CNPC and Sinopec, suggests that Chinese oil companies seek to profit from the building of the pipeline.[162] It has been estimated that oil through the Myanmar-China pipeline will cost $4 per barrel, whereas seaborne oil from the Middle East or Africa to Guangdong costs less than $2 per barrel.[163] However, oil would then have to be piped from Guangdong to southwestern China, which would increase the cost of seaborne crude supplies going to China's southwest. Taken together, the cost of bringing crude oil to southwestern China may be roughly the same for pipeline and seaborne shipments. But if the costs of constructing the pipelines, refineries, and infrastructure are included, it is probably more profitable economically to ship oil on tankers to China's east coast and then to distribute it in the domestic market through existing and future pipeline networks and refinery capacities. A cost-benefit analysis was not likely the primary driver behind the pipeline proposals. Domestic development goals, local provincial government interests, CNPC's scramble for market shares, and even confused considerations regarding the "Malacca dilemma" and the security of supplies were an impetus for the Myanmar-China pipeline option. Nonetheless, it is important also to consider the broader security concerns.

As with China's broader security interests in Central Asia, pursuing a neighborly policy that ensures stable borders is an important objective of Beijing's relations with Myanmar.[164] China shares a 2,200-kilometer border with Myanmar. Occasional fighting between government forces and insurgent groups has occurred in various regions of Myanmar, and Myanmar's repressive policies against ethnic minorities will probably incite more insurgencies in the future.[165] Instability in Myanmar could spill over to China's ethnic groups and to the less-developed provinces in the southwest. It is estimated that the Myanmar-China pipeline will generate $1 billion or more in annual revenue for Myanmar's government, equivalent to about a one-third of the country's existing foreign exchange reserves.[166] This will boost the military junta's ability to maintain control and stability, but the pipeline could be a target for insurgent activity.

Another larger goal behind financing pipeline construction in Myanmar has been to maintain a strategic buffer zone for China and limit the influence other countries might have on Myanmar.[167] There is ongoing competition between China and India for influence in Myanmar, and the increased prospect for political reform in 2012 has invigorated Western interests and activism in Myanmar. The decision to award CNPC equity in Myanmar reserves and to build a pipeline to China dealt a blow to efforts by Indian energy companies to establish a stake in Myanmar. China also undermined India's position and boosted its ambitions to cement its ties with Myanmar through energy deals when, on January 12, 2007, it vetoed a UK/U.S.-backed UNSC draft resolution on Myanmar that called on the country to release all political prisoners, begin dialogue, and end military attacks and human rights abuses against ethnic minorities.[168]

Energy ties have also been an important factor in a broader strategy to prevent Western intrusions into a key neighboring country. Instability in Myanmar, which eventually could bring a friendly Western government to power, might not be conducive to China's interests. Although the pipeline may become a target for separatists and protest movements, the transit fees will generate a stable source of income that will make it more likely that the Myanmar government can withstand domestic protests, Western pressures, and sanctions from the United States and the EU.

Finally, cultivating ties with littoral states around the Indian Ocean provides China with an opportunity to gain access to the SLOCs. Petroleum deals with the Myanmar government can facilitate a Chinese presence in the Bay of Bengal and provide an opportunity to develop supply stations for the PLAN, making it easier to carry out military exercises. Investments in Myanmar therefore constitute a stepping stone, although a small one, in China's ambitions to become a maritime sea power.[169]

CONCLUSION

This chapter has examined China's engagement with neighboring petroleum producers and potential transit states. Breakthroughs in terms of production rights, pipelines, and loan-for-petroleum deals are highlighted to present an overview and an empirical backdrop for an analysis of the drivers behind China's continental petroleum strategy. It is argued that China's continental petroleum strategy provides a hedge against supply disruptions and promotes broader domestic and foreign policy interests.

Not only has China diversified its portfolio of sources of supply and transportation routes by developing a continental petroleum strategy, but it is also hedging its bets for future contracts with the Central Asian states, Russia, and Myanmar. Using future contracts to insure against price risks and volatility is an important characteristic of a hedging strategy and a distinct aspect of China's continental diversification strategy. In other words, China is investing a share of its energy security portfolio in "shorts" while maintaining most of its oil imports through "longs" from the international petroleum spot market. The continental petroleum strategy may work as a protection, or a hedge, if China were to face supply disruptions or security threats that affected affordability, accessibility, and global transportation in the maritime domain.

Noting the differences between a pipeline "market" and the international petroleum market and the distinctions between pipeline and seaborne supplies, it is argued that pipelines provide less security than seaborne supplies during peacetime risks, but they provide more security than seaborne supplies during wartime threats. Hence, mixing pipeline and seaborne supplies is an important hedging strategy that has improved China's continental security of supply.

Taking out land insurance implies that China pays a premium for its pipeline supplies. Developing pipelines as "shorts" in a hedging strategy also makes it difficult to maximize profits. China has chosen to balance profits with security, but CNPC and local governments have benefited from the pipeline projects, and the "losses" or costs for the Chinese government have been compensated for by broader gains to China's national interests and foreign policy objectives, as well as enhancing the central government's domestic development and stability goals. Hedging against peacetime risks and wartime threats globally in the maritime domain and in the continental realm provides China with a combination of insurance and comprehensiveness in its energy security policy.

7

GLOBAL, MARITIME, AND
CONTINENTAL IMPLICATIONS

T HIS CHAPTER IS divided into three sections that analyze the global, maritime, and continental implications of China's energy security policy.
Based on the findings of the previous chapters, we seek to examine
some of the consequences of the strategies and policies that China has developed to hedge against oil dependency and to present the broader implications
of these findings.

Three global implications will be emphasized: first, the more influential
role of China's national oil companies (NOCs) in the international petroleum
market; second, China's growing cooperation with international energy institutions and organizations; and third, the effects of China's expanding global
oil interests on international security.

China's maritime energy security will be analyzed from three perspectives:
first, the implications of China's commercial maritime interests and the buildup
of a state-owned tanker fleet in the international tanker and shipping market;
second, how China's naval buildup, emphasis on protecting sea lines of commu-
nication (SLOCs), and the global commons shape cooperation and conflict at
sea; and third, how the opening of new shipping routes in the Arctic will shape
China's energy security policy as well as its policy toward the Arctic region.

The third section, focusing on the continental aspects of China's energy
security policy, will examine the "energy factor" in Sino-Russian relations and

will analyze the potential for Sino-Russian cooperation, competition, and rivalry in Central Asia and the Far East.

GLOBAL IMPLICATIONS

China's worldwide search for oil, the massive investments by Chinese NOCs in the petroleum sector, and the increasing number of loan-for-oil deals increase competition in the international petroleum market and generate friction and rivalry in international politics. Existing institutions are unlikely to prevent disputes and conflicts of interest. At the same time, a global or regional war over energy that includes China is unlikely. Few states can secure their energy through military means. Instead, energy security remains largely a matter of risk management, and as such China has developed a hedging strategy to minimize the risks that oil supply disruptions will slow down its economic growth or cause domestic instability.

CHINA'S NOCS AND THE INTERNATIONAL PETROLEUM MARKET

The Chinese NOCs have been characterized as hybrid, operating between traditional NOCs and international oil companies (IOCs), and thereby benefiting from both the government's safety net and financing and the business side of the private sector.[1] This characterization suggests that as hybrid NOCs (hNOCs), Chinese NOCs pursue their corporate interests (profits) but also take into account the goals and concerns of the government (security). Although the hNOCs need a "sufficient level of autonomy to devise strategies and conduct their operations," the government retains authority and control over the NOCs through policy making and ownership.[2]

Chinese NOCs are becoming increasingly similar to IOCs, and they are seeking to become internationally competitive. As latecomers to the international petroleum market, their goal is to develop the project management skills possessed by the existing IOCs in order to take on lucrative, large upstream projects. This process involves developing cutting-edge technology, securing and arranging financial backing, managing host-government relations, handling intraconsortium politics, being attentive to environmental impacts, overseeing budgets, and finishing projects on time.[3] Chinese NOCs have increased their expertise and improved their competitiveness through mergers and acquisitions (M&As), by partnering with IOCs for exploration and production (E&P), and by developing more joint ventures with the IOCs.[4] Since mid-2009, Chinese NOCs have focused on building closer ties with the IOCs. Such behavior marks "a maturing of the international strategies of China's oil companies," the key drivers of which have been the Chinese government and the NOCs' desire to gain access to large oil deposits and to enhance access to natural gas in order to diversify China's energy mix.[5] Accordingly, during the past decade, Chinese NOCs have become powerful actors in the energy market.

Iraq is one of the places where Chinese NOCs have been particularly successful. In 2009, China National Petroleum Company (CNPC), in partnership with BP, won the rights to develop the giant South Rumaila oil field, and by the end of 2009 CNPC had signed an agreement to develop the Halfaya oil field in collaboration with Total, Petronas, and Iraq's South Oil Company. China's National Offshore Oil Company (CNOOC) was granted rights to develop the Missan oil field in partnership with the Turkish Petroleum Company. Indeed, by March 2010 estimates suggested that Chinese oil companies had gained access to 18 percent of Iraq's reserves, making the Chinese NOCs the largest external actor in the Iraqi oil market.[6] Meanwhile, because of its involvement in the northern Kurdish region, it seems that Sinopec has been blocked from reaching deals in central and south Iraq.

Although both Chinese NOCs and IOCs can benefit from increased cooperation, Chinese NOCs are becoming stronger competitors and thus may challenge the position of the IOCs in the international oil market. Chinese NOCs have received financial and diplomatic backing from the Chinese government to facilitate the "going abroad" campaign.[7] The Chinese central government and Chinese banks such as the China Development Bank (CDB) provide funding for many of the inducements that the NOCs use to secure overseas deals and M&As. Arms sales, grants, debt relief, subsidies, low-interest loans, and military assistance, together with diplomatic support and high-profile summit meetings, often facilitate important energy deals. In some cases, this has provided a comparative advantage for the Chinese NOCs over the IOCs[8] and has prompted leading executives from the Norwegian oil company Statoil to half-jokingly express frustration with the competition from Chinese NOCs by stating that the Chinese companies simply add an extra zero to their bids, preempting competition from other oil companies for contracts and production rights. Consequently, the emergence of Chinese NOCs has an effect on the dynamics of the international petroleum market.

Thus far, however, Chinese NOCs still are far behind the IOCs when it comes to M&As and overseas oil production, thus "rarely compet[ing] head-to-head with IOCs."[9] Furthermore, the overseas upstream investments and production by Chinese NOCs actually increase the supply of oil in the international petroleum market, as China unlocks reserves from off-limits places such as Sudan. By shipping some of this oil back home, China is competing for less of the "low-risk" oil production in other countries. In contrast to the IOCs, Chinese NOCs are forced to operate with huge losses in the Chinese domestic downstream market. In addition, the "going abroad" campaign, facilitated by government backing, is not much different from the strategy adopted by the major Western oil companies or other Asian oil companies when they established themselves in the international energy market.[10] Economic and diplomatic support from the home country is nothing new, and the Chinese are not the first to cooperate with authoritarian regimes, providing them with arms and overlooking humanitarian, social, and environmental needs in order to advance petroleum interests.[11]

To some extent, Chinese NOCs follow conventional business strategies. As latecomers to the international oil market, they have focused on acquiring energy production and exploration assets overseas, and in certain cases they have paid more than what Western companies would contemplate.[12] During the early phase of gaining access to a market, certain losses can be expected and are acceptable. There are, however, reasons to believe that Chinese NOCs have now reached the next phase, whereby they are becoming increasingly concerned about commercial opportunities, capacity growth, and profits.[13] China's NOCs are market oriented and seek profits,[14] with security and strategic considerations a secondary concern, even though Chairman Wang Yilin of CNOOC stated in a speech to his employees that "large-scale deep-water rigs are our mobile national territory and a strategic weapon." Accordingly, it is pointed out that Wang is running his company as a profit-driven multinational enterprise overseas but promoting it as a political and strategic asset at home.[15] The global search for petroleum by Chinese NOCs may not lead to bellicose actions but instead may present strong competition as well as some opportunities for joint ventures to the IOCs.

CHINA AND GLOBAL ENERGY GOVERNANCE

Energy institutions have proliferated. There are global, regional, and national institutions. Many address only certain types of energy. Others have limited membership. There is as yet no global intergovernmental organization that deals with all types of energy. Even a former head of the International Energy Forum (IEF) has acknowledged that the establishment of a multilateral organization whereby national decision making would be relinquished and replaced by legal binding energy governance seems remote.[16]

The International Energy Agency (IEA), founded in 1974 as an autonomous body of the OECD, is a leading energy organization that acts as energy policy adviser for its twenty-eight member states, with an emphasis on energy security, economic development, and environmental protection. The IEA's cooperation with China is based on a 1996 memorandum of understanding (MOU) and focuses on energy (oil) security, energy statistics, analysis (scenarios and indicators), policy (supply, efficiency, and environment), and technology.[17]

With more than half of the world's energy consumption already taking place in non-OECD countries and with the bulk of oil consumption and the majority of oil consumption moving eastward, Nobuo Tanaka, a former executive director of the IEA, has stated that "in many ways [the Chinese] are already working closely with us. But eventually we wish they would join us."[18] Fearing that the IEA will risk losing its relevance, it has been suggested that China, given its concern about a volatile oil market and oil supply disruptions, cooperate with the IEA in low-key and in-depth discussions on the oil market, conduct a China/IEA simulation exercise, and allow the IEA to assist China's energy policy development.[19] Such discussions have taken place, and China is already regarded as a strategic partner in the IEA.

Despite China's growing cooperation with the IEA and its changing posture in the international community, which has included broader participation in a number of institutions and organizations, China is still not a fully integrated member. On energy issues and collaboration with the IEA, China's lack of or missing oil stock data in terms of oil demand analysis and its continued opaqueness about its national stockpiling policy and implementation are important issues that need to be resolved in order for China to make a contribution to the IEA's objective and mission.[20] Although there are strong incentives for China and the IEA to work together, China remains reluctant to proceed on the membership issue.

According to Chinese experts, membership is not urgent. The ensuing storage and energy-efficiency obligations as an IEA member would present pressure on China, and the nation may put more emphasis on membership when its economic development reaches a certain level.[21] There are, of course, strong incentives for China to maintain its strategic independence, to preserve its freedom to develop and implement its own energy security policy and at the same time "free ride" on the IEA's efforts and ability to stabilize the international petroleum market. At present, China only contributes half-heartedly to international energy institutions.

There are also difficulties to be overcome and adjustments to be made on the part of the IEA before China can become a full member. For example, IEA membership requires OECD membership, which also calls for a commitment to democracy. China's membership in the IEA would also "require a thoroughgoing redistribution of decisionmaking power within the agency, with OECD states giving up decisionmaking power in favor of prospective new members from continental Asia" and a commitment to "review established allocations of voting power."[22] Thus, China's membership in the IEA is affected by the reluctance of some countries to accept the necessary redistribution of vote percentage to reflect China's share of global energy imports. Finally, IEA member states remain sharply divided regarding Chinese membership, and therefore it may be a long time before the world's largest energy consumer and largest oil importer becomes a member of the IEA.

China has used the Shanghai Cooperation Organization (SCO) to engage with the Central Asian states and Russia on energy security issues. Furthermore, in concert with Russia it has sought to limit U.S. influence in Central Asia through regional organizations and bilateral ties with the Central Asian states. However, concerns that the SCO will develop into an energy club or that energy ties will consolidate an anti-Western alliance have been exaggerated. The Central Asian states have been keen to diversify their petroleum exports, China and Russia to a certain extent have also been competing for petroleum contracts in Central Asia, and producer-consumer interests do not always overlap. The SCO was set up to strengthen trust, facilitate the peaceful resolution of border disputes, and improve stability in Central Asia. Thus far it has focused on trade and infrastructure projects and on the three perceived evils of terrorism, separatism, and extremism. Although China has capitalized on these developments to improve its energy security, energy is not the key driver behind China's SCO involvement.

China has promoted its energy interests within the United Nations by protecting client states from sanctions or UN criticism, and in the case of Iran it has pushed for and won exemptions from sanctions in order to continue its oil imports. Again, however, China's interests in the United Nations are much broader than energy, and thus it has simultaneously supported UN Security Council (UNSC) resolutions against Iran, sanctions against Libya, facilitated and participated in the UN peacekeeping operations (UNPKO) in Sudan and Darfur, and participated in UN-mandated antipiracy missions in the Gulf of Aden.

China is engaged in various energy dialogues that seek mutually beneficial solutions to a number of energy challenges, including the Asia Pacific Economic Cooperation (APEC) energy security initiative, the Asian Energy Roundtable, ASEAN+3 common energy goals, and the East Asian summits. The IEF, an intergovernmental organization with the world's largest gathering of energy ministers, reflects the shift of gravity from West to East in terms of energy cooperation. China has also been active in the G20 to address questions of volatility in commodity prices. Finally, some commentators have discussed the possibility of using the Six-Party Talks framework to develop a Northeast Asian multilateral institution to address energy security concerns.

It remains uncertain, however, whether institutions, dialogues, norms, and rules will be able to mitigate energy security competition and potential conflicts over petroleum resources. Adhering to cooperation and mutual gain has often been secondary to great-power politics, the pursuit of national interests, and thinking in terms of relative gain.[23] Competition among states for resources is a natural state of affairs in world politics, but institutions that include China and other major energy consumers can lessen the competition, contribute to a more stable and efficient petroleum market, establish rules of engagement, prevent misunderstandings, enhance trust, and provide information to ensure that states remain sensitive to security dilemmas.

In sum, the implications of China's remarkable economic growth and increasing demand for energy during the last few decades are forcing a shift in the institutionalized world energy order. The new geopolitics and geoeconomics point to an Asia-centered world and question the relevance of institutions that cannot accommodate the role of China, other major energy consumers, and to some extent producers and the oil industry as well. Traditional institutions and old worldviews, in some respect represented by the IEA, are increasingly being replaced by institutions, dialogues, and organizations that embrace the new world order, in particular a more Asia-centered world, as illustrated, for example, by the work of the IEF. This is not to argue that the IEA has become anachronistic. It still encompasses important and valuable knowledge and technological expertise, but as the emerging economies led by China and India will increasingly influence the global energy future, its institutional apparatus seems out of touch with the new realities. A new approach to institutional energy cooperation should capture the new consumption patterns and developments of the new energy era.[24] The extent to which China will contribute remains to be seen, but a more dominant China will hold a stronger stake in the new world energy order.

SAFEGUARDING CHINA'S GLOBAL OIL INTERESTS

It is indisputable that China has become stronger in all aspects of comprehensive national power. At the same time, it has also become more vulnerable.[25] China's continued economic growth will require additional resources. Expanding ties around the world means that China has enlarged its portfolio of interests, and these will need to be sufficiently preserved. China's interests continue to expand as its power expands, but increased relative power brings with it new fears.[26] Thus, China aims to develop new strategies commensurate with its growing international position.

One possibility is to develop powerful military capabilities to protect its interests.[27] But such a policy would likely provoke countermeasures from the United States, India, Japan, and elsewhere, fueling security dilemmas in the region. Another option is to focus on a strategic approach to secure its resource needs. As seen in previous chapters, strategic steps to control equity oil will not be sufficient to "secure" China's energy needs. A final alternative is to rely on market mechanisms. Such an option, however, will leave China vulnerable to U.S. dominance in world energy markets and transportation routes.

Hedging strategies are designed to minimize exposure to unwanted risks by combining security and profit considerations, for example, by insuring that the Chinese economy does not become too dependent on imported oil while at the same time still allowing Chinese NOCs to profit and develop in the increasingly globalized and interdependent world energy market. China's hedging strategies against oil supply disruptions, the insufficient availability of oil, and exceedingly high oil prices are embedded in a mistrust of market mechanisms, a dislike of U.S. preponderance in world affairs, and simultaneously a growing concern about the U.S. ability and willingness to provide common goods beneficial to China's energy security.

China does not fully trust that its interests can be preserved within the international oil market, and therefore it seeks to hedge, for instance, against relying on U.S. maritime dominance and patrol of vital SLOCs, U.S. capabilities to maintain a world energy order, and the continued ability and willingness of the United States to provide stability in key energy regions. In many ways, China would like to insure its interests against the potential negative effects of both U.S. dominance (that is, its control of SLOCs and its security role in oil-rich regions) and the United States' perceived declining power, by developing closer diplomatic ties with oil-exporting and strategically located states, improving its naval capabilities, developing a strong maritime transportation capacity, and trading its way to power by gaining more leverage in neighboring countries and vital energy regions.

China has developed global petroleum interests, and its NOCs are involved in hundreds of global petroleum projects. This gives China influence, but it can also be argued that China has traded its way to power, as a result becoming more exposed and vulnerable. For example, hundreds of thousands of Chinese workers and entrepreneurs are linked to Chinese development work, manufacturing, construction, and agriculture projects in Africa, in addition to substantial Chinese investments on that continent. China became Africa's largest trading partner, with trade of $90 billion in 2009. The United States ranked second, with $86 billion.[28] As pointed out in chapter 4, when conflict erupted in Libya China directed its naval vessels operating in antipiracy missions in the Gulf of Aden through the Suez Canal and into Libyan waters to assist in the evacuation of about 35,000 Chinese workers based in Libya, many of whom were involved in petroleum or pipeline projects.

One can only speculate what the Chinese government will do the next time its workers in an African country are in danger because of conflicts or uprisings or are singled out by a host country eager to end China's increasing dominance by severing ties with the Chinese government and turning instead to other countries for support. Although China has interests and investments in most African countries, particularly South Sudan and Angola, where China has strong petroleum interests, should be followed closely in the years ahead. If conflict in either of these two countries or in other countries where there is a major Chinese presence threatened Chinese interests or endangered the lives of Chinese citizens, there clearly will be calls for the government to rescue Chinese citizens and protect China's interests.

The efficient and well-organized evacuation of Chinese workers from Libya boosted the status and prestige of the Chinese central government and the PLAN, but it simultaneously increased expectations among Chinese that as a great power China should be able to protect its citizens and interests globally. However, will China act unilaterally or multilaterally? Will China wait for a UNSC resolution before it enters a conflict in South Sudan and Angola to protect its interests and citizens? There are no clear answers to such hypothetical questions, but China's growing worldwide presence, driven partly by its increasing need for oil, creates a strong incentive to build up its military power-projection capacity and adjust its "noninterference" policy. Although it remains to be seen whether China will develop a sufficient military capability to protect its investments, oil production, and interests abroad, China is likely to continue to develop and enhance its crisis management plans and capabilities to facilitate evacuations and to safeguard China's regional and global interests. It is common in the West to state that for better or worse, China is changing Africa. However, as one Chinese energy expert succinctly stated in a meeting, Africa is changing China.[29] Indeed, China is gradually shifting its conventional and official policy on sovereignty and noninterference commensurate with its growing global interests and its role as a great power.

Thus, global energy security interests, together with a growing portfolio of worldwide interests, will continue to push China toward a more active role in international affairs. At the same time, China is likely to continue to pursue its hedging strategy, which encompasses a combination of both multilateral efforts and unilateralism, an aggressive "going abroad" strategy and emphasis on joint ventures, and promoting cooperation and simultaneously being fiercely competitive while seeking to manage conflicts of interest and minimize risks.

MARITIME ISSUES

China's ambitions to enhance its energy security have important spillover maritime effects. First, the Chinese plan to build up a large state-owned tanker fleet is contributing to an overcapacity in the tanker market. Second, Chinese tanker companies are likely to maneuver into a stronger position once the tanker market recovers, which will have commercial, strategic, and security implications, since China will then control more of the world's tanker fleet capacity. Third, although China's energy security considerations have not been a key driver behind China's naval modernization, new missions and tasks, including safeguarding its SLOCs and Chinese energy security, are becoming increasingly important, thereby forcing the People's Liberation Army Navy (PLAN) to balance between its traditional war-fighting mission and crisis management tasks encapsulated in military operations other than war (MOOTW). Finally, with a stronger focus on energy security and new energy sources and routes, China will seek a more prominent role in the Arctic as a new petroleum frontier.

MARKET FACTORS

The world fleet of VLCCs consists today of about 600 vessels, and China controls between 10 and 15 percent of the fleet. It is estimated, however, that 160 new ships are on order, with seventy-seven new VLCCs to be delivered in 2011 and fifty-three new vessels to enter the market in 2012.[30] Furthermore, it has been reported that China will build eighty new VLCCs between 2013 and 2016. CNPC and Sinopec have ordered thirty to forty VLCCs, respectively, which will be delivered between 2013 and 2016. The price for these ships is estimated to be $80 million each, or about $34.5 billion for eighty new VLCCs, which is roughly 20 percent lower than the current construction cost for tankers.[31] Shipping analysts are skeptical about the numbers, but they admit that China's ambitions to build a state-owned tanker fleet will be a game changer in the tanker market. According to one analyst, the orderbook shows that seventeen VLCCs will be delivered to Chinese owners, including China Shipping Group, Sinotrans & CSC, CNPC, Grand China Logistics, and China Ocean Shipping Company (COSCO) during the 2011–2013 period. During the same period, twelve VLCCs will be built at Chinese shipyards, but the current owners remain unknown.[32]

There is already an overcapacity in the market: tankers are competing for the same shipments, and freight rates are historically low. Supply is likely to outstrip demand. The tanker fleet is expected to grow 9 percent in 2011, with an added delivery of 36 million dwt of new vessels and further expansion of 29 million dwt, a 7 percent increase in supply, in 2012. This compares to global oil demand growth of between 1–2 percent in 2011 and 2012, according to forecasts by the IEA.[33] Olav Trøim, a CEO and executive at Frontline, the world's largest independent tanker company, operating forty-seven VLCCs, expects that oil production will increase by 6 percent from 2004 to 2013 but that the world tanker fleet will grow by 60 percent during the same period.

This prompted Trøim to conclude that the market since 2008 will be "the worst since the Black Death."[34] In December 2007, the largest tankers could earn close to $300,000 per day operating in the market, but the average earnings in 2011 on the TD3 route, which is the most common trade route into Asia for a VLCC carrying crude from Saudi Arabia, was about $11,000 per day, down more than 65 percent from an average of $32,000 per day in 2010. According to Frontline's cost estimates in August 2011, to break even, a tanker company needs rates at $29,800 per day for its VLCCs and $24,800 per day for its Suezmax vessels.

China has not picked up much of its new tanker capacity in the second-hand market, where many ship owners have been forced to sell their tankers as a result of the weak market. This suggests that the Chinese government values the strategic and security benefits of maintaining and expanding China's own shipbuilding industry, which has important economic spillover effects and implications for China's naval buildup. This also shows that China is seeking long-term security and strategic benefits and is willing to pay the costs of developing a state-owned tanker fleet based on new vessels despite a bearish tanker market, instead of taking advantage of the overcapacity in the market.

More Chinese tankers will increase competition and most likely will postpone any immediate recovery of the tanker market. There might be stronger protests against the Chinese government and Chinese banks subsidizing Chinese shipbuilding and support for a state-owned tanker fleet in a weak market that leads to bankruptcy among some international shipping and shipbuilding companies, which again might trigger a trade conflict if not a trade war. Cheap loans are also a major competitive advantage for Chinese state-owned tankers to survive in the current tanker market. For example, in October 2011 Saga Tankers was forced by the Norwegian bank DNB NOR, the world's largest creditor to international tanker companies, to sell its tankers, as the company was in breach of bank regulations (that is, conditions for the loan, the company's cash on hand, and so forth).[35] This would probably be a less likely outcome for Chinese state-owned tankers.

China's ambitions for a state-owned tanker fleet are not the only or not even the most important factor shaping the bearish tanker market. Ship owners have been "digging their own graves" by ordering more new vessels prior to and in the aftermath of the 2008 financial crisis, thereby leading to today's overcapacity. China's continued economic growth and growing need for petroleum will be a crucial factor in the prospects for a recovery in the tanker market, but as the Chinese economy slows in 2012, China's demand for oil is decreasing. Moreover, global economic turmoil and the economic slowdown in the United States and Europe may overshadow China's growing demand for oil, thereby puncturing any improvements in the tanker market.

One VLCC can supply China with about 2 mbd. A one-way route to China will take about eighteen or nineteen days at fourteen knots from Saudi Arabia, about sixteen days from Iran, about twenty days from Sudan, about twenty-seven days from Angola, and about twenty-seven days from Venezuela through the Panama Canal or thirty-three days around the Cape of Good Hope. If estimating an average overall round trip of fifty days for China's seaborne oil imports, including a two-day turnaround, China will need roughly 150 VLCCs to ship 6 mbd on Chinese tankers.

In 2011, it was reported that China owns thirty-eight VLCCs, with seventy-six more on order. The same report states that "Chinese controlled tonnage accounts for around 12 percent of the total VLCC fleet of 553 double-hull ships currently trading and 19 percent of the global VLCC orderbook of 136 vessels." The numbers, then, do not add up, as China will have more than sixty VLCCs to account for 12 percent of the VLCC fleet and will have more than 50 percent of the orderbook for new vessels.[36] As we have seen, other reports note that China will build eighty new VLCCs between 2013 and 2016 in order to meet the government's target of 50 percent of China's oil imports carried by Chinese-controlled tonnage. More importantly, according to the world fleet register report, in October 2011 China was operating seventy-five tankers.[37] In addition, many PRC tankers are not PRC flagged.

Thus, China is on track to develop a capacity to ship more than half of its oil imports on Chinese tankers. Indeed, shipbuilding and tanker experts argue that the Chinese government has set a new target, as part of its national oil security strategy, that the proportion of imports carried by Chinese-flagged tonnage should be 60 to 70 percent. Similar to how the Chinese NOCs sell their upstream production locally or on the international petroleum market for a profit on a daily basis, Chinese state-owned tankers are likely to be chartered out for profit in their daily business, benefiting countries that depend on imports from the petroleum and shipping market. However, the fact that China can call on its NOCs and state-owned tankers during a crisis has broad strategic and security implications. Such actions may contribute to a limited availability of sufficient oil supplies at affordable prices, as China might "lock up" its equity oil and control a significant percentage of the world's tanker fleet.

China's demand for oil imports coupled with the rapid expansion of domestically built and flagged vessels signal a shift in control of commodity movements and tanker capacities. Subsidizing a state-owned tanker buildup as a result of security concerns and profit considerations will lead to overcapacity in the tanker market. Chinese state-owned companies can more easily absorb a bearish tanker market and may come out of the recession with a stronger market share, giving China control with even more tanker capacity and strategic advantages during any future crisis involving an oil-rich region, short of a major war involving China.

CHINA AND THE MARITIME GLOBAL COMMONS

China's global interests and concern about energy security are becoming important drivers in its quest to develop naval and power-projection capabilities to protect China's vital trade routes, SLOCs, and Chinese citizens and interests abroad. Developing power-projection capabilities commensurate with China's growing power is natural. Indeed, any great power experiencing anything like China's extraordinary economic growth would most likely increase its military power, enhance its global outreach, push for more influence, and demand that its interests be respected. What is exceptional about China's rise, something unprecedented in any historical comparison, is that during the last three decades China has become the dominant power on the Asian mainland, improving both its absolute and relative power formidably and simultaneously solving a number of border disputes peacefully and preventing wars with other major powers.

Such achievements are no guarantee against the fact that a rising China faces the risk of alarming neighboring countries, as well as the United States and the West. If China seeks to protect its interests unilaterally through a military buildup or if, as a traditional land power, it develops its sea power to control maritime Asia, thereby challenging U.S. naval supremacy and East Asian bipolar stability, more confrontation will erupt.[38] China has become more assertive since 2009, but whether the more recent developments suggest a shift from the "peaceful rise" strategy pursued during the last three decades, whereby Chinese leaders have been cautious about taking steps that are overly provocative and in fact have often taken measures aimed at avoiding brinkmanship, remains uncertain.[39]

Rather than seeking to maximize its regional and global presence, China has proceeded with a careful and sophisticated neighborly policy. Despite becoming more reliant on oil imports—and with Chinese commentators calling for a strong navy that can protect China's overseas interests, oil production, and seaborne energy supplies—China's behavior can best be described as showing security dilemma sensibility. Although other great powers and rising powers in the past, notably the Soviet Union and Germany, which were less reliant on seaborne trade than China is today, embarked on an aggressive naval arms race with the dominant sea power, so far China has not followed the tragedy of the past by avoiding alliances, competing in arms races, or engaging in war.[40]

China's security dilemma sensibility can be illustrated in the following ways. First, China has focused on economic growth and domestic stability by channeling its economic surplus into development rather than into expensive weapons programs or a major arms buildup. China's defense spending has averaged more than 10 percent annually over the last two decades and has matched China's GDP growth. But as a percentage of GDP, defense spending has remained in the 2–3 percent range, suggesting that China has not increased its defense budget in order to compete militarily with the United States or to seek expansion through military means. When preparing for war, great powers in the past have spent much more than 2–3 percent of their GDP on defense. Second, China has channeled some of its growing global presence into operations for the common good, thereby seeking to minimize the risk that China's more prominent role in regional and world affairs will stimulate security dilemmas and rivalries.

The PLAN has become increasingly involved in managing risks to China's expanding interests around the world and in safeguarding the global commons that encompasses China's SLOCs. A China more integrated in international affairs includes Chinese participation and partnership on matters for the common good, such as increased participation in UN peacekeeping operations, nonproliferation efforts, stabilizing the international financial system, safeguarding important SLOCs from piracy, preventing terrorist attacks or environmental hazards, and promoting free trade agreements. Efforts on China's part have been productive and welcomed by the international community. Simultaneously, both China's neighbors and the Western countries, as well as China itself, face serious dilemmas as a result of such developments.

The very measures put in place to make China a contributing partner and stakeholder in the international community simultaneously give China clout and valuable experience and earn it international goodwill. Increased Chinese responsibility in due time could possibly challenge U.S. leadership, U.S. regional alliances in Asia, and the U.S.-dominated world (energy)order. Airlift capabilities and naval power will surely facilitate Chinese involvement in missions for the common good, but these capabilities will also strengthen China's power-projection capability and may eventually challenge U.S. predominance at sea. Indeed, the same Chinese naval vessels that have operated in the Gulf of Aden to secure vital SLOCs, protect seaborne oil, and contribute to safeguarding the global commons could also be used to boost China's maritime claims in the South and East China Seas, harass other countries' vessels operating in the area, and eventually be used in naval conflicts or wars. In short, calling for China to become a responsible stakeholder requires that China develops power-projection capabilities to protect the global commons, which in turn will legitimize a stronger Chinese naval buildup that many states fear will provide the means for more aggressive Chinese behavior.

Not only do Chinese leaders need to be sensitive to security dilemmas and counterbalancing against its rising power—and not only do Western states need to be careful about what they wish for—but the new "historical" mission of the PLAN requires a "two-vector navy" that can encompass the twin missions of area denial and peacetime operations, a mix of naval capabilities different from solely focusing on a wartime offshore active defense.[41] Surface ships are important and necessary to peacetime missions, but a submarine force and land-based naval aviation has pride of place in protecting core Chinese interests and national security. Thus, striking the balance between war-fighting capabilities and MOOTW poses a challenge for the PLA.[42]

Chinese naval strategy has prioritized the modernization of China's submarine fleet and access-denial capabilities. The PLAN has focused predominantly on protecting China's coastal waters and on responding to a Taiwan contingency. The more recent ambition is to develop a modern surface fleet that will carry out both traditional war-fighting missions in China's near seas and will participate in MOOTW, including SLOC security and protecting China's energy security interests. This development has been fueled by China's reemergence as a great power with global interests and responsibilities. It will be difficult in the years ahead to develop a first-class modern navy capable of both war fighting in the seas surrounding China if China ends up in a potential conflict and of conducting worldwide peacetime missions to safeguard China's growing global interests.[43] Whether China and the PLAN will manage to balance this ambition will tell us much about the leadership China might play in world affairs in the twenty-first-century and will reveal what kind of great power China might become.

CHINA AND THE ARCTIC

The world is not only becoming more Asia-centric, with more energy moving eastward, but the East and the West are gradually melting together: over the last decade, the Arctic ice cap has been disappearing, shrinking to an area two and a half times that of the United Kingdom. The prospect of the Arctic being partly ice free and navigable during the summer months and the potential for a fully ice free Arctic by midcentury offer opportunities and entail challenges even for the non–Arctic states. The melting Arctic ice cap is of growing interest among Chinese researchers and officials, as it has been increasingly recognized that Arctic petroleum reserves and new transportation routes will offer investment and commercial opportunities for Chinese NOCs and the Chinese shipping sector. The melting of Arctic ice will allow China further to diversify its sources and routes of petroleum imports and trade routes. In short, China's global search for petroleum, its interest in alternative transportation routes, and its new investments will push China to seek a stronger role in shaping the Arctic as a new energy frontier.

China has expanded its polar research capabilities, and its researchers have emphasized that China needs to develop independent research on the Arctic environment instead of relying on data from other countries.[44] China purchased the icebreaker *Xuelong* (*Snow Dragon*) from Ukraine in 1993 to conduct Arctic research and is currently building a new vessel to be used for high-tech polar expeditions. It is also in the process of developing one of the world's strongest polar scientific research programs. Arctic-focused research is growing, more Chinese institutions are focusing on the Arctic, and the region is frequently being discussed in media and publications.

The melting of Arctic ice opens new opportunities for petroleum activity and new routes for maritime transportation. The U.S. Geological Survey (USGS), a frequently used source in the literature on Arctic petroleum reserves, estimates that about 13 percent of the world's total undiscovered oil and about 30 percent of yet-to-be-discovered reserves of natural gas might be found in the Arctic. More than 80 percent of the undiscovered petroleum resources are expected to be offshore, and most of the promising fields are within the littoral states' exclusive economic zones (EEZs) or nondisputed areas in the Arctic Ocean.[45] Indeed, the USGS estimates that about 70 percent of the Arctic resources consist of gas located in western Siberia and the eastern Barents Sea.[46]

High oil prices and energy security concerns are two key factors attracting the petroleum industry and political interest in the Arctic. However, given its challenging, inhospitable, and high-cost environment, which requires state-of-the-art technology, Chinese NOCs are likely to struggle to secure a stake in Arctic E&P. One option is to partner with foreign companies in joint ventures or to invest in companies already operating in the Arctic. In August 2011, China Investment Corporation (CIC) bought 30 percent of GdF Suez E&P, a deal worth about $4 billion, and thereby secured indirect access to E&P and licenses to drill in the Barents Sea.[47] In addition, in 2008 China Oilfield Services bought the Norwegian offshore company Awilco for roughly $2.5 billion. Top officials from the Norwegian Ministry of Petroleum and Energy and the Norwegian petroleum sector anticipate more investments, M&As, and competition for E&P contracts and licenses in the Arctic. They not only expect Chinese companies to participate in this process but also see Chinese involvement and increased competition as "business as usual."[48]

Increased petroleum extraction is expected to add to the emerging interest in new shipping routes between Asia and Europe. With growing demand for petroleum, China will not only be interested in petroleum production in the Arctic but also can be expected to seek more of the increased petroleum exports from the Arctic to Asia. Avoiding maritime piracy and cutting costs with shorter routes between Asia and Europe are often highlighted in reports about the commercial and strategic opportunities presented by a melting Arctic. It is estimated that the maritime route between Asia and Europe could be reduced from 15,000 miles to under 8,000 miles. Using the Northern Sea route, the sailing distance from Yokohama to Rotterdam can be reduced by 40 percent, meaning that shipping companies will save time and reduce fuel and maintenance costs. The Arctic route is even shorter for ships that are unable to transit the Suez Canal because of their size or overcapacity in the canal. In addition, there seems to be no short-term solution to the increase in piracy in the Gulf of Aden, which is driving up the cost of insurance for ships traveling via the Gulf of Aden toward the Suez Canal and thus making other routes more attractive.[49]

Currently, the Northern Sea route at best is only usable during the summer season, and even then ships have to contend with floating ice, a lack of infrastructure, and limited search-and-rescue capabilities. Thus, several factors are working against the commercial viability of the Arctic route. Darkness, strong currents, the narrow passage, and shallow waters make it challenging for VLCCs to transit the Bering Strait or make full usage of the Northern Sea route.[50] For many years in the future, icebreakers will be needed when transiting Arctic waters, even in the summer months, thereby driving up costs. Shipping companies will also have to pay a premium for insurance when sailing the Northern Sea route. But even with insurance, bankruptcy is a likely outcome were there to be an accident or oil spill in the fragile Arctic environment.[51] Hence, many large and established shipping companies are currently not willing to take the risk of sending ships through the Arctic Ocean. Major bulk carriers such as Maersk are instead developing business models that emphasize reliability and punctuality, which thus far the Arctic sea routes cannot provide.

Nonetheless, there are several shipping companies willing to take the risk and potentially profit from a new shipping route between Europe and Asia. Russian vessels have operated the Northern Sea route for many years, but in 2009 the Beluga Group was the first Western company to sail the Northern Sea route from Asia, and other companies, such as Tschudi Shipping, are also beginning to operate the route. Both the Russian shipping company Sovcomflot and the Russian oil company Novatek are shipping oil to the Far East via the Northern Sea route. Although it is expected that about 1.15 million tons will be shipped in 2011, this is only a tiny fraction of the 500 million tons handled at world ports annually. The true promise of the Arctic lies in its fivefold growth rate from 2010 to 2011 and its huge future.

Some Chinese and Western researchers and commentators are anticipating confrontation and conflict as states scramble for resources and control of sea lanes in the Arctic. Additionally, some Chinese pundits are questioning the legal framework for the region.[52] Rear Admiral Yin Zhou's comment in March 2010 that "the North Pole and surrounding areas are the common wealth of the world's people and do not belong to any one country" is often cited in the literature assessing Chinese perspectives on the Arctic. With no authoritative Arctic strategy published by the central government, it appears that alarmist voices have been allowed to shape China's Arctic policy.[53]

China may not have an official public policy on the Arctic, but contrary to much of the alarmist writing, Assistant Foreign Minister Hu Zhengyue has clearly stated that China supports both the legal framework of the Arctic and the cooperation promoted by the Arctic Council. In July 2009, Hu stated that "international law, including a series of conventions, has provided the basic legal framework for addressing Arctic affairs." Although Hu simultaneously noted that "full consideration should be given to the fact that the outer continental shelf is linked to the International Seabed Area which is the common heritage of mankind to ensure a balance between the rights and interests of the coastal states and the common interests of the international community," he stressed that "China supports the current legal order in the Arctic, including the law of the sea . . . [and] China respects the sovereignty of Arctic states, and their sovereign rights and jurisdiction in accordance with international law." Thus, sovereignty over the EEZs and the littoral states' delimited continental shelf, where most of the hydrocarbon reserves and the important sea routes are located, are not questioned by China officially, and Chinese UNCLOS experts at the China Institute of Marine Affairs (CIMA), an institute under the State Oceanic Administration (SOA), have confirmed this position.[54]

China is pursuing a hedging strategy to improve its energy security by searching for global petroleum deals, seeking out worldwide energy sources, and diversifying its routes to import petroleum. China is likely to draw more attention to and to put more resources into promoting its interests in the Arctic in the years ahead, as that area represents a new frontier. As Europe and Asia literally melt together in the future, there will be commercial, environmental, strategic, and military implications for China, Asia, and the Arctic, even as the long-term consequences remain uncertain. Currently, the situation in the Arctic is characterized by cooperation and benign relations. Littoral states and other actors have more common than diverging interests. Other areas of the world remain more important to the great powers, and the stakes in the Arctic are not high enough to warrant confrontation or conflict.

THE ENERGY FACTOR IN SINO-RUSSIAN RELATIONS

The Sino-Russian relationship is at a critical juncture. The border disputes have been resolved, cooperation is maintained within the SCO, and the bonds that kept the two countries together—including their shared interest in combating the "three evils" (separatism, religious extremism, and terrorism), their common interests to oppose U.S. primacy and presence in Central Asia, and Russian arms sales to China—are gradually diminishing. As China grows stronger, Russia, as a junior partner in the bilateral relationship, is becoming more concerned about China's dominance in Asia.

After more or less a decade of negotiations, the Eastern Siberia–Pacific Ocean (ESPO) pipeline spur from Skovordinov to Daqing became operational in 2011, linking the world's biggest oil producer, Russia, and the world's largest oil importer, China. The financial crisis in 2008 and 2009 became a game changer and facilitated a record loan-for-oil deal of $25 billion in return for the delivery of 300,000 b/d over a twenty-year period. Negotiations for a gas pipeline are continuing, and despite the setbacks and the obstacles that need to be overcome, one cannot rule out that more petroleum deals will be signed in the future. This could strengthen Sino-Russian cooperation and their strategic partnership.

Some researchers have argued that the two important pillars of the Sino-Russia relationship—arms sales and energy cooperation—are crumbling. China has not placed a significant order for Russian arms since 2005 and buys only a fraction of its energy imports from Russia. Indeed, Russia and China are faced with a number of challenges in maintaining or advancing their strategic partnership, and they are likely to "continue to be pragmatic partners of convenience."[55] Nonetheless, China's oil imports from Russia, standing at roughly 7 percent of the total and with Russia consistently being among China's top five suppliers, is not insignificant, and despite the remaining challenges in developing petroleum reserves in Siberia and the lack of infrastructure in the Far East, growing willingness in Moscow to boost the Russian presence in the Far East could stimulate further energy cooperation between China and Russia. Furthermore, neither China or Russia has much to gain from moving away from important energy ties and toward a more confrontational relationship, unless China becomes so dominant in Asia that Russia decides to counterbalance China's supremacy.[56] Sino-Russian arms sales did pick up again in 2012, after Russia decided to sell forty-eight Su-35 fighters and four Lada-class diesel-electric submarines, and bilateral diplomatic ties were strengthened through China and Russia's joint opposition to UNSC action against Syria.

CONCLUSION

China's hedging strategy against oil supply disruptions, which combines security and profit considerations, has global, maritime, and continental implications. China's profit-driven NOCs have become more active and important players internationally and are contributing to a better-supplied world oil market. The NOCs' growing presence makes the international petroleum market more competitive, as they are supported by the Chinese government and Chinese state-owned banks or have developed cutting-edge technology or business strategies to contend with the IOCs and other NOCs in the international marketplace.

Chinese tankers are also profit driven and are mostly chartered out, thereby enhancing capacity in the tanker market, which today has an overcapacity and historically low freight rates. Profit considerations are also partially driving the ambition to develop a state-owned tanker fleet. Shipbuilding boosts the entire economic chain and creates a large number of jobs, and China will end up controlling more of the tanker fleet and the tanker market, including freight rates. Nonetheless, Chinese tankers are most likely to continue to be chartered out during normal market conditions.

One important consequence of China's hedging strategy is that during a crisis or conflict in an oil-rich region the Chinese government can instruct its NOCs and state-owned tankers to ship their overseas oil production back home. Hence, less oil may be available in the market, more oil-importing states might be dependent on the oil located in the crisis or conflict area, and there may be less tanker capacity in the international market and fewer tankers able to operate in a war-exclusion zone. Although a crisis in the international petroleum market can be expected to hurt the Chinese economy, China might be relatively better off than other major oil importers as a result of the hedging strategies it has developed in the international petroleum and tanker markets.

Both China's domestic public and the international community are calling for China to take up global leadership roles commensurate with its growing power. China combines these incentives into a strategy that emphasizes both security and profit. Military power and crisis management capabilities are developed to secure China's global interests, advance energy security, and protect Chinese citizens abroad. Simultaneously, China seeks to profit from its role as a responsible stakeholder that safeguards the global commons and vital SLOCs, which is conducive to its energy security interests and legitimizes its military and naval buildup, reflecting its great-power status, leadership, and more prominent role in international affairs. China's worldwide search for new sources of petroleum and transportation routes has motivated China to develop a more comprehensive Arctic policy. This is expected to continue, and China will be an important competitor for a larger stake in the Arctic energy frontier and will take advantage of new sea routes. However, China is not likely to compromise its longstanding adherence to the principles of sovereignty, and China's interpretation of the UNCLOS strengthens the jurisdiction of the Arctic coastal states and limits the potential for conflict in the Arctic Ocean.

Despite the ups and downs in their energy relationship, Sino-Russian relations and energy ties remain strong. Fearful of China's dominance over continental Asia, Russia's position as an energy superpower will be one of the few factors that will reinforce close Sino-Russian relations and will represent an area where Russia can still negotiate from a position of strength without alarming China. Thus, energy ties will be crucial to manage Sino-Russian conflicts of interest and will be an important factor shaping the prospects for either continued cooperation or eventual rivalry and conflict.

8
CONCLUSION

C HINA'S ENERGY SECURITY policy is characterized by the interaction of security/strategic and profit/market elements. Currently, there is a void in the literature regarding how strategic and market approaches can be combined theoretically with respect to China's energy security policy. This book establishes a new theoretical framework based on the concept of hedging, linking market and strategic elements, or what has been referred to as the interrelationship between profit and security, in order to analyze China's energy security policy. It has revealed how and why market and strategic factors are linked in a number of hedging strategies that the Chinese government has developed to reduce exposure to potential oil supply disruptions, to insure against sudden price rises, and to adapt to the increasing net oil import gap.

Although some studies mix market and strategic approaches to analyze China's energy security policy, they do so relatively loosely. In contrast, the new hedging framework established here, by systematically showing how market and strategic factors interact, better explains important aspects of China's energy security policy. The hedging framework shifts the analysis from wartime threats to peacetime risks, from securing to insuring an adequate oil supply, and from controlling to managing any disruptions to the sea lines of communications (SLOCs). China's energy security cannot be secured, but it can be insured based on the ability of the government to develop hedging strategies. This more comprehensive understanding provides two important nuances to much of the existing literature.

First, contrary to a strategic approach, a hedging framework is more attentive to the distinction between threats and risks and between wartime and peacetime scenarios. Because it disregards this important distinction, the strategic approach neglects, or cannot sufficiently examine, how China's energy security policy predominantly addresses peacetime risks rather than unlikely wartime threats. Although risks cannot be eliminated, they can be managed. As the International Energy Agency (IEA) has noted, "energy security in practice is best seen as a problem of risk management, that is, reducing to an acceptable level the risks and consequences of disruption and adverse long-term market trends."[1] A hedging strategy can be applied either to minimize or to manage risks. This study has shown how China has developed hedging strategies against oil supply disruptions, rather than taking an alarmist strategic view of how China can, has, or should[2] "secure" its overseas sources of oil, "control" transportation routes, and "lock up" energy supplies.

Second, China's energy security policy is pursued according to many of the principles advocated by a market approach, but the security considerations are often ignored and replaced by market factors, the autonomy of the national oil companies (NOCs), and profit considerations. Instead of pursuing the most profitable and cost-effective market mechanisms to safeguard sufficient oil supplies at affordable prices (longs), China has invested in a number of more costly insurances (shorts), corresponding to different hedging strategies. For example, it has invested in expensive pipelines, even though they are not the most cost-effective strategy, and it has focused on developing a large state-owned tanker fleet, despite the overcapacity in the tanker market.

We have stressed that China's energy security policy cannot be approached from an "either/or" perspective, that is, emphasizing either a strategic, mercantilist, or state-controlled energy security policy in the pursuit of national interests or focusing on market mechanisms, weak governmental institutions, and powerful NOCs that pursue profits and commercial interests. Instead, a "more or less" perspective captures both market and strategic considerations. As the hedging framework suggests, such an approach systematically demonstrates how in some cases the pursuit of profits on the part of NOCs is a driving factor, whereas in other cases, government concern about security of supply is the main driving factor.

The theoretical framework is outlined in chapter 2, which draws a distinction between risks and threats. Whereas threats are already established and can be identified—for example, a blockade of China's oil supply—risks are dangers in the offing whose potential costs and opportunities are unknown—for example, a potential conflict between the United States and Iran. The ways to cope with threats and risks differ. In principle, threats can be eliminated; risks must be managed. By adapting the view of the IEA regarding energy security as a problem of risk management, it is argued that hedging is a comprehensive strategy that can both minimize and manage risks.

It is then argued that the Chinese government has developed a number of hedging strategies to deal with energy security uncertainties. Unsure whether a market approach or a strategic approach will best enhance China's interests, hedging, and the combination of security and profit considerations, is the preferred strategy, rather than choosing one approach at the expense of the other.

China's hedging strategy was developed by combining the government's concern about security of supply with the NOCs' search for profits, that is, by combining shorts and longs. As an investor seeking to hedge against a volatile market through derivative products, future contracts, and short selling, China bets against cooperation in the international petroleum market and buys insurance "outside" this market. Such a hedging strategy will likely leave China reasonably well off during a period of crisis or during a war in an oil-rich region.

Uncertainty in the decision-making process facilitates a hedging policy in sensitive cases, such as in Sudan and Iran, where to some extent China's national and diplomatic interests clash with the commercial interests of the NOCs. Although few researchers have access to discussions at the highest level of the Chinese leadership, the Politburo strives to reach a consensus among the NOCs, the People's Liberation Army (PLA), the Ministry of Foreign Affairs (MFA), and government institutions such as the National Development Reform Commission (NDRC) and seeks to mediate differing interests. In the case of Iran, the outcome of this pluralistic decision-making process can be described as China's "dual game."[3] However, it can also be seen as a hedging policy that mixes cooperation with the West (that is, longs, or betting on U.S./Western abilities to maintain stability and to prevent disruptions in oil-rich regions, emphasizing the importance of U.S.-China relations and China's status and prestige in the international community), probably advocated by the MFA and the NDRC, and confrontation (that is, shorts, or supporting Iran and safeguarding China's unilateral energy interests and challenging U.S. interests), probably advocated by the PLA and to some extent the NOCs.

China does not trust that the cooperation in the international petroleum market, or longs, will safeguard China's energy interests. Hence, its short strategies seek to insure against supply disruptions, high prices, and instability in the international market. A hedging strategy depends on developing both longs and shorts. An investor that only sells shorts is a speculator who is seeking to profit when the market goes down but who will lose money when the market goes up. A hedger is more risk averse and does not seek to maximize profit; rather, he seeks to minimize risks, by investing in both longs and shorts so as to be reasonably well off irrespective of whether the market goes up or down. China can profit from both longs (i.e., seaborne oil supplies) and shorts (i.e., pipelines). Relying only on seaborne oil would probably be most cost effective. However, China would lose out by following such a strategy if seaborne oil supplies were to be disrupted. Accordingly, there are costs attached to a hedging strategy, just as buying insurance is a more costly option.

Finally, hedging goes beyond diversification. Although diversification is part of a hedging strategy, it is only one aspect. As an illustration, a hedger will not only diversify an investment portfolio by buying different stocks, commodities, properties, and currencies, but simultaneously she will short sell a percentage of her investment portfolio. How and when the portfolio is shorted, for example, through derivative products or future contracts, will depend on the hedger's risk profile. These measures provide a hedger with more tools to limit and manage risks than an investor, who only diversifies. China's cross-border pipelines diversify China's energy sources and routes. However, the pipeline "market" differs from the international petroleum market, the seaborne and the pipeline routes offer different security, and the development of cross-border pipelines provide China with broader economic, strategic, and development opportunities. Hence, China's cross-border pipelines, encompassed in its hedging strategy, represent more than mere diversification.

EMPIRICAL FINDINGS

The preceding chapters in this volume focus on how the central government controls the NOCs and how this authority gives the government the means to develop and implement hedging strategies, especially when it comes to long-term planning or during a crisis. After describing what the government is hedging against in chapters 3 through 6, the respective analyses in each chapter examine how China hedges in the domestic, global, maritime, and continental aspects of its energy security policy.

The central government has several mechanisms to control the NOCs: appointing NOC leaders (the nomenklatura system); setting up party groups within the NOCs through its role as lawmaker, as regulator, and law enforcer; by collecting dividend payments; by setting domestic oil product prices; through taxation; by approving investments; by endorsing credits; and by developing overall planning and granting rights and licenses for exploration. But the central government cannot control the entire Chinese energy sector, which is growing at an unprecedented pace. The NOCs have become powerful and autonomous actors, especially in their daily business and in their pursuit of profits at home and abroad. The government has difficulties overseeing all the deals, investments, production, and exploration plans of the NOCs. At the same time, controlling all aspects of the NOCs' activities is probably not the main aim of the government.

NOC autonomy does not override the central government's authority, and autonomous does not mean independent. Corporate interests cannot undermine the overarching objective of the government to provide oil supply security, which is linked to China's economic growth, domestic stability, and CCP legitimacy. The NOCs may invest in countries without the approval of the central leadership, they may sell more of their oil on the international market for profit instead of shipping it back to China, they may seek to manipulate the domestic downstream market in order to pressure the central government to raise oil product prices, and at times their presence in certain countries may weaken China's diplomatic status. Nonetheless, the government maintains a number of mechanisms to control the NOCs, and it has the authority to mandate that the NOCs provide for China's oil supply security in times of an emergency or a crisis. This is facilitated by the hedging strategies developed by the central government.

The NOCs are involved in the planning and implementation of the five-year plans, thereby shaping strategies and policies based on their own interests. Nonetheless, the NOCs were initially reluctant to participate in some of the core hedging strategies that have been outlined here, such as the decision to develop cross-border pipelines or to build up strategic petroleum reserves (SPRs), and they were not the key players driving the goal to develop a large state-owned tanker fleet. In addition, maintaining the coal sector as the bedrock of China's energy security, emphasizing energy efficiency and conservation, and seeking to develop and increase the use of alternative sources of energy, all strategies highlighted in the various long-term plans, are goals that are not directly linked to the interests of the NOCs.

More importantly, whether the five-year plans or the long-term strategies coincide with the NOCs' interests is not the core issue. The fact is that roughly all the strategies put forward by the central government and by energy institutions in the early 2000s, including a willingness to hedge to reduce the risks and consequences of oil supply disruptions, sudden price rises, and the growing net oil import gap, have been developed and advanced. This provides evidence that the central government has been a key driver behind China's energy security policy and has successfully implemented a number of hedging strategies stipulated in the five-year plans, such as maintaining a strategically favorable energy mix, building an SPR, increasing refinery capacity, investing in long-term future contracts, securing overseas equity oil production, building a large state-owned tanker fleet, developing cross-border pipelines, and diversifying China's sources, routes, and energy mix.

Despite a recognition that China's growth model is not beneficial to the environment and is unsustainable, China's central leadership has emphasized that coal will remain a crucial factor in China's energy security policy. Maintaining a favorable energy mix, whereby roughly 90 percent of China's energy demand is provided for by domestic sources, reflects China's continued emphasis on self-sufficiency but also demonstrates that the central government is willing to pay the costs of insuring against oil supply disruptions and to hedge against the availability of sufficient oil supplies at affordable prices.

China is not significantly vulnerable to oil supply disruptions or to a blockade in a wartime contingency. It has a domestic oil production of about 4 mbd and a favorable energy mix, and automobile use will largely drive future oil imports. Personal automobile use is not strategic and can be curtailed. Many interviewees in China have pointed out that the Chinese will be able to "tighten their belts" if China becomes involved in an unlikely war with the United States, the only great power capable of blockading or severely disrupting China's oil imports.

To hedge against a crisis in the international petroleum market or other emergencies short of a war involving China, measures have been taken to develop an SPR to insure against oil supply disruptions and price volatility. Refinery capacity has been expanded, mainly because of the NOCs' profit considerations but in part to enhance import flexibility and security of supply by expanding China's ability to refine crude oil from more suppliers.

The Chinese NOCs' global search for equity oil production, marketing rights, and growing importance in the world energy market is hard to overstate. It is estimated that Chinese NOCs spent at least $47.59 billion to acquire oil and gas assets worldwide from January 2009 to December 2010, spent about $48 billion on M&As from 2009 through 2010, and during the same period were involved in twelve loan-for-oil deals in the same period worth $77 billion. The NOCs have invested in about fifty countries and have a stake in roughly two hundred upstream projects, with an overseas equity production accounting for approximately 1.4 mbd.

The NOCs have been profit driven in their "going abroad" strategies since the 1990s. However, their commercial interests are beneficial and conducive to China's oil supply security. More importantly, their overseas equity investments, production, and marketing rights provide China with a hedge against risks in the international petroleum market. Having the opportunity and authority to mandate that the NOCs ship more of their overseas oil production back home during a crisis in the international petroleum market allows China to be relatively better off than other major oil consumers during periods of crisis.

This strategy is reinforced by the fact that one way China hedges against exceedingly high oil prices in the international market during a crisis or a war in an oil-rich region through the buildup of a large state-owned tanker fleet, which makes it possible to ship China's NOCs' overseas oil production back to China on Chinese tankers, thereby with lower freight costs than if China were to rely on chartering international tankers during a crisis. A state-owned tanker fleet thus also becomes a hedge against disruptions in the availability of oil during a crisis, since the central government can force Chinese tankers to operate, despite the high risks, in a conflict area, in order to access oil terminals where other shipping companies may not be willing to operate. Finally, by having its own tankers and its own capacity, China can self-insure and avoid high insurance premiums. The self-insurance strategy can also be applied short of a wartime scenario, for example, if one of China's oil suppliers are sanctioned, embargoed, or blockaded, as was the case regarding China's oil trade with Iran in 2012 and 2013.

These hedging strategies are developed to manage risks in the international petroleum market but not to eliminate threats to China's energy security. For example, China's overseas oil production cannot be protected or shipped back home if China finds itself at war with the United States. Neither can China's tankers and vital SLOCs be safeguarded by the PLAN. Indeed, a state-owned tanker fleet will be counterproductive if China becomes involved in a war with the United States, since it will be much easier to identify which tankers are carrying oil to China. Accordingly, the distinction between peacetime risks and wartime contingencies is crucial to understanding key aspects of China's energy security policy and how and why China adopts the hedging strategies it has.

In addition, the cross-border pipelines that China has developed act as a hedge, or as "short" strategy, that provides protection if China's reliance on "longs," or seaborne petroleum supplies and the international oil market, is disrupted. Since the security of pipelines and seaborne supply routes differ and the pipeline "market" and the international petroleum market remain distinct, in the case of China's energy security, pipelines are considered more vulnerable during peacetime risks and safer during wartime threats, whereas seaborne energy supplies are safer during peacetime risks and more vulnerable during wartime threats. Thus, combining pipeline and seaborne supply routes to hedge against supply disruptions strengthens China's risk management capabilities.

IMPLICATIONS

China's powerful NOCs are playing an increasingly influential role and are becoming strong competitors in the international petroleum market. Backed by the Chinese government and Chinese banks, the NOCs will continue to seek out global investments, M&As, and loan-for-oil deals. Regardless of whether the NOCs develop into hybrid oil companies or come to resemble IOCs, at the end of the day they will be controlled by the central government and will serve China's national interests when they are called upon by the Chinese leadership. This will have direct implications for the international oil market.

The international oil market is akin to a large pool of oil. If China's NOCs were to ship all their oil back home, then China will buy less elsewhere, which means there will be more oil available to other consumers.[4] However, if China's NOCs' equity oil, which is normally sold on the international petroleum market for profit, is shipped home during a crisis or a war in an oil-rich region and replaces, for example, disrupted Iranian oil that China normally imports, then there will be less oil available to all other consumers.[5] Such a "lock-up" scenario is only likely to occur when China hedges against supply disruptions in the international market during a crisis that falls short of a war involving China. During "normal" market conditions, the NOCs will be in the driver's seat, and a large percentage of their overseas oil production will be sold locally or on the international market for profit, whereas during a wartime scenario involving China it will be difficult for China to ship equity oil back to China.

Oil will be available in the market, but probably at an exceedingly high price. China will be reasonably well off compared to other major oil importers during such a situation, as Chinese leaders can call on its state-owned tankers to ship oil back to China and can self-insure its tankers operating in areas of conflict. Again, this will affect the petroleum and tanker market since some of the Chinese tankers that normally are chartered out will be directed to serve Chinese interests, while at the same time some of the remaining tankers in the world tanker fleet may be reluctant to enter an area of conflict or be prohibited because of sanctions and embargo, thereby limiting tanker capacity and driving up freight rates. This situation is likely to be reinforced in the years ahead: China's state-owned tankers can be expected to play a more prominent role in the tanker market, both as a result of the tanker buildup and the fact that government-backed state-owned tankers are more likely to survive the current bearish tanker market.

Chinese NOCs' global search for petroleum investment opportunities and oil production deals confers on both the NOCs and the Chinese government a more prominent role in the petroleum market and world affairs. The Chinese government, banks, and the PLA often facilitate the NOCs' "going abroad" strategy by a number of sweeteners, including diplomatic support, loans to host countries, and arms deals. When the NOCs operate more or less as autonomous actors, Chinese diplomatic, strategic, political, economic, military, and security interests will be affected. China's energy security policy and the activities of the NOCs signal that China has become a powerful actor in international politics.

This corresponds with a geopolitical trend whereby China's interests expand as its power increases. A more dominant China acquires a stronger global stake as its interests are spread throughout the world. Simultaneously, as China expands its worldwide interests, they will need to be protected. However, this is a delicate and sensitive task. More Chinese military power, intervention, policing, diplomatic engagement, assertiveness, and influence could trigger Chinese security dilemmas, arms races, and counterbalancing or containment against China. More involvement often leads to further engagement and expansion, and these in turn produce new areas or interests that need to be protected.[6] China's growing involvement in UN peacekeeping operations (UNPKO), its participation in constabulary tasks and patrolling of SLOCs, its shifting but cautious stance on sovereignty and "noninterference," and its need to balance between developing war-fighting access-denial platforms and more power-projection capabilities show how China's growing global energy interests are shaping broader Chinese foreign, security, and defense policies.

Commensurate with China's rising power is also its growing importance in international energy institutions, and these institutions and organizations recognize that cooperation with China is of critical importance. China is also becoming more involved in Arctic affairs and is seeking to shape this new energy frontier in ways conducive to China's interests. Finally, great-power relations will be crucial in promoting cooperation and managing conflicts of interest related to energy security.

China and the United States are interacting globally on energy-related questions. The United States remains the key guarantor of security for the Middle East, but at the same time China's presence in the region is growing, and it is expected that China's oil imports from the Middle East will increase in the future. U.S. naval supremacy and control of SLOCs, which are vital to China's trade and oil imports, also are cause for concern in China. Simultaneously, China's naval ambitions and growing presence in distant waters are bound to alarm policy makers in the United States. Hence, energy security can either bring the United States and China closer together through their roles as the world largest oil consumers or fuel rivalry and future conflicts of interest.

Although Russia and China eventually managed to construct a cross-border pipeline, seal a major oil-for-loan deal, and keep the potential for future energy cooperation sound, the overall Sino-Russian strategic and security partnership remains fragile. Their common interest in challenging the U.S. presence in Asia is eroding as the United States withdraws most of its troops from the region. Instead, China has become the dominant power on the Asian mainland; Russia is falling into a junior position in their bilateral relationship. While there remain several obstacles to improving Sino-Russian energy relations, which are largely tied to uncertainty about Russian petroleum reserves in eastern Siberia, the lack of Russian infrastructure in Siberia and the Far

East, the difficult investment conditions in the Russian petroleum sector, and disputes about contracts and prices, energy ties are one of the few areas where the strategic partnership potentially might be revitalized and prevent the Sino-Russian relationship from eroding.

In under twenty years, China has evolved from being a net oil exporter to becoming the world's largest oil importer. The Chinese central government has been successful in carrying out most of China's long-term energy strategies. It has addressed many of the challenges to China's energy security that have been identified since the early 2000s, and it has improved China's risk management capacity through a number of hedging strategies. It may be that the hedging framework gives the impression that the different elements in China's approach to energy security are more coordinated, integrated, and carefully planned than they are in reality. This is undoubtedly a difficult issue to research. However, the Chinese government has the authority to control the NOCs and the means to develop and execute its hedging strategies. The five-year plans and other long-term plans emphasize a preference for hedging, and Chinese behavior in important aspects of its energy security policy fits a hedging pattern.

NOTES

1. INTRODUCTION

1. Philip Andrews-Speed and Roland Dannreuther, *China, Oil, and Global Politics* (London: Routledge, 2011); Carrie Liu Currier and Manochehr Dorraj, eds., *China's Energy Relations with the Developing World* (New York: Continuum, 2011); Bo Kong, *China's International Petroleum Policy* (Santa Barbara, Calif.: Praeger Security International, 2010); and Erica Downs, "China's Energy Security," Ph.D. diss., Princeton University, 2004.

2. The Chinese government generally refers to the central party-state apparatus, for example, the Chinese Communist Party (CCP) Politburo, the State Council, the National Energy Administration (NEA), and the National Development and Reform Commission (NDRC). Because the central and local governments may have conflicting interests in the energy sector as well as in terms of energy security policy, I will explicitly refer to the provincial or local governments when discussing their roles.

3. Profits do not simply refer to money or profitability but also include corporate aggrandizement. In addition, since the major NOCs are state-owned companies, those who stand to benefit from the NOCs profits are not only shareholders but also the Chinese government and the CCP. It is beyond the scope of this book to examine how the NOCs use their profits to buy "political capital" or how NOC revenues fund the different sections of the Chinese government and the factions within the CCP.

4. Throughout this work, "hedging" is sometimes referred to as a "framework," an "analytical tool," a strategy, a policy, an explanatory factor, or a concept that enhances understanding of China's energy security policy. The meaning of "hedging" and its applicability to China's energy security policy will be demonstrated in the following chapters. It is not uncommon or contradictory that a theory or framework, such as the balance of power theory or rational

choice, predicts, explains, or provides analytical tools and understanding of behavior but also is simultaneously a strategy, a policy objective, and a tool of statecraft.

5. The United States, for the most of the post–Cold War period, has been uncertain as to whether China represents a threat that warrants a balancing strategy. Instead, China has been perceived as a risk: it might become an aggressive peer competitor in the future. Hedging was therefore seen as an appropriate strategy. Recently, it seems the uncertainty about China's rise and whether it represents a threat to the United States is no longer as profound. Indeed, China is seen as increasingly challenging U.S. interests, especially in Asia. The United States is therefore more willing to balance instead of hedging against China.

6. Peter C. Fusaro, ed., *Energy Risk Management: Hedging Strategies and Instruments for the International Energy Markets* (New York: McGraw-Hill, 1998); Peter C. Fusaro and Tom James, *Energy Hedging in Asia: Market Structure and Trading Opportunities* (New York: Palgrave Macmillan, 2005); Michael Wesley, ed., *Energy Security in Asia* (London: Routledge, 2007); and Downs, "China's Energy Security," 49, 89.

7. But see Øystein Tunsjø, "Hedging Against Energy Dependency: New Perspectives on China's Energy Security," *International Relations* 24, no. 1 (2010): 25–45; and Øystein Tunsjø, "Zhongguo nengyuan anquan de dui chong zhanlüe" (China hedges its energy security bets), *Shijie jingji yu zhengzhi* (World economics and politics) (August 2008): 42–51.

8. Mikkal Herberg and David Zweig, "China's 'Energy Rise': The U.S. and the New Geopolitics of Energy," Pacific Council on International Policy, April 2010, http://www.pacificcouncil.org/document.doc?id=159.

9. See chapter 2 for a discussion of these two approaches.

10. For example, a terrorist attack is a risk that cannot be eliminated, but it can be prevented, and the risk can be managed. If a suicide bomber enters a building, plane, or ship and the person is observed by security guards, this situation will pose a threat, but theoretically this threat can be eliminated by the security guards before the terrorist will cause harm. However, it is impossible to eliminate the danger or risk that another suicide bomber will seek to attack the building, plane, or ship the next day; it is possible to manage and prevent this risk from materializing into a threat through intelligence, surveillance, and increased security measures.

11. See the excellent study by Gabriel B. Collins and William S. Murray, "No Oil for the Lamps of China?" *Naval War College Review* 61, no. 2 (Spring 2008): 79–95.

12. Instead, a maritime blockade, in theory, could be enforced without having to attack a sovereign state. See Downs, "China's Energy Security," 112.

13. Of course, the United States would face the risk of escalation if the U.S. Navy were ordered to enforce a limited maritime blockade, but that risk is greater if the United States were to bomb Chinese territory than if the United States were to board a Chinese-flagged tanker on the high seas. Again, both scenarios are unlikely because they could quickly escalate into a major war and so have a limited effect on China's energy security policy.

14. Important contacts were established when the author co-organized the Energy Security in Asia conference, Beijing, May 2009, with participation by leading energy experts from

the NDRC, NOCs, PLA, think tanks, and universities. At the conference, Xu Dingming, State Council counselor, chairman of the National Energy Expert Advisory Committee, and former director of the National Energy Bureau, presented a keynote speech. A number of experts from the International Energy Agency (IEA), the United States, Japan, South Korea, and Europe, international oil companies, and the shipping and insurance sectors, together with representatives from the diplomatic community in Beijing, participated in the conference.

15. The oil import dependency rate now stands at about 55 percent.

16. The overseas production of Chinese NOCs is likely about 2 mbd, with equity oil production accounting for approximately 1.4 mbd. China's total oil imports currently stand at close to 6 mbd.

17. See chapter 4. In short, it would be extremely difficult, short of a major war, to identify tankers going to China. They may be parceled out, rerouted to China after inspection, or rerouted through other straits. The blockading power would suffer diplomatically because the tankers and their crews are multinational and their cargo is sold on the international market. The blockading power would also undermine diplomatic ties with its allies because of supply disruptions and rising prices. For these and other reasons, a blockade of China's oil supply is highly unlikely short of a major war.

18. Daniel Yergin, *The Quest: Energy, Security, and the Remaking of the Modern World* (London: Allen Lane, 2011), 196. China's oil imports amount to 10 percent of China's energy mix. Even during a war it would be difficult to blockade completely all of China's oil imports. More importantly, domestically China produces more than 4 mbd and in a wartime contingency nonstrategic or military oil consumption could be curtailed. In fiscal year 2004, when the U.S. military was engaged in Iraq and Afghanistan, it used approximately 395,000 bpd of oil. Thus, domestically, China produces more than ten times the amount that the United States needs to maintain its global military activities. See Collins and Murray, "No Oil for the Lamps of China?" 79–95.

19. Robert Ross, "China's Naval Nationalism," *International Security* 34, no. 2 (Fall 2009): 46–81.

2. CHINA'S ENERGY SECURITY: A NEW FRAMEWORK FOR ANALYSIS

1. Philip Andrews-Speed, "China's Energy Policy and Its Contribution to International Stability," in Marcin Zaborowski, "Facing China's Rise: Guidelines for an EU Strategy," *Chaillot Paper* 94 (December 2006): 73; and Philip Andrews-Speed, Xuanli Liao, and Roland Dannreuther, "The Strategic Implications of China's Energy Needs," *Adelphi Paper* 346 (Oxford: Oxford University Press, 2002), 19.

2. François Godement, Francois Nicolas, and Taizp Yakushiji, "An Overview of Options and Challenges," in *Asia and Europe: Cooperating for Energy Security*, ed. François Godement, Françoise Nicolas, and Taizo Yakushiji (Paris: Centre Asia, Institut français des relations internationales, 2004), 20; and Christian Constantin, "Understanding China's Energy Security," *World Political Science Review* 3, no. 3 (2007): 5.

3. Ziad Haider, "Oil Fuels Beijing's New Power Game," *YaleGlobal* (March 11, 2005). For an in-depth analysis of relative gains in international politics, see Kenneth Waltz, *Theory of International Politics* (New York: Random House, 1979), 105; and Joseph M. Grieco, *Cooperation Among Nations: Europe, America, and Non-Tariff Barriers to Trade* (Ithaca, N.Y.: Cornell University Press, 1990), esp. chaps. 3, 10.

4. Cherie Canning, "Pursuit of the Pariah: Iran, Sudan, and Myanmar in China's Energy Security Strategy," *Security Challenges* 3, no. 1 (February 2007): 51; Andrews-Speed, "China's Energy Policy and Its Contribution to International Stability," 73; Flynt Levrett, "The Geopolitics of Oil and America's International Standing," statement before the Committee on Energy and Natural Resources, U.S. Senate, January 10, 2007.

5. Statements by former U.S. Deputy Secretary of State Robert Zoellick and former head of the China Desk in the U.S. State Department, Randall Schriver, in September 2005. See "U.S. Warns China on Iran Oil," *Reuters* (September 7, 2005), http://www.financialexpress.com/old/latest_full_story.php?content_id=101772. See also Senator Joseph Lieberman, "China-US Energy Policies: A Choice of Cooperation or Collision," Council on Foreign Relations, November 30, 2005; and Michael T. Klare, *Rising Powers, Shrinking Planet: The New Geopolitics of Energy* (New York: Metropolitan, 2008).

6. Gal Luft, "U.S., China Are on a Collision Course Over Oil," *Los Angeles Times* (February 2, 2004); Gal Luft, "Fuelling the Dragon: China's Race Into the Oil Market," Institute for the Analysis of Global Security, http://www.iags.org/china.htm; and Peter Hatemi and Andrew Wedeman, "Oil and Conflict in Sino-American Relations," *China Security* 3, no. 3 (Summer 2007): 110.

7. At the conference on Energy Security in Asia, Beijing, May 21–22, 2009, leading Chinese energy security experts emphasized the importance of market mechanisms.

8. Hu Jintao, "G8 Written Statement," St. Petersburg, July 17, 2006, http://www.uofaweb.ualberta.ca/chinainstitute/nav03.cfm?nav03=48114&nav02=43884&nav01=43092; Hu Jintao, "An Open Mind for Win-Win Cooperation," speech at the Asia Pacific Economic Cooperation (APEC) CEO summit, Busan, Republic of Korea, November 17, 2005; Angie Austin, "Energy and Power in China: Domestic Regulation and Foreign Policy," Foreign Policy Centre, April 2005, http://fpc.org.uk/fsblob/448.pdf; and Eurasia Group, "China's Overseas Investment in Oil and Gas Production," report for the U.S.-China Economic and Security Review Commission, October 16, 2006.

9. U.S. Department of Energy (DOE), *Energy Policy Act 2005 Section 1837: National Security Review of International Energy Requirements* (February 2006), 3.

10. See Gary Dirks, "China's Energy: Challenges and Implications," speech at the German Council of Foreign Policy, Berlin, September 13, 2007. Trevor Houser and Roy Levy argue that overseas oil production by Chinese companies totaled 620,000 bpd in 2007, only half of which was shipped back to China and accounting for less than 10 percent of the 3.25 mbd of Chinese imports that year. Trevor Houser and Roy Levy, "Energy Security and China's UN Diplomacy," *China Security* 4, no. 3 (Summer 2008): 70.

11. Julie Jiang and Jonathan Sinton, "Overseas Investments by Chinese National Oil Companies: Addressing the Drivers and Impacts," Information Paper (IEA) (February 2011), 17.

12. Mikkal E. Herberg, statement before the U.S.-China Economic and Security Review Commission hearing on "China's Energy Consumption and Opportunities for U.S.-China Cooperation to Address the Effects of China's Energy Use," June 14–15, 2007.

13. Uwe Nerlich, "Energy Security or a New Globalization of Conflicts? Oil and Gas in Evolving New Power Structures," *Strategic Insight* 7, no. 1 (February 2008), http://www .isn.ethz.ch/isn/Digital-Library/Publications/Detail/?ots591=0c54e3b3-1e9c-be1e -2c24a6a8c7060233&lng=en&id=48295.

14. Kenneth Lieberthal and Mikkal Herberg, "China's Search for Energy Security: Implications for U.S. Policy," *NBR Analysis* 17, no. 1 (2006).

15. Henry Lee and Dan A. Shalmon, "Searching for Oil: China's Oil Initiatives in the Middle East," BCSIA Discussion Paper (Cambridge, Mass.: Belfer Center for Science and International Affairs, Kennedy School of Government, Harvard University, January 2007), 8. See also Cindy Hurst, "China's Oil Rush in Africa," *IAGS Energy Security* (July 2006); Esther Pan, "China, Africa, and Oil," Council on Foreign Relations, *Backgrounder* (January 26, 2007); and Michael Klare and Daniel Volman, "The African 'Oil Rush' and U.S. National Security," *Third World Quarterly* 27, no. 4 (May 2006): 622.

16. Robert O. Keohane, *Neorealism and Its Critics* (New York: Columbia University Press, 1986); Joseph S. Nye, "Neorealism and Neoliberalism," *World Politics* 40, no. 2 (1988): 235–251; David Baldwin, ed., *Neorealism and Neoliberalism: The Contemporary Debate* (New York: Columbia University Press, 1993); Øystein Tunsjø, "Zhongguo nengyuan anquan de dui chong zhanlüe" (China hedges its energy security bets), *Shijie jingji yu zhengzhi* (World economics and politics) (August 2008): 42–51.

17. International Crisis Group (ICG), "China's Thirst for Oil," *Asia Report* 153 (June 9, 2008); Bernard Cole, *Sea Lines and Pipe Lines: Energy Security in Asia* (Westport, Conn.: Praeger, 2008); Bo Kong, "An Anatomy of China's Energy Insecurity and Its Strategies," Pacific Northwest Center for Global Security, December 2005; Zha Daojiong, "Energy Interdependence," *China Security* 2, no. 2 (Summer 2006): 2–16; Linda Jakobson and Zha Daojiong, "China and the Worldwide Search for Oil Security," *Asia-Pacific Review* 13, no. 2 (2006): 60–73; James Tang, *With the Grain or Against the Grain? Energy Security and Chinese Foreign Policy in the Hu Jintao Era* (Washington, D.C.: The Brookings Institution, 2006); Xuecheng Liu, "China's Energy Security and Its Grand Strategy," *Policy Analysis Brief* (The Stanley Foundation) (September 2006); and Peter Cornelius and Jonathan Story, "China and Global Energy Markets," *Orbis* 51, no. 1 (Winter 2007): 5–20.

18. Carrie Liu Currier and Manochehr Dorraj, eds., *China's Energy Relations with the Developing World* (New York: Continuum, 2011); Michal Meidan, ed., *Shaping China's Energy Security: The Inside Perspective* (Paris: Asia Centre, 2007); Lieberthal and Herberg, "China's Search for Energy Security"; Michael Wesley, ed., *Energy Security in* Asia (London: Routledge, 2007); Constantin, "Understanding China's Energy Security"; Erica Strecker Downs, "China's Energy Security," Ph.D. thesis, Princeton University, 2004; and Andrews-Speed, Liao, and Dannreuther, "The Strategic Impact of China's Energy Needs."

19. Richard Weitz, "Why US Keeps Hedging over China," *The Diplomat* (January 11, 2011); Greg Sheridan, "Asia Primed for a New Cold War," *The Australian* (February 2, 2011); John J. Tkacik, "Panda Hedging: Pentagon Urges New Strategy for China," *Web Memo* (Heritage Foundation) 1093 (May 24, 2006); Robert Sutter, "Why Rising China Can't Dominate Asia," *Glocom Platform* (September 11, 2006); David Shambaugh, "Asia in Transition: The Evolving Regional Order," *Current History* (April 2006): 153–160; Donald G. Gross, "Transforming the U.S. Relationship with China," *Global Asia* 2, no. 1 (Spring 2007); Hugh White, "Why War in Asia Remains Thinkable," *Survival* 50, no. 8 (2008): 99; Carin Zissis, "Crafting a U.S. Policy on Asia," Council on Foreign Relations, April 10, 2007; Sharman Katz and David Stewart, "Hedging China with FTAs," *Asia Times* (October 1, 2005); and Graeme Dobell, "Correspondent Report: US Details its Hedging Policy Towards China," *ABS Online* (April 1, 2007).

20. Rosemary Foot, "Chinese Strategies in a U.S.-Hegemonic Global Order: Accommodating and Hedging," *International Affairs* 82, no. 1 (2006): 77–94; Evelyn Goh, "Meeting the China Challenge: The U.S. in Southeast Asian Regional Security Strategies," *Policy Studies* 16 (Washington, D.C.: East-West Center, 2005), 1–57; Robert J. Art, "Europe Hedges Its Security Bets," in *Balance of Power: Theory and Practice in the Twenty-First Century*, ed. T. V. Paul, James J. Wirtz, and Michel Fortman (Stanford, Calif.: Stanford University Press, 2004), 179–213; Øystein Tunsjø, "Geopolitical Shifts, Great Power Relations, and Norway's Foreign Policy," *Cooperation and Conflict* 46, no. 1 (March 2011): 60–77; Evan S. Medeiros, "Strategic Hedging and the Future of Asia-Pacific Stability," *Washington Quarterly* 29, no. 1 (2005–2006): 145–167; and Øystein Tunsjø, *U.S. Taiwan Policy: Constructing the Triangle* (London: Routledge, 2008), 109–119.

21. The term frequently occurred in documents from the Bush administration: U.S. Department of Defense, *National Defense Strategy* (Washington, D.C., June 2008); U.S. Department of Defense, *Quadrennial Defense Review Report* (Washington, D.C., February 6, 2006); The White House, *The National Security Strategy of the United States of America* (Washington, D.C., March 2006); and U.S. Department of Defense, *Annual Report to Congress on the Military Power of the People's Republic of China* (Washington, D.C., 2007), executive summary.

22. Tunsjø, "Geopolitical Shifts."

23. Goh, "Meeting the China Challenge"; and Evelyn Goh, "Understanding 'Hedging' in Asia-Pacific Security," *Pacific Forum* (CSIS) (August 31, 2006).

24. Goh, "Meeting the China Challenge," 2–3.

25. Richard J. Samuels, *Tokyo's Grand Strategy and the Future of East Asia* (Ithaca, N.Y.: Cornell University Press, 2007), 8.

26. See note 10 in chapter 1.

27. International Energy Agency, *2007 World Energy Outlook*, 161, http://www.iea.org/textbase/nppdf/free/2007/weo_2007.pdf.

28. Jon Elster, *Nuts and Bolts for the Social Sciences* (Cambridge: Cambridge University Press, 1989), 26.

29. Tunsjø, "Geopolitical shifts," 2011; Brock Tessman and Wojtek Wolfe, "Great Powers and Strategic Hedging: The Case of Chinese Energy Security Strategies," *International Studies*

Review 13 (2011): 214–240; and Brock Tessman, "System vStructure and State Strategy: Adding Hedging to the Menu," *Security Studies* 21, no. 2 (2012).

30. Tessman and Wolfe, "Great Powers."

31. Tunsjø, "Geopolitical Shifts."

32. Foot, "Chinese Strategies"; Tunsjø, "Geopolitical Shifts."

33. Tessman, "System Structure."

34. Some residual, or basic, risk is intrinsic, and a hedging strategy is contingent on the risk profiles of the analysts, firms, or state stakeholders: some are risk takers; others are more risk averse. See Darrell Duffie, *Future Markets* (Englewood Cliffs, N.J.: Prentice-Hall, 1989); James Dow, "Arbitrage, Hedging, and Financial Innovation," *Review of Financial Studies* 11, no. 4 (Winter 1998): 739–755; and Richard A. Brealey and Stewart C. Myers, *Financing and Risk Management* (New York: McGraw-Hill, 2003), 319–325.

35. China's SPR can also be a "long" strategy if China coordinates its releases with the International Energy Agency (IEA). However, it remains to be seen if China will closely cooperate with the IEA in its energy security policy.

36. Kenneth Waltz, *Man, the State, and War: A Theoretical Analysis* (New York: Columbia University Press, 1959), 168.

37. Kenneth Waltz, *Theory of International Politics* (New York: Random House, 1979), 105.

38. Downs, "China's Energy Security," 49, 89; Erica Downs, "The Chinese Energy Security Debate," *China Quarterly* 177 (2004): 41; Wesley, ed., *Energy Security in Asia*, 1, 8; William Tow, "Strategic Dimensions of Energy Competition in Asia," in Wesley, ed., *Energy Security in Asia*, 165; and Zha Daojiong, "Oiling the Wheels of Foreign Policy? Energy Security and China's International Relations," Asia Security Initiative Policy Series Working Paper 1 (March 2010), on the use of the term hedging.

39. Brealey and Myers, *Financing and Risk Management*, 319–325.

40. Zhang Wenmu, "Sea Power and China's Strategic Choices," *China Security* 3 (Summer 2006), 17–31.

41. For various views on China's assertive behavior see, among others, Thomas J. Christensen, "The Advantages of an Assertive China," *Foreign Affairs* 90, no. 2 (2011): 54–67; June Teufel Dreyer, "Grimm Foreign Policy," *The Diplomat* (February 12, 2011), http://the-diplomat.com/2011/02/12/grimm-foreign-policy/; Minxin Pei, "China's Bumpy Ride Ahead," *The Diplomat* (February 16, 2011), http://the-diplomat.com/whats-next-china/chinas-bumpy-ride-ahead/; Kerry Brown and Loh Su Hsing, "Trying to Read the New 'Assertive' China," *Asia Program Paper* (Chatham House) 2 (January 2011), http://www.chathamhouse.org/sites/default/files/public/Research/Asia/0211brownhsing_pp.pdf; Wang Jisi, "China's Search for a Grand Strategy," *Foreign Affairs* 90, no. 2 (2011): 68–79; and Michael D. Swaine and Taylor M. Fravel, "China's More Assertive Behavior. Part Two: The Maritime Periphery," *China Leadership Monitor* 35 (Summer 2011).

42. Robert D. Kaplan, "Center Stage for the Twenty-First Century, Power Plays in the Indian Ocean," *Foreign Affairs* 88, no. 2 (March/April 2009): 16–32.

43. Chris Devonshire-Ellis, "China's String of Pearls Strategy," *China Briefing* (March 18, 2009), http://www.china-briefing.com/news/2009/03/18/china%E2%80%99s-string-of-pearls

-strategy.html. On overseas bases, see "China's National Defense White Paper," issued by the Information Office of the State Council.

44. Taylor Fravel, *Strong Borders, Secure Nation: Cooperation and Conflict in China's Territorial Disputes* (Princeton, N.J.: Princeton University Press, 2008).

45. Ken Booth and Nicholas J. Wheeler, *The Security Dilemma: Fear, Cooperation, and Trust in World Politics* (New York: Palgrave Macmillan, 2008).

46. Various views and approaches also shaped discussions among leading Chinese, Western, and Asian experts at the conference on Energy Security in Asia, Beijing, May 21–22, 2009.

47. Zha Daojiong is a representative of the market, or interdependent, view, whereas Zhang Wenmu argues that China's increasing dependence on oil imports is a strategic threat. Several articles in *China Security* 3 (Summer 2006) explore this debate.

48. Zha, "Energy Interdependence," 8.

49. Daniel Yergin, "Ensuring Energy Security," *Foreign Affairs* 85, no. 2 (March–April 2006): 69–82.

50. See Tunsjø, *U.S. Taiwan Policy*, 107–119, for the linkage between hedging and risk management in explaining U.S. China policy; and Øystein Tunsjø, "Zhongguo nengyuan anquan de dui chong zhanlüe" (China hedges its energy security bets), *Shijie jingji yu zhengzhi* (World economics and politics) (August 2008): 42–51, which also links hedging to risk management and differentiates this approach from securitization.

51. Ulrich Beck, *Risk Society: Towards a New Modernity*, trans. Mark Ritter (Cambridge: Polity, 1992); Anthony Giddens, *Modernity and Self-Identity: Self and Society in the Late Modern Age* (Cambridge: Polity, 1991); and Michael Foucault, "Governmentality," in *Essential Works of Michael Foucault, 1954–1984*, vol. 3: *Power*, ed. James D. Faubion (New York: The New Press, 2000), 201–222.

52. Christopher Coker, *War in an Age of Risk* (Cambridge: Polity, 2009); Yee-Kuang Heng, *War as Risk Management: Strategy and Conflict in an Age of Globalised Risks* (London: Routledge, 2006); Mikkel Vedby Rasmussen, *The Risk Society at War: Technology and Strategy in the Twenty-First Century* (Cambridge: Cambridge University Press, 2006); Louise Amoore and Mareke de Goede, eds., *Risk and the War on Terror* (London: Routledge, 2008); and Tunsjø, *U.S. Taiwan Policy*, 109–119.

53. Heng, *War as Risk Management*, 10.

54. Mathias Albert, "From Defending Borders Towards Managing Geopolitical Risks? Security in a Globalized World," *Geopolitics* 5, no. 1 (2001): 64.

55. Øystein Tunsjø, "Hedging Against Oil Dependency: New Security Perspectives on Chinese Energy Security Policy," *International Relations* 24, no. 1 (March 2010): 32; IEA, *World Energy Outlook*, 2007.

56. Yergin, *The Quest: Energy, Security, and the Remaking of the Modern World* (London: Allen Lane, 2011), 213.

57. Gabriel B. Collins and William S. Murray, "No Oil for the Lamps of China?" *Naval War College Review* 61, no. 2 (Spring 2008): 81.

58. Yergin, *The Quest*, 196.

3. CHINA'S DOMESTIC ENERGY SECTOR

1. Philip Andrews-Speed, Xuanli Liao, and Roland Dannreuther, "The Strategic Implications of China's Energy Needs," *Adelphi Paper* 346 (Oxford: Oxford University Press, 2002), 100; Bo Kong, "Institutional Insecurity," *China Security* 3 (Summer 2006): 64–88; and see parts 1 and 2 in Michal Meidan, ed., *Shaping China's Energy Security* (2007). Christian Constantin, "Understanding China's Energy Security," *World Political Science Review* 3, no. 3 (2007): 1–30. Although the model is slightly outdated, this decision-making environment has also been described as a "fragmented authoritarianism." See Kenneth Lieberthal and Michel Oksenberg, *Policy Making in China: Leaders, Structures, and Processes* (Princeton, N.J.: Princeton University Press, 1988); Susan Shirk, *The Political Logic of Economic Reform in China* (Berkeley: University of California Press, 1993); Erica Strecker Downs, "China's Energy Security," Ph.D. thesis, Princeton University, 2004.

2. Erica Downs, "Who's Afraid of China's Oil Companies?" in *Energy Security: Economics, Politics, Strategies, and Implications*, ed. Carlos Pascual and Jonathan Elkind (Washington, D.C.: Brookings Institution Press, 2010), 76–77; Bo Kong, "An Anatomy of China's Energy Insecurity and Its Strategies," Pacific Northwest Center for Global Security (Seattle), 2005; Zha Daojiong and Hu Weixiong, "Promoting Energy Partnership in Beijing and Washington," *Washington Quarterly* 30, no. 4 (Autumn 2007): 105–115; Zha Daojiong, "Energy Interdependence," *China Security* 3 (Summer 2006).

3. Kenneth Lieberthal and Mikkal Herberg, "China's Search for Energy Security: Implications for U.S. Policy," *NBR Analysis* 17, no. 1 (2006); Bo Kong, *China's International Petroleum Policy* (Santa Barbara, Calif.: Praeger Security International, 2010); Mikkal J. Herberg, "The Rise of Energy and Resource Nationalism in Asia," in *Strategic Asia 2010–11: Asia's Rising Power and America's Continued Purpose*, ed. A. J. Tellis, A. Marble, and T. Tanner (Seattle: National Bureau of Asian Research, 2010), 125, http://www.nbr.org/publications/strategic_asia/pdf/Preview/SA10/SA10_Energy_preview.pdf.

4. Su Shilin is now the governor of Fujian province. It was long predicted that Jiang Jiemin would become the governor of Yunnan province in late August 2011. Li Jiheng was appointed acting governor. See "New Acting Governor of SW China's Yunnan Province Announced," *Xinhua* (August 30, 2011), http://www.globaltimes.cn/NEWS/tabid/99/ID/673251/New-acting-governor-of-SW-Chinas-Yunnan-Province-announced.aspx; Willy Lam, "The Rise of the Energy Faction in Chinese Politics," *China Brief* (April 22, 2011), http://www.jamestown.org/programs/chinabrief/single/?cHash=f436aaedf32f78d903120b09a3a3ad10&tx_ttnews%5BbackPid%5D=25&tx_ttnews%5Btt_news%5D=37836.

5. Erica Downs, "China's 'New' Energy Administration," *China Business Review* 35, no. 6 (November–December 2008): 42–45.

6. Cited in Julian Wong, http://greenleapforward.com/2010/02/04/the-national-energy-commission-myth-busting-the-new-energy-super-ministry/.

7. Lieberthal and Herberg, "China's Search," 17; and Downs, "China's 'New' Energy Administration."

8. Erica S. Downs, "China," *Energy Security Series, The Brookings Foreign Policy Studies* (Washington, D.C.: Brookings Institution, December 2006); Downs, "China's 'New' Energy Administration"; Downs, "Who's Afraid of China's Oil Companies?"

9. Downs, "China's Energy Security"; Downs, "China's 'New' Energy Administration."

10. Mikkal E. Herberg, statement before the U.S.-China Economic and Security Review Commission hearing on "China's Energy Consumption and Opportunities for U.S.-China Cooperation to Address the Effects of China's Energy Use," June 14–15, 2007, 126.

11. Philip Andrews-Speed and Roland Dannreuther, *China, Oil, and Global Politics* (London: Routledge, 2011), 47.

12. Kong, *China's International Petroleum Policy*; Downs, "China's 'New' Energy Administration"; Erica Downs, "Business Interest Groups in Chinese Politics: The Case of the Oil Companies," in *China's Changing Political Landscape: Prospects for Democracy*, ed. Cheng Li (Washington, D.C.: Brookings Institution Press, 2008), 123–124. Interview with the president of one of the IOCs operating in China and representatives from other IOCs present in China, Beijing, December 2010.

13. Downs, "Who's Afraid of China's Oil Companies?" 75–76; Downs, "China's 'New' Energy Administration"; Downs, "Business Interest Groups in Chinese Politics," 123–124.

14. Kong, *China's International Petroleum Policy*, 25.

15. Emphasized in interviews with government representatives in China.

16. Kong, *China's International Petroleum Policy*, 26; Downs, "Who's Afraid of China's Oil Companies?"; Barry Naughton, "SASAC and Rising Corporate Power in China," *China Leadership Monitor* 24 (Spring 2008), http://falcon.arts.cornell.edu/am847/pdf/SASAC1.pdf.

17. Naughton, "SASAC and Rising Corporate Power in China"; Andrews-Speed and Dannreuther, *China, Oil, and Global Politics*, 81.

18. IEA, "Oil Market Report," January 16, 2009, 15; and Kong, *China's International Petroleum Policy*, 24–25.

19. Bo Kong, "China's Energy Decision Making—Becoming More Like the United States?" *Journal of Contemporary China* 18, no. 62 (November 2009): 798–812; Bo Kong, "China Overhauls Natural Resource Tax," *Caixin Online* (May 25, 2010), http://www.energychinaforum.com/news/35788.shtml.

20. Downs, "Who's Afraid of China's Oil Companies?" 76.

21. This control mechanism in recent years has not always been successful, and the NDRC has been "forced into issuing a directive requiring the NOCs to form consortia rather than bidding against each other." See Andrews-Speed and Dannreuther, *China, Oil, and Global Politics*, 84; and conversation with Erica Downs, Washington, D.C., 2010.

22. Conversely, it has been shown that the China Development Bank (CDB) at times acts independently and promotes overseas investments, M&As by NOCs, and loan-for-oil deals in the pursuit of commercial interests. See Erica Downs, "Inside China, Inc.: China Development Bank's Cross-Border Energy Deals," John L. Thornton China Center Monograph Series 3 (March 2011), http://www.brookings.edu/~/media/Files/rc/papers/2011/0321_china_energy_downs/0321_china_energy_downs.pdf. Still, it is reasonable to suggest that the CCP Politburo can direct the CDB whether or not to provide loans to certain countries

or companies, just as the central government can control and direct the NOCs on other important sensitive issues.

23. Deloitte, "A New Era Is Dawning: The Rise of the Hybrid National Oil Company," 2010, 4, http://www.deloitte.com/view/en_GX/global/industries/energy-resources/2e4745a204368210VgnVCM200000bb42foaRCRD.htm.

24. Emphasized in an interview with the president of an IOC operating in China, Beijing, December 2010.

25. Michal Meidan, Philip Andrews-Speed, and Ma Xin, "Shaping China's Energy Policy: Actors and Processes," *Journal of Contemporary China* 18, no. 61 (September 2009): 609. See also Philip Andrews-Speed, "China's Ongoing Energy Efficiency Drive: Origins, Progress and Prospects," *Energy Policy* 37, no. 4 (April 2009): 1331–1344.

26. Philip Andrews-Speed, "The Institutions of Energy Governance in China," *IFRI* (January 2010): 40–42, http://www.ifri.org/?page=contribution-detail&id=5842&id_provenance=97; and Andrews-Speed and Dannreuther, *China, Oil, and Global Politics*, 54.

27. Deloitte, "A New Era is Dawning," 4.

28. I thank Erica Downs for pointing this out.

29. Meidan, Andrews-Speed, and Ma, "Shaping China's Energy Policy," 597.

30. Li Peng, "China's Policy on Energy Resources," *Xinhua* (May 28, 1997), appearing as "Li Peng on Energy Policy," *World News Connection* (June 23, 1997).

31. Kong, *China's International Petroleum Policy*, 47.

32. Ibid., 48.

33. Ibid., 56–57.

34. Ibid., 57.

35. Andrew-Speed and Dannreuther, *China, Oil, and Global Politics*; Richard Lester and Edward Steinfeld, "China's Real Energy Crisis," *Harvard Asia Review* 9, no. 1 (Winter 2007); and Downs, "Business Interest Groups in Chinese Politics."

36. Lester and Steinfeld, "China's Real Energy Crisis," 3–4; Andrews-Speed and Dannreuther, *China, Oil, and Global Politics*.

37. Edward Cunningham, "China's Energy Governance: Perception and Reality," *Audits of the Conventional Wisdom* (MIT Center for International Studies, March 2007), 1; Kong, "Institutional Insecurity," 65.

38. The electricity crisis in California in 2000–2001 is one example.

39. See, among others, Yergin, *The Quest: Energy, Security, and the Remaking of the Modern World* (London: Allen Lane, 2011), chaps. 9, 10.

40. Lester and Steinfeld, "China's Real Energy Crisis"; Cunningham, "China's Energy Governance."

41. Yergin, *The Quest*, 221.

42. However, as will be seen in chapter 6, local and provincial governments can play an important role in driving China's energy security policy, especially in the case of the Myanmar-China pipelines.

43. Downs, "China's Energy Security"; Andrews-Speed, Liao, and Dannreuther, "The Strategic Implications of China's Energy Needs."

44. Kong, *China's International Petroleum Policy*, 27, 58–60.

45. Andrews-Speed and Dannreuther, *China, Oil, and Global Politics*; Linda Jakobson and Dean Knox, "New Foreign Policy Actors," *Sipri Policy Paper* 26, http://books.sipri.org/files/PP/SIPRIPP26.pdf.

46. For example, Chinese bulk carriers were requisitioned by the Chinese government to ensure coal shipments to China's domestic market as coal prices spiraled in 2008. See "Coal Prices Jump, Hit by the Perfect Storm," *SeaTrade Asia* (January 30, 2008), http://www.seatradeasia-online.com/print/2264.html. This will be further explored in the following chapters.

47. Xiaojie Xu, *Petro-Dragon's Rise: What It Means for China and the World* (Florence: European Press Academic Publishing, 2002). Analysts and researchers at China's NOCs have emphasized how path dependency and self-reliance have shaped China's energy security policy. Interviews in Beijing, China, May 2009 and December 2010.

48. Andrews-Speed and Dannreuther, *China, Oil, and Global Politics*, 36–42; Downs, "China's Energy Security"; Fred C. Bergsten et al., *China's Rise: Challenges and Opportunities* (Washington, D.C.: Peterson Institute for International Economics, October 2009), 137–168.

49. The U.S.-led embargo from 1950 to 1970 prevented China from exporting oil to the international market and isolated the Chinese oil industry. The Sino-Soviet split had a large impact because China was dependent on the USSR for virtually all of the refined products used by its military.

50. Andrews-Speed, "The Institutions of Energy Governance," 20; Andrews-Speed and Dannreuther, *China, Oil, and Global Politics*, 36–42.

51. Ibid., 39.

52. Constantin, "Understanding China's Energy Security."

53. Jakobson and Knox, "New Foreign Policy Actors." However, the authors do not argue that the government or state energy institutions cannot control the NOCs, especially in times of crisis or emergency. Conversation with Jakobson and Knox, Beijing, November/December 2010.

54. Downs, "Who's Afraid of China's Oil Companies?" 77; Deloitte, "A New Era Is Dawning"; interviews in China and the United States.

55. Kat Cheung, *Integration of Renewables: Status and Challenges in China*, Working Paper, IEA (2011), 7.

56. Xiaoli Liu and Xinmin Jiang, "China's Energy Security Situation and Countermeasures," *International Journal of Energy Sector Management* 3, no. 1 (2009): 87–88; Edward S. Steinfeld, Richard K. Lester, and Edward A. Cunningham, "Greener Plants, Greyer Skies? A Report from the Front Lines of China's Energy Sector," *Energy Policy* 37, no. 5 (2009): 1811.

57. In 2006 alone, 102 GW of new generating capacity was added, an increment substantially larger than the United Kingdom's entire electric power system. See Steinfeld, Lester, and Cunningham, "Greener Plants," 1809. Bergsten et al., *China's Rise*, 151, note that in 2006 and 2007, 200 GW of new capacity was added to the grid, more than the entire installed base of Germany and Italy combined.

58. Steinfeld, Lester, and Cunningham, "Greener Plants," 1810, 1817–1818.

59. Andrews-Speed and Dannreuther, *China, Oil, and Global Politics*, 21. Shipments of coal to China were boosted by 31 percent to a record 165 million tons in 2010, but in 2011 imports slowed after surges in global prices. See "China's Coal Import to Slow as Floods Boost Prices," *Bloomberg News* (February 1, 2011), http://www.bloomberg.com/news/2011-02-01/china-s-coal-import-growth-rate-to-slow-as-australian-floods-boost-prices.html.

60. "China's Crude Oil Output up 6.9 pct—Stats Bureau," *Reuters* (January 20, 2011), http://www.reuters.com/article/2011/01/20/china-crude-output-idUSBJI0025412011010120.

61. Andrews-Speed and Dannreuther, *China, Oil, and Global Politics*.

62. The president of an IOC operating in China estimated that China has roughly 1 mbd in offshore production. Interview, Beijing, China, December 2010.

63. Yergin, *The Quest*, 210.

64. "China Net Crude Imports Rise as Factories Boost Fuel Demand," *Bloomberg News* (May 10, 2011), at http://www.bloomberg.com/news/2011-05-10/china-s-april-net-crude-oil-imports-rise-as-factories-increase-fuel-demand.html.

65. International Energy Agency (IEA), *2010 World Energy Outlook*, http://www.iea.org/weo/2010.as.

66. Liu and Jiang, "China's Energy Security Situation," 87; IEA, *2010 World Energy Outlook*.

67. It should be noted that roughly 15 to 20 percent of the imported oil is transported by cross-border pipelines from Kazakhstan and Russia.

68. IEA, "Oil Market Report," August 10, 2012; Andrews-Speed and Dannreuther, *China, Oil, and Global Politics*, 16–17; Bergsten et al., *China's Rise*, 139–140.

69. Cheung, *Integration of Renewables*, 17.

70. Sabine Johnson-Reiser, "China's Hydropower Miscalculation," *China Brief* 12, no. 11 (May 25, 2012); Huw Pohlner, "Chinese Dam Diplomacy: Leadership and Geopolitics in Continental Asia," *East Asia Forum* (August, 19, 2010).

71. Andrews-Speed and Dannreuther, *China, Oil, and Global Politics*, 28–29.

72. http://www.eia.doe.gov/cabs/China/NaturalGas.html.

73. Graham Cunningham and Hanzhi Ding, "The Asia Petrolizer," *Citigroup Global Markets* (November 9, 2009): 3.

74. For example, it will be argued in chapter 6 that the Myanmar pipeline is less secure than the Russian or Central Asian pipelines.

75. Correspondence with Mikkal Herberg, October 24, 2011.

76. Julie Jiang and Jonathan Sinton, "Overseas Investments by Chinese National Oil Companies: Addressing the Drivers and Impacts," *Information Paper* (IEA), February 2011, 10; Andrews-Speed and Dannreuther, *China, Oil, and Global Politics*, 26; Philip Andrews-Speed, "China's Booming Gas Sector: Threat or Promise?" *Oil, Gas, and Energy Law* (February 11, 2011).

77. As a comparison, in 2011 Russia piped gas to Germany through the Nord Stream with a price of five hundred dollars per thousand cubic meters.

78. Philip Andrews-Speed, "China-Russia Energy Relations: Which Party Holds the Stronger Hand?" *Oil, Gas, and Energy Law* (July 2011). As one expert succinctly puts it, "the Chinese gas boat has already sailed." Conversation with Mikkal Herberg, Oslo, May 2011.

79. Neil Beveridge and Angus Chan, "Bernstein Asia-Pac Energy: Coal Bed Methane: Is It Time for China's Unconventional Gas Revolution?" *Bernstein Research* (May 18, 2010).

80. Andrews-Speed and Dannreuther, *China, Oil, and Global Politics*, 139.

81. James T. Areddy, "China's Nuclear Energy Officials Watch Japan," *Wall Street Journal* (March 14, 2011), http://blogs.wsj.com/chinarealtime/2011/03/14/chinas-nuclear-energy -officials-watch-japan/.

82. Philip Andrews-Speed, "China's Nuclear Power Plans After Fukushima," *Oil, Gas, and Energy Law* (April 2011); Wang Ying, "China to Build 28 More Nuclear Power Reactors by 2020," *Bloomberg News* (March 23, 2010), http://www.businessweek.com/news/2010-03 -23/china-to-build-28-more-nuclear-power-reactors-by-2020-update1-.html; Keiko Yoshi-oka, "China to Build 60 Nuclear Reactors Over Next Decade," *Asahi* (March 11, 2011), http://www.asahi.com/english/TKY201103100233.html.

83. Wang Haibin, "Profits Outweigh Self-Reliance as China's Nuclear Industry Expands," *Uranium Intelligence Weekly* 44, no. 4 (November 1, 2010): 9.

84. James T. Areddy and Brian Spegele, "China, Also on Fault Lines, Faces New Atomic Scrutiny," *Wall Street Journal* (March 16, 2011), http://online.wsj.com/article/SB1000142 4052748703566504576202512705277514.html. See also Andrews-Speed and Dannreuther, *China, Oil, and Global Politics*, 29; Andrews-Speed, "China's Nuclear Power Plans."

85. Zhen Li, "China's Utility Powerhouses Push Into Nuclear," *Uranium Intelligence Weekly* 44, no. 4 (November 1, 2010): 3–4.

86. Silvia Yu, "China Doubling Down on Domestic Uranium Exploration Efforts," *Uranium Intelligence Weekly* 44, no. 4 (November 1, 2010): 4–5.

87. "China Not to Change Plan for Nuclear Power Projects: Government," *Xinhua* (March 12, 2011), http://news.xinhuanet.com/english2010/china/2011-03/12/c_13774519.htm; Areddy, "China's Nuclear Energy Officials Watch Japan."

88. "Nuclear Safety Should be Priority," *China Daily* (March 22, 2011), http://www.chinadaily .com.cn/opinion/2011-03/22/content_12206416.htm.

89. Statements by Wang Haibin, Zhao Hongtu, and Mikkal Herberg at the Fueling China's Rise seminar, Norwegian Institute for Defense Studies, Oslo, May 9, 2011; Andrews-Speed, "China's Nuclear Power Plans." Wang Haibin has recently pointed out that on May 31, 2012, indications emerged that the central government might resume steps toward nuclear expansion. See the roundtable debate on "The Nuclear Approach to Climate Risk," *Bulletin of the Atomic Scientists* (August 30, 2012), http://www.thebulletin.org/web-edition/ roundtables/the-nuclear-approach-to-climate-risk#.

90. Daniel Rosen and Trevor Houser, *China Energy: A Guide for the Perplexed* (Washington, D.C.: Peterson Institute for International Economics, 2007), 17.

91. Andrews-Speed and Dannreuther, *China, Oil, and Global Politics*, 29.

92. "Wind Power Installation Slowed in 2010, Outlook for 2011 Stronger: AWEA," *Platts* (January 24, 2011), http://www.platts.com/RSSFeedDetailedNews/RSSFeed/ElectricPower/6773195.

93. http://en.wikipedia.org/wiki/Solar_power_in_China.

94. Rosen and Houser, *China Energy*, 8; Bergsten et al., *China's Rise*, 141.

95. Guy C. K. Leung, "China's Oil Use, 1990–2008," *Energy Policy* 38, no. 2 (2010): 932–944.

96. From 1990 to 2007, the average annual growth rate of oil demand in the various sectors was as follows: industry, 5.8 percent; the household sector, 13.4 percent; transport, 12.7 percent; the commercial sector, 8.6 percent; and agriculture, 4.5 percent. Ibid., 936–937.

97. Ibid., 940; Andrews-Speed and Dannreuther, *China, Oil, and Global Politics*.

98. Rosen and Houser, *China Energy*, 8. Kong maintains that the automobile sector is one of the primary contributors to China's surging oil demand and that automobiles consume 75 percent of the country's imported oil. See Kong, "China's Energy Decision Making," 792. Austin argues that the "increase in the demand for oil in China in the last two years has been driven by a sharp increase in use of motor vehicles." See Angie Austin, "Energy and Power in China: Domestic Regulation and Foreign Policy," Foreign Policy Centre, April 2005, http://fpc.org.uk/fsblob/448.pdf.

99. Leung, "China's Oil Use," 940; Rosen and Houser, *China Energy*, 14.

100. Bergsten et al., *China's Rise*, 158–159.

101. "China Ends U.S.'s Reign as Largest Auto Market," *Bloomberg News* (January 11, 2010), http://www.bloomberg.com/apps/news?pid=newsarchive&sid=aE.x_r_l9NZE; "China Overtakes US as World's Biggest Car Market," *Guardian* (January 8, 2010), http://www.guardian.co.uk/business/2010/jan/08/china-us-car-sales-overtakes; Leung, "China's Oil Use," 942.

102. Yergin, *The Quest*, 217.

103. Leung, "China's Oil Use," 943.

104. Andrews-Speed, "The Institutions of Energy Governance in China," 32; Andrews-Speed, "China's Ongoing Energy Efficiency," 1338.

105. Bergsten et al., *China's Rise*, 155.

106. Andrews-Speed and Dannreuther, *China, Oil, and Global Politics*, 29–33.

107. Cheung, *Integration of Renewables*, 8.

108. Austin, "Energy and Power in China," 4; Andrews-Speed, "China's Ongoing Energy Efficiency," 1335.

109. Bergsten et al., *China's Rise*, 141.

110. Yergin, *The Quest*, 220.

111. Key measures have been taken in the industrial sector through the energy-pricing policy to set targets for local governments and to reduce waste in buildings and transport through public awareness. See Andrew-Speed, "China's Ongoing Energy Efficiency," 1338–1341. For different views on whether China is on track to meet its target, see Li Jing, "Fight Hard and Long to Save Environment, Wen Tells Nation," *China Daily* (March 6, 2010), http://www.chinadaily.com.cn/china/2010npc/2010-03/06/content_9546849.htm; Stephen Howes, "China's Energy Intensity Target: On Track or Off?" *East Asian Forum* (March 31, 2010), http://www.eastasiaforum.org/2010/03/31/chinas-energy-intensity-target-on-track-or-off/.

112. "Key Targets of China's 12th Five-Year Plan," *Xinhua* (March 5, 2011), http://www.chinadaily.com.cn/china/2011npc/2011-03/05/content_12120283.htm.

113. Leung, "China's Oil Use," 943.

114. Gabriel Collins, "China Fills First SPR Site, Faces Oil, Pipeline Issues," *Oil and Gas Journal* (August 20, 2007).

115. Kang Wu and Liutong Zhang, "China's Strategic Reserves Capacity to Double by 2011," *Oil and Gas Journal* (September 21, 2009).

116. Kang Wu and Liutong Zhang, "China Works to Double SPR Capacity by 2013," *Oil and Gas Journal* (October 4, 2010); Wu and Zhang, "China's Strategic Reserves Capacity to Double"; Collins, "China Fills First SPR Site"; Daniel Nieh et al., "Study Examines Chinese SPR Growth Alternatives," *Oil and Gas Journal* (July 23, 2007); Andrews-Speed and Dannreuther, *China, Oil, and Global Politics*, 23.

117. IEA, "Oil & Gas Security Emergency Response of IEA Countries: People's Republic of China," 2012, http://www.iea.org/publications/freepublications/publication/China_2012 .pdf.

118. Collins, "China Fills First SPR Site."

119. Wu and Zhang, "China Works to Double SPR Capacity; Feng Li, "Work on Tianjin Crude Reserve Base Starts," *China Daily* (May 13, 2010), http://www.chinadaily.com.cn/ bizchina/2010-05/13/content_9843659.htm.

120. Wu and Zhang, "China's Strategic Reserves Capacity to Double by 2011."

121. Collins, "China Fills First SPR Site."

122. Zhou Yan, "CNPC Plans to Extend Pipeline Network," *China Daily* (February 22, 2011), http://www.chinadaily.com.cn/bizchina/2011-02/22/content_12056227.htm.

123. Wu and Zhang, "China Works to Double SPR Capacity; Nieh et al., "Study Examines Chinese SPR Growth Alternatives."

124. Nieh et al., "Study Examines Chinese SPR Growth Alternatives"; Collins, "China Fills First SPR Site."

125. Nieh et al., "Study Examines Chinese SPR Growth Alternatives."

126. Collins, "China Fills First SPR Site."

127. Leung, "China's Oil Use," 934.

128. Fayen Wong and Judy Hua, "Analysis-China Sharpens Axe to Cull 'Teapot' Refineries," *Reuters* (April 29, 2011), http://www.reuters.com/article/2011/04/29/businesspro-us-china -energy-refining-idUSTRE73S2DA20110429.

129. Gabriel Collins, "China's Refining Expansion to Reshape Global Oil Trade," *Oil and Gas Journal* (February 18, 2008).

130. Wong and Hua, "Analysis-China Sharpens Axe to Cull 'Teapot' Refineries."

131. Collins, "China's Refining Expansion."

132. In 2006, only 15 percent of China's refining capacity could handle the kind of sour crude oil that comes from Iran, Saudi Arabia, Venezuela, and other core suppliers. To put this in perspective, 82 percent of U.S. refineries can handle sour crude oil, with a lower percentage able to process a large proportion of high-acid crudes. Since 2008, China has boosted its refinery capacity, which increased more than 10 percent in 2010, and improved its ability to handle sour crude oil. See Collins, "China's Refining Expansion."

133. Leung, "China's Oil Use," 935.

134. Andrews-Speed and Dannreuther, *China, Oil, and Global Politics*, 137.

135. Leung, "China's Oil Use," 934.

136. Andrews-Speed and Dannreuther, *China, Oil, and Global Politics*, 137.

137. Exxon Mobil is not a national oil company, but it is a major oil producer and possesses technical and managerial expertise.

138. "Exxon, Saudi Aramco in $3.5B China Refinery Deal," *FoxNews* (July 8, 2005), http://www.foxnews.com/story/0,2933,161945,00.html.

139. "Update 1—Exxon Mobil, Aramco's China Refinery Starts Up-Source," *Reuters* (May 7, 2009), http://www.reuters.com/article/2009/05/07/sinopec-exxonmobil-idUSPEK29391720090507.

140. "Exclusive-Update 2-China, Venezuela Refinery Gets Beijing's Nod," *Alibaba.com* (January 21, 2010), at http://news.alibaba.com/article/detail/energy/100236217-1-exclusive-update-2 -china%252C-venezuela-refinery-gets.html.

141. "China's Sinopec, Kuwait Petroleum to Begin Building $9 Billion Refining Complex in Q1 2012" (May 23, 2011), http://www.hydrocarbonprocessing.com/Article/2835204/ Latest-News/Chinas-Sinopec-Kuwait-Petroleum-to-begin-building-9-billion-refining -complex-in-Q1-2012.html.

142. Anna Shiryaevskaya and Maria Kolsenikova, "Russia, China to Build $5 Billion Refinery, Extend Gas Talks," *Bloomberg News* (September 22, 2010), http://www.bloomberg.com/ news/2010-09-21/rosneft-cnpc-may-build-refinery-in-china-in-two-years-ria-novosti -says.html.

143. Collins, "China's Refining Expansion."

144. Downs, "Inside China, Inc."

145. José Orozco, "China's Buying Spree: Venezuela—A Match Made in Globalization," *Latin Trade* (February 7, 2011), http://latintrade.com/2011/02/china%E2%80%99s-buying-spree -venezuela-a-match-made-in-globalization; Mamdouh G. Salameh, "China's Oil 'Adventure' Into Venezuela," *International Association for Energy Economics*, 2nd quarter 2011, http://www.iaee.org/en/publications/fullnewsletter.aspx?id=18.

146. "Venezuela Increases Oil Exports to China by 21 pct," *Reuters* (April 13, 2010), http:// in.reuters.com/article/2010/04/13/venezuela-oil-china-idINN1219684120100413.

147. James Fowler, "Exports to China Continue Surge—JP Morgan—Venezuela," *Business News Americas* (December 21, 2010), http://www.bnamericas.com/news/oilandgas/Exports-to -China-continue-surge---JP-Morgan.

148. "Venezuela Increases Oil Exports to China by 21 pct"; Salameh, "China's Oil 'Adventure' Into Venezuela"; "China and the Politics of Venezuelan Oil," *Financial Sense* (June 9, 2011), http://www.financialsense.com/contributors/oil-price/2011/06/09/china-and-the-politics -of-venezuelan-oil. The importance of the two-way trade was emphasized by tanker bro-kers, and it was estimated that this development would boost China's oil import from Ven-ezuela. These brokers also argued that oil shipment from Columbia to China is also likely to increase. Interview, Beijing, November, 28, 2012.

149. Collins, "China's Refining Expansion"; Salameh, "China's Oil 'Adventure' Into Venezuela."

150. Salameh, "China's Oil 'Adventure' Into Venezuela."

151. "China Mulls Allowing Oil Companies to Set Oil Prices—Media," *Reuters* (June 14, 2011), http://www.reuters.com/article/2011/06/14/china-oil-pricing-idUSL3E7HE1JY20110614.

152. Bergsten et al., *China's Rise*, 146.

153. IEA, "Oil Market Report," 2009, 15, http://omrpublic.iea.org/omrarchive/16jan09dem.pdf.

154. Andrews-Speed and Dannreuther, *China, Oil, and Global Politics*, 33, 46; Andrews-Speed, "The Institutions of Energy Governance in China," 38.

155. "China Mulls Allowing Oil Companies to Set Oil Prices—Media."

156. Kong, *China's International Petroleum Policy*; Andrews-Speed and Dannreuther, *China, Oil, and Global Politics*, 31–33; Collins, "China's Refining Expansion."

157. "Platts Report: China's March Oil Demand Stays Robust at 9.2 Mil. Barrels per Day," *Platts* (April 21, 2011).

158. Collins, "China's Refining Expansion."

4. THE GLOBAL SEARCH FOR PETROLEUM

1. Daniel Yergin, *The Quest: Energy, Security, and the Remaking of the Modern World* (London: Allen Lane, 2011), 202.

2. Kenneth Waltz, *Theory of International Politics* (New York: Random House, 1979), 105.

3. "Platts Report: China's December Oil Demand at Record High 9.6 mil b/d," *Platts* (January 31, 2011), http://www.platts.com/PressReleases/2011/12111.

4. Energy Information Agency (EIA), November 2010.

5. International Energy Agency (IEA), *2009 World Energy Outlook 2009*, http://www.iea.org/weo/2009.asp. In 2007, the IEA projected that China's oil imports would jump from 3.5 mbd in 2006 to 13.1 mbd in 2030, and the import share of demand was forecasted to rise from 50 percent to 80 percent. IEA, *2007 World Energy Outlook: China and India Insights*, 44–45, http://www.iea.org/weo/2007.asp.

6. IEA, *2010 World Energy Outlook*, http://www.iea.org/weo/2010.asp.

7. Han Wenke, presentation at the Energy Security in Asia conference, Beijing, May 21–22, 2009.

8. Information Office of the State Council of the People's Republic of China, *China's National Defense in 2006* (Beijing: Foreign Language Press, 2006), 4; Hu Jintao, "G8 Written Statement," St. Petersburg, July 17, 2006, http://www.uofaweb.ualberta.ca/chinainstitute/nav03.cfm?nav03=48114&nav02=43884&nav01=43092. See also David Zweig and Bi Jinhai, "China's Global Hunt for Energy," *Foreign Affairs* 84, no. 5 (September–October 2005): 25–38; Erica S. Downs, "China," *Energy Security Series, The Brookings Foreign Policy Studies* (Washington, D.C.: Brookings Institution, December 2006), 14; Xu Yi-Chong, "China's Energy Security," *Australian Journal of International Affairs* 60, no. 2 (June 2006): 268; Yang Yi, "Engagement, Caution," *China Security* 3, no. 4 (Autumn 2007): 29–39; Zha Daojiong and Hu Weixing, "Promoting Energy Partnership in Beijing and Washington," *Washington Quarterly* 30, no. 4 (Autumn 2007): 107; Amy Myers Jaffe and Steven W. Lewis, "Beijing's Oil Diplomacy," *Survival* 44, no. 1 (Spring 2002): 115–134.

9. A history and a chronology of the major NOC investments, mergers, and acquisitions can be found at the companies' webpages. For CNPC, see http://www.cnpc.com.cn/en/aboutcnpc/companyprofile/history/default.htm; for CNOOC, see http://en.cnooc.com.cn/data/html/english/channel_114.html; and for Sinopec, see http://www.sinopecgroup.com/english/business/Pages/default.html.

10. Bo Kong, *China's International Petroleum Policy* (Santa Barbara, Calif.: Praeger Security International, 2010), 63–67; John Mitchell and Glada Lahn, "Oil for Asia," Briefing Paper (London: Chatham House, March 2007).

11. Wan Xu, "CNPC: Expects 2010 Overseas Oil, Gas Output of 85 Mln Tons Equivalent," *Dow Jones International News* (December 7, 2010).

12. "China: Energy Companies Change Strategic Tack," *Oxford Analytica*, Daily Brief (April 26, 2010). In addition to the foreign upstream projects, the service arms of China's NOCs have extended their presence throughout the world. See Kong, *China's International Petroleum Policy*. According to a 2011 IEA report on overseas investments by Chinese NOCs, "Chinese oil companies are now operating in 31 countries and have equity production in 20 of these countries, though their equity shares are mostly located in four countries: Kazakhstan, Sudan, Venezuela and Angola." See Julie Jiang and Jonathan Sinton, "Overseas Investments by Chinese National Oil Companies: Addressing the Drivers and Impacts," *Information Paper* (IEA) (February 2011): 7, 10.

13. Kong, *China's International Petroleum Policy*, 67. In June 2010, it was reported that "state-run Chinese companies spent a record $32 billion last year [2009] acquiring energy and resources assets overseas." See Rakteem Katakey and John Duce, "India Loses to China in Africa-to-Kazakhstan-to-Venezuela Oil," *Bloomberg News* (June 29, 2010), http://www.bloomberg.com/news/2010-06-30/india-losing-to-china-in-africa-to-kazakhstan-to-venezuela-oil-purchases.html.

14. Wood Mackenzie, "Chinese NOCs International Expansion," *Corporate Service Insight* (June 2010). Wood Mackenzie notes that two of the deals, worth in total $7.75 billion (Sinopec in Canada buying from ConocoPhillips and CNOOC in Argentina buying from Bridas Holdings) were based on announcements. These are now completed. See "Update 1—Canada Clears Sinopec to Buy Syncrude Stake," *Reuters* (June 25, 2010), http://www.reuters.com/article/2010/06/25/sinopec-conocophillips-idUSN2516802520100625; "CNOOC LTD, Bridas Energy Holdings Finish JV Deal," *AsiaInfo Services* (May 5, 2010), http://www.highbeam.com/doc/1P1-179531949.html.

15. Jiang and Sinton, "Overseas Investments," 10. Others, drawing to some extent on memoranda of understanding (MOU) and "future investments"—or deals that have been announced but are yet to be completed or where the Chinese oil companies may not have put money on the table or the Chinese government has not agreed to the MOUs—arrive at much higher investment numbers. See International Crisis Group, "The Iran Nuclear Issue: The View from Beijing," *Asia Briefing* 100 (February 17, 2010), appendix C; Paula Dittrick, "Chinese Oil Companies Invest Heavily Abroad," *Oil and Gas Journal* (February 8, 2010).

16. According to Downs's excellent study on the involvement of the CDB in China's cross-border energy deals, "in 2009 and 2010 CDB agreed to provide $65 billion in credit to national energy companies and government entities in Russia, Brazil, Venezuela, Turkmenistan, and Ecuador, often on terms that the borrowers would have had trouble obtaining elsewhere. The only major energy-backed loan brokered by Chinese firms that did not involve the CDB was a $1 billion loan extended by PetroChina to PetroEcuador in 2009 as

pre-payment for oil deliveries of 96,000 b/d over two years. Although the $10 billion that CNPC and the China Eximbank loaned to Kazakhstan's national oil company and development bank in 2009 are often grouped together with CDB's energy loans in the media, the lines of credit extended to Kazakhstan are not backed by an energy supply contract." See Erica Downs, "Inside China, Inc.: China Development Ban's Cross-Border Energy Deals" (Washington, D.C.: John L. Thornton China Center, Brookings Institution, 2011), 38. In September 2009, CNPC received a $30 billion five-year loan from the CDB at a discounted interest rate to buy energy resources. "China's CNPC Gets $30 Billion Loan to Support Global Expansion" (September 10, 2009), http://goliath.ecnext.com/coms2/gi_0199-11393952/China-s-CNPC-gets-30.html. See also Wood Mackenzie, "Overseas Investments by Chinese National Oil Companies."

17. Kong, *China's International Petroleum Policy*, 69; see also "China Seals Oil Deals," Center for American Progress (April 27, 2010), http://www.americanprogress.org/issues/2010/04/china_oil_map.html; and Michael T. Klare, "China's Global Shopping Spree," http://www.countercurrents.org/klare010410.htm.

18. "Venezuela Says China Offers $20 Billion in Financing," *Reuters* (April 17, 2010), http://www.reuters.com/article/idUSTRE63H02M20100418.

19. Jiang and Sinton, "Overseas Investments," 10.

20. "China Seals Oil Deals," Center for American Progress (April 27, 2010), http://www.americanprogress.org/issues/2010/04/china_oil_map.html.

21. Interviews with experts in the United States and China.

22. Dan Molinski, "Chávez, Shunning IMF, Gushes About China Oil-for-Credit Deal," *Wall Street Journal* (May 4, 2010); Wood Mackenzie, "Overseas Investments by Chinese National Oil Companies." However, as pointed out in chapter 3, crude exports from Venezuela to China are picking up and are set to increase substantially in the years ahead; these have been facilitated and fueled by investment in refining capacity. Reports from 2012 claim that the Venezuelan state-owned oil company PDVSA already has to send 430,000 b/d of crude oil and products to cover its debts with China. See Brian Ellsworth and Marianna Parraga, "Venezuela Expands China Loan-for-Oil deal to $8 Billion," *Reuters* (May 22, 2012).

23. Vladimir Soldatkin, "Russia in Milestone Oil Pipeline Supply to China," *Reuters* (January 1, 2011).

24. Erica Downs, "Inside China Inc.," 41–45.

25. "Argus Briefing on ESPO," http://web04.us.argusmedia.com/ArgusStaticContent/snips/sectors/pdfs/ArgusESPO.pdf.

26. China will most likely pay market price for the crude oil piped to China through the Myanmar-China pipeline since it will be shipped on tankers to Myanmar. Little oil will be produced domestically in Myanmar, hence it will be difficult to strike a long-term deal similar to the CNPC deal in Russia.

27. Daniel Rosen and Trevor Houser, *China Energy: A Guide for the Perplexed* (Washington, D.C.: Peterson Institute for International Economics, 2007), 17.

28. Wood Mackenzie, "Overseas Investments by Chinese National Oil Companies."

29. Mikkal J. Herberg, "The Rise of Energy and Resource Nationalism in Asia," in *Strategic Asia 2010–11: Asia's Rising Power and America's Continued Purpose*, ed. Ashley J. Tellis et al. (Seattle: National Bureau of Asian Research, 2010), 129.

30. "CNPC Clinches Deal to Develop Iranian Gas Field," *China Daily* (February 11, 2010), http://www.chinadaily.com.cn/bizchina/2010-02/11/content_9461870.htm.

31. According to a January 2010 brief from FACTS Global Energy, based in Singapore, net oil production abroad in 2008 by CNPC and its subsidiary, PetroChina Co. Ltd., was only 612.000 b/d, indicating an even larger gap in the numbers. See Dittrick, "Chinese Oil Companies Invest Heavily Abroad."

32. Wood Mackenzie, "Chinese NOCs Step up International Expansion."

33. Wan, "CNPC: Expects 2010." According to the 2011 IEA report on overseas investments by China's NOCs, "successful acquisition allowed China's NOCs to expand their overseas equity shares from 1.1 mbd in 2009 to 1.36 mbd in the first quarter of 2010." These numbers are much higher than the FACTS Global Energy estimates and slightly higher than the Wood Mackenzie data but lower than the overseas production announced by CNPC.

34. Erica Downs, "The Facts and Fiction of Sino-Africa Energy Relations," *China Security* 3, no. 3 (Summer 2007): 65 n. 12; Rosen and Houser, *China Energy*, 32 n.78; and Bo Kong, *China's International Petroleum Policy*, 92.

35. Rosen and Houser, *China Energy*, 32–33.

36. *Oil and Gas Journal* (February 8, 2010). In the references above, both the Wood Mackenzie and the FACTS Global Energy reports refer to overseas production. Still, their numbers are lower than the IEA report referring to overseas equity production.

37. Jiang and Sinton, "Overseas Investments."

38. Wood Mackenzie, "Overseas Investments by Chinese National Oil Companies."

39. Interview with a top executive from an IOC operating in China, Beijing, December 2010. Another expert noted in an interview in Washington in May 2010 that some Chinese oil companies might have an interest in manipulating production numbers in order to boost the stock prices of their listed subsidiaries. While such a proposition is controversial, a number of other leading energy experts interviewed in the United States and China did not disregard it.

40. Interview, Beijing, December 2010; http://www.eppo.go.th/ref/UNIT-OIL.html.

41. Jiang and Sinton, "Overseas Investments," 7, 10.

42. Andrew Monaghan, "Russia and the Security of Europe's Energy Supplies: Security in Diversity?" Conflict Studies Research Centre, Special Series, Defence Academy of the United Kingdom (January 2007), 8.

43. Yergin, quoted in Linda Jakobson and Zha Daojiong, "China and the Worldwide Search for Oil Security," *Asia-Pacific Review* 13, no. 2 (2006): 62. See also Daniel Yergin, "Ensuring Energy Security," *Foreign Affairs* 85, no. 2 (March–April 2006): 69–82; Ambassador Arne Walther, "The Geopolitics of Oil and Global Energy Security," paper presented at the Sixteenth Annual Energy Conference, The Oil Era: Emerging Challenges, Abu Dhabi, November 8–10, 2010.

44. Downs, "China," 30.

45. Vlado Vivoda, "Diversification of Oil Import Sources and Energy Security: A Key Strategy or an Elusive Objective?" *Energy Policy* 37, no. 11 (2009): 4615–4623.

46. Mikkal Herberg and David Zweig, "China's "Energy Rise": The U.S. and the New Geopolitics of Energy," Pacific Council on International Policy, April 2010, http://www .pacificcouncil.org/document.doc?id=159. See also "Chinese Oil Import from Middle East Down," *China Daily* (January 17, 2008), http://www.china.org.cn/english/ business/239753.htm; Vivoda, "Diversification of Oil," 4621. When China's dependence on imported oil grew from 30 percent in 2002 to just under 50 percent in 2006, China accounted for 36 percent of the growth in global oil demand, compared with the United States at 14 percent. See Fred C. Bergsten et al, *China's Rise: Challenges and Opportunities* (Washington, D.C.: Peterson Institute for International Economics, October 2009), 151.

47. Herberg and Zweig, "China's 'Energy Rise,'" 73–74, appendix 1. See also figure 2, based on the *China Customs Statistics Yearbook* available through 2007, in John Garver, Flynt Leverett, and Hillary Mann Leverett, "Moving (Slightly) Closer to Iran: China's Shifting Calculus for Managing Its 'Persian Gulf Dilemma,'" Johns Hopkins School of Advanced International Studies (SAIS), October 2009, 12.

48. "China's Crude Oil Import Dependence Exceeds 50%, Raises Security Concerns," *Global Times* (March 29, 2010), http://business.globaltimes.cn/china-economy/2010-03/517059 .html; Jon B. Alterman and John W. Garver, *The Vital Triangle: China, the United States, and the Middle East* (Washington, D.C.: Center for Strategic and International Studies, 2008), 6–7, present lightly higher numbers for the period until 2007.

49. The drop in Indonesian exports to China is, of course, mainly a result of declining production and disappointing oil exploration results. Indonesia is now importing light and sweet Nile Blend crude from Sudan. "Indonesia to Import 600,000 Oil Barrels from Sudan," *Sudan Tribune* (February 6, 2010). Arif Sharif and Rob Verdonck, "Saudi Aramco CEO Says China Overtakes U.S. as Largest Customer," *Bloomberg News* (January 28, 2010).

50. Kenneth Lieberthal and Mikkal Herberg, "China's Search for Energy Security: Implications for U.S. Policy," *NBR Analysis* 17, no. 1 (2006); Kong, *China's International Petroleum Policy*, 12; and Bo Kong, "An Anatomy of China's Energy Insecurity and its Strategies," report for the Pacific Northwest Center for Global Security, 2005, 13–14. Kong's projections differ from those of Downs. See Downs, "China," 32.

51. For example, Alterman and Garver, *The Vital Triangle*, 4–8.

52. Edward L. Morse, "Energy 2020: North America, the New Middle East?" CITI GPS (March 20, 2012), http://csis.org/files/attachments/120411_gsf_MORSE_ENERGY_2020_North _America_the_New_Middle_East.pdf.

53. IEA, *World Energy Outlook*, 2010.

54. David Victor and Linda Yueh, "The New Energy Order: Managing Insecurities in the Twenty-first Century," *Foreign Affairs* 89, no. 1 (2010): 61–73; Robert D. Kaplan, "Center Stage for the 21st Century, Power Plays in the Indian Ocean," *Foreign Affairs* 88, no. 2

(March/April 2009): 16–32; Klare, "China's Global Shopping Spree"; and China Seals Oil Deals," Center for American Progress (April 27, 2010), http://www.americanprogress.org/issues/2010/04/china_oil_map.html.

55. Herberg, "The Rise of Energy and Resource Nationalism in Asia," 125.

56. Jaffe and Lewis, "Beijing's Oil Diplomacy," 116.

57. Richard J. Ellings, "Foreword," in Lieberthal and Herberg, "China's Search for Energy Security"; Joe Barnes and Amy Myers Jaffe, "The Persian Gulf and the Geopolitics of Oil," *Survival* 48, no. 1 (Spring 2006): 154; Victor and Yueh, "The New Energy Order," 61.

58. Cindy Hurst, "China's Oil Rush in Africa," *IAGS Energy Security* (July 2006): 14; Jakobson and Zha, "China and the Worldwide Search for Oil Security," 65; Downs, "The Facts and Fiction of Sino-Africa Energy Relations," 42–68.

59. Theodore H. Moran, *China's Strategy to Secure Natural Resources: Risks, Dangers, and Opportunities*, Policy Analyses in International Economics 92 (Washington, D.C.: Peterson Institute for International Economics, 2010), 2.

60. Rosen and Houser, *China Energy*, 33.

61. U.S. Department of Energy, *Energy Policy Act 2005, Section 1837, National Security Review of International Energy Requirements* (February 28, 2006), 28.

62. Daniel Yergin, statement before the Committee on Energy and Commerce, U.S. House of Representatives (May 4, 2004).

63. Hu Jintao, "G8 Written Statement," St. Petersburg, July 17, 2006, http://www.uofaweb.ualberta.ca/chinainstitute/nav03.cfm?nav03=48114&nav02=43884&nav01=43092; Hu Jintao, "An Open Mind for Win-Win Cooperation," speech at the Asia Pacific Economic Cooperation (APEC) CEO summit, Busan, Republic of Korea, November 17, 2005; Angie Austin, "Energy and Power in China: Domestic Regulation and Foreign Policy," Foreign Policy Centre, April 2005, http://fpc.org.uk/fsblob/448.pdf; Eurasia Group, "China's Overseas Investment in Oil and Gas Production," report for the U.S.-China Economic and Security Review Commission, October 16, 2006.

64. Zha Daojiong, "Energy Interdependence," *China Security* 2, no. 2 (Summer 2006): 8; Zha Daojiong, "Oiling the Wheels of Foreign Policy? Energy Security and China's International Relations," Asia Security Initiative Policy Series Working Paper 1 (March 2010).

65. Leading Chinese energy security experts emphasized the importance of market mechanisms at the Energy Security in Asia conference, Beijing, May 21–22, 2009.

66. Walther, "The Geopolitics of Oil," 4.

67. This theme was emphasized by a number of energy experts, including representatives from the Chinese government and the IEA during the Energy Security in Asia conference, Beijing, May 21–22, 2009.

68. Herberg, "The Rise of Energy and Resource Nationalism in Asia," 125.

69. Xiaojie Xu, *Petro-Dragon's Rise: What It Means for China and the World* (Florence: European Press Academic Publishing, 2002); Downs, "Who's Afraid of China's Oil Companies?"; Kong, *China's International Petroleum Policy*; and Andrews-Speed and Dannreuther, *China, Oil, and Global Politics*.

70. Kong, *China's International Petroleum Policy*, 91.

71. Interview with energy expert, Beijing, November 2007. See also Kong, *China's International Petroleum Policy*, 38.

72. Xu, "China's Energy Security," 276.

73. Ibid.

74. Kong, *China's International Petroleum Policy*, 60.

75. Zha, "Oiling the Wheels," 3–4.

76. Kong, *China's International Petroleum Policy*, 46–56.

77. Ibid., 59–60.

78. After a referendum in January 2011, South Sudan became independent from North Sudan on July 9, 2011. This section refers to Chinese NOCs investments in equity oil production, pipelines, and service sector and port facilities in South Sudan and Sudan.

79. Trade between China and Africa averaged more than 30 percent annual growth from 2000 to 2008, with China emerging as Africa's largest trading partner. From 2002 to 2003, trade between China and Africa doubled to $18.5 billion; by 2005 it stood at $32.17 billion; by 2007, it had reached $73 billion; and by the end of November 2010 bilateral trade had increased to $114.8 billion, surpassing the pre-financial-crisis level of $106.8 billion for all of 2008. It is important to note that China's trade with Africa has been largely balanced. Exports to Africa totaled $15.25 billion, while imports from Africa were $16.92 billion in 2005, compared with a surplus of $940 million in 2007 and a deficit of $5.16 billion in 2008. See, among others, "China-Africa Trade Hits Record High," *China Daily* (December 24, 2010), http://www.chinadaily.com.cn/china/2010-12/24/content_11748176.htm; Stephanie Hanson, "China, Africa, and Oil," Council on Foreign Relations Backgrounder, June 6, 2008, http://www.cfr.org/publication/9557/china_africa_and_oil.html; "China-Africa Trade Jumps by 39%," *BBC News* (January 6, 2006), http://news.bbc.co.uk/2/hi/business/4587374.stm; "Huge Surge in China-Africa Trade" (December 23, 2010), http://www.fin24.com/Economy/Huge-surge-in-China-Africa-trade-20101223.

80. Xu, *Petro-Dragon's Rise*, 46; Jakobson and Zha, "China and the Worldwide Search for Oil Security," 60–73.

81. International Crisis Group (ICG), "China's Thirst for Oil," *Asia Report* 153 (June 9, 2008): 23; Chen Zhu, "CNPC Braces for Changes in Sudan's Oil Fields," *Caixin Online* (March 11, 2011), http://english.caing.com/2011-03-11/100235332.html.

82. Margaret Cornish, "Behaviour of Chinese SOEs: Implications for Investment and Cooperation in Canada," Canadian Council of Chief Executives, February 2012; Downs, "Who's Afraid."

83. "Chinese Foreign Minister to Meet Sudan's Bashir," *Reuters* (August 8, 2011), http://www.reuters.com/article/2011/08/08/us-sudan-china-idUSTRE7773CT20110808; Henry Lee and Dan A. Shalmon, "Searching for Oil: China's Oil Initiatives in the Middle East," *BCSIA Discussion Paper* (Cambridge, Mass.: Belfer Center for Science and International Affairs, Kennedy School of Government, Harvard University, January 2007), 24; Frederik Balfour, "China: Spielberg's Olympic-Sized Snub," *Bloomberg News* (February 13, 2008), http://www.businessweek.com/print/globalbiz/content/feb2008/gb20080213_989993.htm.

84. Ian Taylor, "China's Oil Diplomacy in Africa," *International Affairs* 82, no. 5 (2006): 942; William Engdahl, "Darfur? It's the Oil, Stupid," *Geopolitics-Geoeconomics* (May 20, 2007).

85. Downs, *China's Energy Security*, 37–41.

86. Human Rights Watch, "China's Involvement in Sudan: Arms and Oil" (November 24, 2003); Ian Taylor, "Beijing's Arms and Oil Interests in Africa," *China Brief* 5, no. 21 (October 13, 2005); Chris Alden, *China in Africa* (London: Zed, 2007); "The New Colonialists, China's Hunger for Natural Resources Is Causing More Problems at Home than Abroad," *The Economist* (March 13, 2008).

87. Eurasia Group, "China's Overseas Investments," 21.

88. Chen, "CNPC Braces for Changes in Sudan's Oil Fields."

89. Interview with Zha Daojiong, Beijing, November 2007. Although Houser and Levy support Zha's findings from 2006, their statistics show that Sudanese oil exported in 2007 was shipped back to China. Trevor Houser and Roy Levy, "Energy Security and China's UN Diplomacy," *China Security* 4, no. 3 (Summer 2008): 70.

90. Chen, "CNPC Braces for Changes in Sudan's Oil Fields."

91. Nicholas D. Kristof, "China's Genocide Olympics," *New York Times* (January 24, 2008), http://www.nytimes.com/2008/01/24/opinion/24kristof.html.

92. Jakobson and Zha, "China and the Worldwide Search," 67; ICG, "China's Thirst."

93. Houser and Levy, "Energy Security."

94. Downs, "The Facts and Fiction of Sino-Africa Energy Relations," 58.

95. Erica Downs, "China's 'New' Energy Administration," *China Business Review* 35, no. 6 (November–December 2008): 43.

96. Chin-Hao Huang, "China's Evolving Perspective on Darfur: Significance and Policy Implications," *PacNet Newsletter* 40 (July 25, 2008).

97. Bates Gill, Chin-hao Huang, and Stephen J. Morrison, "Assessing China's Growing Influence in Africa," *China Security* 3, no. 3 (Summer 2007): 15.

98. Kristof, "China's Genocide Olympics."

99. ICG, "China's Thirst," 22–32; Houser and Levy, "Energy Security," 70.

100. ICG, "China's Thirst."

101. Herberg, "China's Energy Consumption and Resource Nationalism in Asia," 126.

102. Lee and Shalmon, "Searching for Oil."

103. Ibid., 10.

104. Ibid., 24.

105. Chris Buckley, "Sudan's Bashir Warns, Reassures China on South Split," *Reuters* (June 27, 2011), http://af.reuters.com/article/energyOilNews/idAFL3E7HR03S20110627.

106. "Sudan's Bashir Arrives a Day Late in China," *South China Morning Post* (June 29, 2011).

107. "China 'Helpful' on South Sudan: US," *Straits Times* (July 2, 2011), http://www.straitstimes.com/BreakingNews/Asia/Story/STIStory_686397.html.

108. Kathrine Dennys, "'A Foreign Policy Migraine': Assessing China's Role in the Conflict Between Sudan and South Sudan," *Consultancy Africa Intelligence* (June 18, 2012), http://

www.consultancyafrica.com/index.php?option=com_content&view=article&id=1040
:a-foreign-policy-migraine-assessing-chinas-role-in-the-conflict-between-sudan-and
-south-sudan-&catid=57:africa-watch-discussion-papers&Itemid=263.

109. Chen, "CNPC Braces for Changes in Sudan's Oil Field"; Eoin O'Cinneide, "South Sudan 'Picks Kenya Pipeline,'" *UpstreamOnline* (July 6, 2011), http://www.upstreamonline.com/live/article265666.ece.

110. "Sudan's Bashir Arrives a Day Late in China," *South China Morning Post* (June 29, 2011).

111. In comparison, statistics from the EIA show that Saudi Arabia was the world's top oil producer in 2008, with production of 10,782 mbd; that Russia was the second largest, with production of 9,790 mbd; and that the United States was the third largest, with production of 8,514 mbd. According to the EIA, Iran produced 4,174 mbd in 2008. Iran was also the world's fourth-largest net exporter of oil, exporting 2,342 mbd. This was behind Saudi Arabia's exports of 8,030 mbd, Russia's exports of 7,017 mbd, and the UAE's exports of 2,474 mbd. See http://www.eia.doe.gov/country/index. The CIA *World Factbook*, however, states that Iran's oil exports total 2.8 mbd, the third largest in the world. See Garver et al., "Moving (Slightly) Closer to Iran," 25.

112. Sanam Vakil, "Iran: Balancing East against West," *Washington Quarterly* 29, no. 4 (2006): 55; Lee and Shalmon, "Searching for Oil," 19; ICG, "China's Thirst"; Christina Lin, "The Caspian Sea: China's Silk Road Strategy Converges with Damascus," *Pitts Report* (The Jamestown Foundation) (September 7, 2010), http://www.pittsreport.com/2010/09/the-caspian-sea-china%E2%80%99s-silk-road-strategy-converges-with-damascus/.

113. Saira H. Basit, "The Iran-Pakistan-India Pipeline Project. Fuelling Cooperation?" *Oslo Files* 4 (Oslo: Norwegian Institute for Defense Studies, 2008).

114. Garver et al., "Moving (Slightly) Closer to Iran," 14; Willem van Kemenade, "Iran's Relations with China and the West: Cooperation and Confrontation in Asia," Netherlands Institute of International Relations, Clingendael, November 2009, 121–122.

115. Andrew S. Erickson and Gabriel B. Collins, "China's Oil Security Pipe Dream: The Reality, and Strategic Consequences, of Seaborne Imports," *Naval War College Review* 63, no. 2 (Spring 2010): 88–111; Interviews with Zha Daojiong, Beijing, 2008 and 2009.

116. Erickson and Collins, "China's Oil Security Pipe Dream," 101–104. Whether or not a pipeline is built, China's close ties with Pakistan facilitated the construction of a deep-water port in Gwadar. The Pakistanis have proposed building liquefied natural gas (LNG) terminals in Gwadar to transform the piped natural gas from Iran into LNG to be exported to China, thereby avoiding the Strait of Hormuz for shipments from Iran to China.

117. Garver et al., "Moving (Slightly) Closer to Iran," 32; Lin, "The Caspian Sea."

118. ICG, "The Iran Nuclear Issue," 6. Bilateral trade between China and Iran at $21.2 billion in 2009 made China Iran's "most significant trade partner." See Ariel Farrar-Wellman and Robert Frasco, "China-Iran Foreign Relations," *IranTracker* (July 13, 2010), http://www.irantracker.org/foreign-relations/china-iran-foreign-relations; "Iran, China Explore Using Yuan to Settle Trade: Rpt," *China Daily* (June 23, 2010), http://www.chinadaily.com.cn/bizchina/2010-07/23/content_11043011.htm.

119. Garver et al., "Moving (Slightly) Closer to Iran," 27, 15.

120. According to Garver et al., the "buyback" system requires that "the contracting company typically funds all investments to develop a new or existing field and is compensated through a share of the field's production that is supposedly sufficient to give the contract a guaranteed rate of return on its investments. Once the contract is completed, operation of the field passes back to the NOC." Garver et al., "Moving (Slightly) Closer to Iran," 26–27, n. 27.

121. "China and Iran Signs Biggest Oil & Gas Deal," *China Daily* (October 31, 2004), http://www.chinadaily.com.cn/english/doc/2004-10/31/content_387140.htm.

122. I thank Erica Downs for pointing this out. See also Jiang and Sinton, "Overseas Investments," 42.

123. Sanchez Wang, "Iran, China Sign $5 Billion Contract on South Pars Gas Field," *Bloomberg News* (June 9, 2010), http://www.bloomberg.com/apps/news?pid=newsarchive&sid=aWV nwjppjOXI; Reza Derakhshi, "Sinopec in $6.5 Billion Iran Refinery Deal: Iranian Media," *Reuters* (November 25, 2010), http://www.reuters.com/article/idUSTRE5AO20C20091125; Jiang and Sinton, "Overseas Investments," 42.

124. Lin, "The Caspian Sea."

125. Pomfret and Lynch, "U.S. Criticized on Iran Sanctions"; John Pomfret, "European Oil Companies Pledge to End Investment in Iran Over Nukes Program," *Washington Post* (September 30, 2010), at http://www.washingtonpost.com/wp-dyn/content/article/2010/09/30/AR2010093006452.html. As an illustration to how the numbers are thrown around in many of the reports, and $20 billion is not a small number, this same report notes that China's NOCs have "commitments" of $80 billion in Iran's energy sector without clarifying what the distinction is between "commitments" and "signed memorandums committing to invest more than $100 billion." See also Herberg, "The Rise of Energy and Resource Nationalism in Asia," 129.

126. Press TV, "China Boosts Investment in Iran's Energy" (July 31, 2010), http://www.infowars.com/china-boosts-investment-in-irans-energy/; Sara Bulley, "Sanctions Give Boost to Chinese Investments in Iran," *CSIS* (August 2, 2010), http://csis.org/blog/sanctions-give-boost-chinese-investments-iran.

127. ICG, "The Iran Nuclear Issue"; Kong, *China's International Petroleum Policy*.

128. Erica Downs, "China-Gulf Energy Relations," in *China and the Gulf: Implications for the United States*, ed. B. Wakefield and S. L. Levenstein (Washington, D.C.: Woodrow Wilson International Center for Scholars, 2011), http://www.brookings.edu/articles/2011/0801_china_gulf_energy_downs.aspx.

129. Justin Li, "Chinese Investment in Iran: One Step Forward and Two Steps Backward," *East Asian Forum* (November 3, 2010), http://www.eastasiaforum.org/2010/11/03/chinese-investment-in-iran-one-step-forward-and-two-steps-backward/.

130. "CNPC begins South Pars Phase II Project," *Platts* (October 19, 2010); Benoit Faucon, "CNPC Nears End of Iran Gas Study But No Drilling Yet," *Dow Jones Newswires* (October 18, 2010).

131. Downs, "China-Gulf Energy Relations."

132. Chinese NOCs such as Sinochem and Nanjing Tankers are also charging big premiums (rates are doubled) for chemical tanker trips from Iran as sanctions have come into effect.

133. "Sinopec Ships Rare Gasoline Cargo Spore-Iran—Trade," *Reuters* (April 14, 2010), http://uk.reuters.com/article/2010/04/14/sinopec-iran-gasoline-idUKLDE63DoGS20100414.

134. Lin, "The Caspian Sea."

135. Downs, "China-Gulf Energy Relations."

136. In the first quarter of 2009, China was Iran's second-largest buyer of crude oil, purchasing 484,093 b/d. Japan was Iran's top buyer at 519,518 b/d, and India with 426,360 b/d and South Korea with 244,989 in third and fourth place respectively. "Iran's Major Oil Customers and Energy Partners," *Alexander's Gas and Oil Connections* (June 5, 2009), http://www.gasandoil.com/goc/news/ntm93113.htm; Garver et al., "Moving (Slightly) Closer to Iran," 24.

137. "FACTBOX—Ties Binding China and Iran," *Reuters* (April 2, 2010), http://in.reuters.com/article/idINIndia-47399820100402?sp=true; International Crisis Group (ICG), "The Iran Nuclear Issue: The View from Beijing," *Asia Briefing* 100 (February 17, 2010): 5.

138. ICG, "The Iran Nuclear Issue," 12.

139. The full text of the resolution: http://www.un.org/News/Press/docs/2010/sc9948.doc.htm.

140. Neil MacFarquhar, "U.N. Approves New Sanctions to Deter Iran," *New York Times* (June 9, 2010), http://www.nytimes.com/2010/06/10/world/middleeast/10sanctions.html?_r=1.

141. John W. Garver, "Is China Playing a Dual Game in Iran?" *Washington Quarterly* 34, no. 1: 75–88.

142. Øystein Tunsjø, "Hedging Against Oil Dependency: New Security Perspectives on Chinese Energy Security Policy," *International Relations* 24, no. 1 (March 2010): 37. The ICG calls this a hedging strategy. See "The Iran Nuclear Issue," 10. For a thorough analysis of China's balancing act, see Garver et al., "Moving (Slightly) Closer to Iran."

143. President Barack Obama, "Remarks by President Obama to the Australian Parliament," The White House, Office of the Press Secretary (November 17, 2011), http://www.whitehouse.gov/the-press-office/2011/11/17/remarks-president-obama-australian-parliament; Hillary Clinton, "America's Pacific Century," *Foreign Policy* (November 2011), http://www.foreignpolicy.com/articles/2011/10/11/americas_pacific_century.

144. Mriganka Jaipuriyara, "China Cuts Iranian Crude Oil Imports by 30 Percent in H1 2010," *Platts* (July 26, 2010), http://www.platts.com/RSSFeedDetailedNews.aspx?xmlpath=RSSFeed/HeadlineNews/Oil/8941642.xml; Emran Hussain, "Iranian Oil Export to China Down 30 Percent in 1st Half," *ArabianOilandGas.com* (August 2, 2010), http://www.arabianoilandgas.com/article-7639-iranian-oil-exports-to-china-down-30-in-1st-half/.

145. Jaipuriyara, "China Cuts Iranian Crude Oil." Downs supports this view, arguing that uncompetitive Iranian crude prices were the main reason for the fall in Iranian export to China. See Downs "China-Gulf Energy Relations."

146. Edward Morse, "Pincer Movement," *Fixed Income Research* (Credit Suisse) (April 19, 2010); "China's Oil Demand Increase 'Astonishing,' Says IEA," *BBC* (March 12, 2010), http://news.bbc.co.uk/2/hi/business/8563985.stm.

147. Jaipuriyara, "China Cuts Iranian Crude Oil"; Hussain, "Iranian Oil Exports."

148. Andrew Jacobs and Mark Landler, "Strains Easing, Obama Talks with Chinese Leader," *New York Times* (April 2, 2010), http://www.nytimes.com/2010/04/03/world/asia/03china.html?_r=1. Of course, the renminbi issue is about more than simply sanc-

tions. U.S.-China economic issues are complicated, and if China was accused of currency manipulations the costs to the United States would be high, given the prospect of possible Chinese retaliation.

149. Andrew Jacobs, "Israel Makes Case to China for Iran Sanctions," *New York Times* (June 8, 2010), http://www.nytimes.com/2010/06/09/world/middleeast/09israel.html?_r=1; Chris Zambelis, "Shifting Sands in the Gulf: The Iran Calculus in China-Saudi Arabia Relations," *China Brief* 10, no. 10 (May 13, 2010): 5.

150. ICG, "The Iran Nuclear Issue"; "Saudi, UAE Ready to Press China on Iran Sanctions: US" (March 11, 2010), http://www.france24.com/en/20100311-saudi-uae-ready-lobby-china -iran-sanctions-us.

151. ICG, "The Iran Nuclear Issue," 14. Dennis Ross, a senior National Security Council adviser at the White House, allegedly proposed in early 2009 that if China severed its energy ties with Iran, Saudi Arabia would provide for all of China's incremental oil demand in the future. Others have noted that U.S. Secretary of State Hillary Clinton tried to enlist Saudi Arabia and Qatar to support persuading China to back U.S.-led efforts to impose sanctions on Iran. See Garver et al., "Moving (Slightly) Closer to Iran," 53; Zambelis, "Shifting Sand in the Gulf," 5.

152. The "priority" formula refers to a discount on transportation costs for oil shipped to the United States from the kingdom. Interview, Washington, D.C., May 2010.

153. Consequently, the U.S.-Saudi strategic partnership suffered as the kingdom sought to retain Iraq as a balance against Iran. Interview, Washington, D.C., May 2010.

154. Morse, "Pincer Movement."

155. Zambelis, "Shifting Sands," 6.

156. Morse, "Pincer Movement."

157. Aizhu Chen and Judy Hua, "Iran-China Oil Trade Runs Smoothly—Beijing Sources," *Reuters* (July 25, 2011); "Iran's Oil Exports to China up 46 Percent, Jan-Jul 2011," *Moj News* (August 23, 2011), http://www.payvand.com/news/11/aug/1219.html.

158. Ibid.

159. Judy Hua and Fayen Wong, "China Iran Oil Imports Recover, Recoup Earlier Fall," *Reuters* (June 21, 2012), http://www.reuters.com/article/2012/06/21/us-china-oil-iran -idUSBRE85K0L020120621.

160. Many PRC tankers are not PRC flagged and may have managed to get waivers from U.S. for a limited period. Some Iranian-owned tankers have also shifted names and flagged out. But the extraterritorial arm of the United States is getting very long, and many countries and companies do not want to get caught importing or shipping Iranian oil in breach of U.S. sanctions.

161. Downs, "China-Gulf Energy Relations."

162. Duc Huynh statement taken from "China's Oil Companies Play Cool Hand in Iran" (August 19, 2012), http://china-wire.org/?p=5863.

163. In October 2008, CNOOC announced a $1.1 billion shale gas deal with Chesapeake Energy, an independent U.S. firm. Pending final U.S. government approval, this was CNOOC's first deal since its 2005 bid for the U.S. firm Unocal was blocked by U.S. regulators and

politicians. "Exclusive—China Slows Iran Oil Work as U.S. Energy Ties Warm," *Markets News & Commentary* (October 28, 2010).

164. Cen Aizhu, "Exclusive-China Slows Iran Oil Works as U.S. Energy Ties Warm," *TD Waterhouse, Market News & Commentary* (October 28, 2010).

165. Chen Aizhu, "Sinopec Turns Down Cut-Price Iran Crude: Source," *Reuters* (June 12, 2012), http://www.reuters.com/article/2012/06/12/us-iran-oil-sinopec-idUSBRE85B06720120612.

166. Garver, "Is China Playing a Dual Game in Iran?"

167. ICG, "The Iran Nuclear Issue," 16.

168. For an interesting assessment of the potential role played by these various actors in China's Iran policy and the CCP's leaders balancing and mediation among these divergent points of view, see Garver, "Is China Playing a Dual Game in Iran?" 84–86.

169. Garver et al., "Moving (Slightly) Closer to Iran," 2–3, 53–54.

170. Lee and Shalmon, "Searching for Oil," 11.

5. SAFEGUARDING CHINA'S SEABORNE PETROLEUM SUPPLIES

1. Robert Ross, "China's Naval Nationalism," *International Security* 34, no. 2 (Fall 2009): 46–81.

2. Wu Lei and Shen Qinyu, "Will China Go to War Over Oil?" *Far Eastern Economic Review* (April 2006): 39; see also Wu Kang and Ian Storey, "Energy Security in China's Capitalist Transition: Import Dependence, Oil Diplomacy, and Security Imperatives," in *China's Emergent Political Economy: Capitalism in the Dragon's Lair*, ed. Christopher A. McNally (London: Routledge, 2008), 200; Philip Andrews-Speed and Roland Dannreuther, *China, Oil, and Global Politics* (London: Routledge, 2011), 110–111.

3. This section draws on the work by leading U.S. experts on the PLA Navy and conversations with experts at the U.S. Naval War College/China Maritime Studies Institute, the Center for Naval Analysis/China Studies, and the National Defense University. However, the author has also benefited from conversations and presentations by leading Chinese naval experts at the China Foundation for International and Strategic Studies, the China Institute for Marine Affairs, the Academy of Military Science, the National Defense University, and the PLA Naval Institute. In addition, leading experts in Japan, India, and Europe have participated in a number of book projects, conferences, workshops, and seminars that I have co-organized with U.S., Asian, and European partners since 2007. See the following publications: Peter Dutton, Robert Ross, and Øystein Tunsjø, eds., *Twenty-First-Century Seapower: Cooperation and Conflict at Sea* (London: Routledge, 2012); Saira Basit and Øystein Tunsjø, "Emerging Naval Powers in Asia: China and India's Quest for Seapower," Oslo Files 2 (Oslo: Norwegian Institute for Defense Studies, 2012), http://ifs.forsvaret.no/publikasjoner/oslo_files/OF_2012/Sider/OF_2_2012.aspx; the contributions to Bjørn Terjesen and Øystein Tunsjø, eds., "The Rise of Naval Powers in Asia and Europe's Decline," Oslo File 6 (Oslo: Norwegian Institute for Defense Studies, 2012), http://ifs.forsvaret.no/publikasjoner/oslo_files/OF_2012/Sider/OF_6_2012.aspx; and the Inter-

national Order at Sea Workshop Series webpage, http://ifs.forsvaret.no/publikasjoner/ios/Sider/default.aspx. For the traditionally defensive orientation of the PLAN, see Michael McDevitt, "The Strategic and Operational Context Driving PLA Navy Building," in *Right Sizing the People's Liberation Army: Exploring the Contours of China's Military*, ed. R. Kamphausen and A. Scobell (Carlisle Barracks, Penn.: U.S. Army War College, 2007), 494.

4. Nan Li, "The Evolution of China's Naval Strategy and Capabilities: From 'Near Coast' and 'Near Seas' to Far Seas," *Asian Security* 5, no. 2 (2009): 144–169.

5. For more on the evolution and modernization of the PLAN see, among others, Li, "The Evolution," 148–149; Peter Howarth, *China's Rising Sea Power: The PLA Navy's Submarine Challenge* (London: Routledge, 2006); Bernard D. Cole, *Sea Lanes and Pipelines: Energy Security in Asia* (London: Praeger Security International, 2008), 138–139; Bernard D. Cole, *The Great Wall at Sea: China's Navy in the Twenty-First Century*, 2nd ed. (Annapolis, Md.: Naval Institute Press, 2010), 86–95.

6. McDevitt, "The Strategic and Operational Context," 487–488; Bernard D. Cole, "Right-Sizing the Navy: How Much Naval Force Will Beijing Deploy?" in *Right Sizing the People's Liberation Army: Exploring the Contours of China's Military*, ed. R. Kamphausen and A. Scobell (Carlisle Barracks, Penn.: U.S. Army War College, 2007), 552.

7. Li, "The Evolution," 155.

8. Ibid.

9. McDevitt, "The Strategic and Operational Context," 490.

10. James R. Holmes and Toshi Yoshihara, *China's Naval Strategy in the Twenty-First Century: The Turn to Mahan* (London: Routledge, 2008); Li, "The Evolution"; Cole, *The Great Wall at Sea*.

11. Li, "The Evolution," 155.

12. McDevitt, "The Strategic and Operational Context," 490.

13. Ross, "China's Naval Nationalism," 58–60. An access denial strategy can be linked to Posen's concept of a "contested zone." See Barry Posen, "Command of the Commons: The Military Foundations of U.S. Hegemony," *International Security* 28, no. 1 (2008): 5–46.

14. See Ronald O'Rourke, "China Naval Modernization: Implications for U.S. Navy Capabilities—Background and Issues for Congress," Congressional Research Service (April 22, 2011); McDevitt, "The Strategic and Operational Context," 492–496.

15. Office of the U.S. Secretary of Defense, *Annual Report to Congress: Military Power of the People's Republic of China* (Washington, D.C., 2009), 22, 48. Associate Professor Andrew Erickson at the U.S. Naval War College has set up a blog that provides a good starting point for those seeking information on China's antiship ballistic missile plans and the U.S. response. See http://www.andrewerickson.com/2011/07/china%E2%80%99s-anti-ship-ballistic-missile-asbm-reaches-equivalent-of-%E2%80%9Cinitial-operational-capability%E2%80%9D-ioc%E2%80%94where-it%E2%80%99s-going-and-what-it-means/.

16. David A. Shlapal et al., *A Question of Balance, Political Context, and Military Aspects of the China-Taiwan Dispute* (Santa Monica, Calif.: RAND Corporation, 2009).

17. The PLAN has about three or four Yuan, twelve Kilo, fourteen Song, and twenty-one Ming SS and about a total of seventy-seven principal surface combatants, including twenty-six destroyers (four Sovremenny, two Luzhou, four Luyang, one Luhai, two Luhu, and thirteen

Luda) and fifty-one frigates (eight Jiankai, fourteen Jiangwei, and twenty-nine Jianghu). See Cole, *The Great Wall at Sea*, 93.

18. Bernard Cole from the National Defense University (NDU) notes that "the PLAN seems to be investing little in mine warfare," whereas Andrew Erickson from the U.S. Naval War College writes that China has devoted "considerable attention to . . . the sea mine." See Cole, *Sea Lanes and Pipelines*, 140; Andrew S. Erickson, "Can China Become a Maritime Power?" in *Asia Looks Seaward: Power and Maritime Strategy*, ed. T. Yoshihara and J. R. Holmes (London: Praeger Security International, 2008), 77.

19. Despite the progress, China's naval modernization has weaknesses on two fronts. First, PLAN shipbuilding is still dependent on foreign design in almost all areas. Second, the PLAN "has yet to demonstrate the command and control capability necessary successfully to conduct net-centric operations in a twenty-first-century maritime battlespace." Cole, *The Great Wall at Sea*, 102–103, 113.

20. However, as Michael McDevitt notes, "the PLAN has *not* played a central role" in deterring or potentially punishing Taiwan if the island declared independence. To "'reach out and touch Taiwan' in a way that was not possible in earlier decades" has largely been the task of the Second Artillery's hundreds of ballistic missiles and the PLAAF's tactical aircraft systems. McDevitt, "The Strategic and Operational Context," 490, 497.

21. Li, "The Evolution," 150–151; O'Rourke, "China's Military Modernization."

22. Eric A. McVadon, "China's Navy Today: Looking Toward Blue Water," in *China Goes to Sea: Maritime Transformation in Comparative Historical Perspective*, ed. Andrew Erickson et al. (Annapolis, Md.: Naval Institute Press, 2009), 373–400.

23. Cole, *The Great Wall at Sea*, 190.

24. Ross, "China's Naval Nationalism," 59–60; Erickson, "Can China Become a Maritime Power?" 108.

25. Li, "The Evolution," 158, 160.

26. Ibid.; McDevitt, "The Strategic and Operational Context," 506; Erickson, "Can China Become a Maritime Power?" 74, 101.

27. Hu Jintao, written statement at the G8 meeting in St. Petersburg (July 17, 2006), http://www.uofaweb.ualberta.ca/chinainstitute/nav03.cfm?nav03=48114&nav02=43884&navoi=43092. See also Amy Myers Jaffe and Steven W. Lewis, "Beijing's Oil Diplomacy," *Survival* 44, no. 1 (Spring 2002): 115–134.

28. Cited in Xu Yi-Chong, "China's Energy Security," *Australian Journal of International Affairs* 60, no. 2 (June 2006): 268.

29. Information Office of the State Council of the People's Republic of China, *China's National Defense in 2006* (Beijing: Foreign Language Press, 2006), 4.

30. Andrew Erickson and Lyle Goldstein, "Gunboats for China's New 'Grand Canals'? Probing the Intersection of Beijing's Naval and Oil Security Policies," *Naval War College Review* 62, no. 2 (Spring 2009): 48, 71 n. 32. The full text of the 2010 *Defense White Paper* is available at http://news.xinhuanet.com/english2010/china/2011-03/31/c_13806851.htm.

31. Erica S. Downs, "China," in *Energy Security Series: The Brookings Foreign Policy Studies* (Washington, D.C.: Brookings Institution, December 2006), 14; Cole, "Right-Sizing the

Navy," 545; David Zweig and Bi Jinhai, "China's Global Hunt for Energy," *Foreign Affairs* 84, no. 5 (September/October 2005): 34; Wu and Storey, "Energy Security in China's Capitalist Transition"; Ji Xiaohua, "It Is Not Impossible to Send Troops Overseas to Fight Terrorism," *Sing Tao Jih Pao* (Hong Kong) (June 17, 2004), A27, in Foreign Broadcast Information Service CPP20040617000054.

32. Wu and Storey, "Energy Security in China's Capitalist Transition"; *Wall Street Journal* (October 7, 2005).

33. "Chinese President Calls for Strengthened, Modernized Navy," *People's Daily* (December 27, 2006), http://english.peopledaily.com.cn/200612/27/eng20061227_336273.html.

34. Andrew Erickson and Lyle Goldstein, "Gunboats for China's New 'Grand Canals'?" *Naval War College Review* 62, no. 2 (2009): 48.

35. Shi Hongtao, "China's 'Malacca Straits,'" *Qingnian bao* (June 15, 2004), in Foreign Broadcast Information Service CPP20040615000042.

36. Ibid., 52; Ross, "China's Naval Nationalism," 70.

37. See Ross, "China's Naval Nationalism," 69–71, for a survey of alarmist Chinese voices.

38. James R. Holmes and Toshi Yoshihara, "China's 'Caribbean' in the South China Sea," *SAIS Review* 26, no. 1 (Winter/Spring 2006): 79–92. The authors note: "China sees economic development and in turn energy security as a keystone of state power—giving it dual commercial and political motives in the South China sea similar to those Mahan claimed for the United States in the Caribbean basin" (52). See also Robert D. Kaplan, "Center Stage for the 21st Century: Power Plays in the Indian Ocean," *Foreign Affairs* 88, no. 2 (March/April 2009): 16–32.

39. Dan Blumenthal, "Concerns with Respect to China's Energy Policy," in *China's Energy Strategy: The Impact on Beijing's Maritime Policies*, ed. Gabriel B. Collins et al (Annapolis, Md.: U.S. Naval Institute Press, 2008), 418–436; James Boutilier, "Great Nations, Great Navies: Looking for Sea Room in Asia," presentation at the NATO-Russia Council: From Vladivostok to Vancouver conference, Vladivostok, May 11, 2006, 12. See also Tarique Niazi, "The Ecology of Strategic Interests: China's Quest for Energy Security from the Indian Ocean and the South China Sea to the Caspian Sea Basin," *China and Eurasia Forum Quarterly* 4, no. 4 (2006): 97–116; Ian Storey, "China's 'Malacca Dilemma,'" *China Brief* (The Jamestown Foundation) 6, no. 8 (April 12, 2006); Arnold Gay, "Oil Fueling China's Naval Build-up," *Strait Times* (May 2, 2005).

40. Scott Bray, "Seapower Questions on the Chinese Submarine Force," U.S. Navy, Office of Naval Intelligence (December 20, 2006), http://www.fas.org/nuke/guide/china/ONI2006 .pdf.

41. You Yi, "Dealing with the Malacca Strait Dilemma: China's Efforts to Enhance Energy Transportation Security," *EAI Background Brief* 329 (April 12, 2007).

42. Wu and Shen, "Will China Go to War Over Oil?" 40. One writer paraphrases Mackinder's often-quoted dictum stating that "he who controls not just the production of oil and gas but also the supply, and has discovered substitutes, will rule the world." Indeed, this observer goes so far as to argue that oil security is "the single most important determinant of China's foreign policy," and he reminds us that the U.S.-imposed oil blockade of Japan in

December 1941 led to Japan's attack on Pearl Harbor. See Mohan Malik, "The Geopolitics of Energy Security," paper presented at the Twenty-First Asia-Pacific Roundtable, Kuala Lumpur, June 4–8, 2007.

43. Erickson and Goldstein's excellent survey of the Chinese literature on this issue finds that those "who believe that greater reliance on the international oil market is the best path to oil supply security have gained strength over the past several years," but they also note that "the mercantilists still exert significant influence." See Erickson and Goldstein, "Gunboats for China's New 'Grand Canals'?" 51.

44. Xiaoli Liu and Xinmin Jiang, "China's Energy Situation and Countermeasures," *International Journal of Energy Sector Management* 3, no. 1 (2009): 87.

45. Ibid., 92.

46. "Chinese President Calls for More Efforts in Energy Saving," *People's Daily* (December 27, 2006), http://english.people.com.cn/200612/26/eng20061226_336015.html. See also chapter 3.

47. Interviews with Zha Daojiong and Zhao Hongtu, Beijing, August 2009, 2010, and 2011. On the important linkage between energy security and domestic measures, see, among others, Erica S. Downs, "China's Energy Security," Ph.D. thesis, Princeton University (2004), 164–223; Erica S. Downs, "The Chinese Energy Security Debate," *China Quarterly* 177 (March 2004): 21–41; Xu Yi-chong, "China's Energy Security," in *Energy Security in Asia*, ed. Michael Wesley (London: Routledge, 2007), 42–67; Kong Bo, "Institutional Insecurity," *China Security* 2, no. 2 (Summer 2006): 64–88; Wang Qingyi, "Energy Conservation as Security," *China Security* 2, no. 2 (Summer 2006): 89–105.

48. Interview with Zhao Hongtu.

49. McDevitt, "The Strategic and Operational Context," 508.

50. Cole, *The Great Wall at Sea*, 190.

51. Interviews, Beijing, 2007–2011.

52. Zha Daojiong, "Oiling the Wheels of Foreign Policy? Energy Security and China's International Relations," Asia Security Initiative Policy Series Working Paper 1 (March 2010), 4–5.

53. Interviews in Beijing, Shanghai, and Hong Kong, 2007–2011.

54. Zha, "Oiling the Wheels of Foreign Policy?" 6.

55. A more assertive stand emerged in response to the *Impeccable* incident in March 2009.

56. While rejecting that the PLAN will challenge the U.S. Navy in the foreseeable future, Rear Admiral McDevitt (ret.) notes that this "is not to say that maintaining distant squadrons lacks high utility in peacetime and in periods of crisis." See McDevitt, "The Strategic and Operational Context."

57. Zha, "Oiling the Wheels of Foreign Policy?" 6; interviews in Beijing and Washington, 2007–2011.

58. One Western observer argues that China is developing an "indirect strategy," which assumes that "China's prospective enemies, finding themselves encircled or obstructed by powers aligned with Beijing, will be unable to envision a military campaign to deny China oil at an acceptable level of cost. They will, therefore, be deterred from threatening China, e.g., by interrupting its oil supplies." See Jacqueline Newmyer, "Oil, Arms, and Influence:

The Indirect Strategy Behind Chinese Military Modernization," *Orbis* (Spring 2009): 205–219. The phrase "string of pearls" was first used in a report entitled "Energy Futures in Asia" by the defense contractor Booz-Allen-Hamilton to describe China's emerging maritime strategy. This report was commissioned in 2005 by the Office of Net Assessment of the U.S. Department of Defense. See also Christopher J. Pehrson, "String of Pearls: Meeting the Challenge of China's Rising Power Across the Asian Littoral," *U.S. Army War College* (July 2006), http://www.strategicstudiesinstitute.army.mil/pdffiles/pub721.pdf.

59. Cole, *The Great Wall at Sea*, 191.

60. Ross, "China's Naval Nationalism," 80; Andrew F. Diamond, "Dying with Eyes Open or Closed: The Debate Over a Chinese Aircraft Carrier," *Korean Journal* 13, no. 1 (Spring 2006): 35–58.

61. Ross, "China's Naval Nationalism," 69; Erickson and Goldstein, "Gunboats for China's New 'Grand Canals'?" 47.

62. Ross, "China's Naval Nationalism," 47. See also Diamond, "Dying with Eyes Open or Closed," 35–58.

63. Ross, "China's Naval Nationalism," 69; Erickson and Goldstein, "Gunboats for China's New 'Grand Canals'?" 46.

64. Quoted in Ross, "China's Naval Nationalism," 70.

65. Ibid., 69.

66. Ibid.

67. This section draws heavily on the excellent study by Gabriel Collins and William Murray, "No Oil for the Lamps of China?" *Naval War College Review* 61, no. 2 (2008): 79–95.

68. Ibid., 87.

69. McDevitt, "The Strategic and Operational Context," 505.

70. Collins and Murray, "No Oil for the Lamps of China?" 86.

71. McDevitt, "The Strategic and Operational Context," 505. To counter any circumnavigation, the United States could impose a more distant blockade at another choke point, such as the Strait of Hormuz. This might interrupt oil shipments from the Middle Eastern region, which exports the most oil to China, but such a blockade would not affect oil coming from Africa or Latin America. More importantly, short of a major war between the United States and China, a U.S. blockade of oil shipments to China in the Strait of Malacca and Hormuz would be unfeasible and highly unlikely.

72. Collins and Murray, "No Oil for the Lamps of China?" 84–85.

73. Ibid., 84. See also Andrew S. Erickson and Gabriel Collins, "Beijing's Energy Security Strategy: The Significance of a Chinese State-Owned Tanker Fleet," *Orbis* 51, no. 4 (Fall 2007): 681.

74. Collins and Murray, "No Oil for the Lamps of China?" 88.

75. Ibid.

76. McDevitt, "The Strategic and Operational Context," 507. See also Zha, "Oiling the Wheels of Foreign Policy?"

77. Collins and Murray, "No Oil for the Lamps of China?" 83–84.

78. Ibid., 85.

79. Ibid., 89–92.

80. Ibid., 90. Bud Cole estimates a shorter transit time between Singapore and South Korea and Japan. See Cole, *Sea Lanes and Pipelines*, 18. In addition, Chinese shipbuilders argue that their newest ULCC (ultra-large crude carrier), capable of carrying more than 300,000 tons of oil and traveling at a speed of 30 km per hour, can make the route from Guangzhou to terminals in the Middle East in twenty days. "China's Largest Oil Tanker Set to Sail," *China Daily* (January 14, 2010), http://www.chinadaily.com.cn/china/2010-01/14/content_9317494.htm.

81. Collins and Murray, "No Oil for the Lamps of China?" 90.

82. Ibid., 91.

83. Ibid., 81.

84. McDevitt, "The Strategic and Operational Context," 505.

85. Gabriel Collins, "China Seeks Oil Security with New Tanker Fleet," *Oil and Gas Journal* (October 9, 2006); McDevitt, "The Strategic and Operational Context," 505.

86. Andrews-Speed and Dannreuther, *China, Oil, and Global Politics*, 82.

87. "Coal Prices Jump, Hit by the Perfect Storm," *SeaTrade Asia* (January 30, 2008), http://www.seatradeasia-online.com/print/2264.html. Emphasis added.

88. Interviews, 2007–2011, in Beijing, Hong Kong, Washington, Boston, and Oslo with the chairman of one of China's largest shipping companies, top executives of IOCs operating in China, a managing director of a shipping insurance company in Hong Kong, and government representatives and energy experts in China and the United States.

89. China was not as dependent on overseas oil imports in 2002 as it is today or will be in the years ahead, and China did not possess a large state-owned tanker fleet prior to the Iraq War. The current intention to develop such a fleet points to the fact that it might be called upon as an important hedging strategy if another conflict in which China is not involved were to erupt in the Middle East.

90. Interviews with executives from shipping companies, a managing director from a shipping insurance company, an analyst in the tanker and petroleum market, and an analyst at one of the world's largest shipping banks: Beijing, May 2009; Hong Kong, August 2009; Oslo, October 2009; Beijing, March 2011; and Oslo, September 2011.

91. Wood Mackenzie, "Chinese NOCs Step-up International Expansion," *Corporate Service Insight* (June 2010).

92. EIA, "US, China Oil," country analysis briefs at http://www.eia.gov/.

93. Julie Jiang and Jonathan Sinton, "Overseas Investments by Chinese National Oil Companies: Addressing the Drivers and Impacts," *Information Paper* (IEA) (February 2011): 7.

94. China imported more than 5 mbd in 2011, and it is likely to import close to 6 mbd by the end of 2012. China imported 4.7 mbd in 2010. However, it should be noted that China's NOCs have production in Kazakhstan and imports from that country, which provide a hedge against seaborne oil disruptions.

95. Interviews in China, 2007–2011.

96. "China's Refinery Growth to Outpace Oil Demand," *Alexander's Gas and Oil Connections* 14, no. 1 (January 29, 2009); "China to Boost Oil Refinery Capacity," *China Daily* (March 17, 2006).

97. But by combining overseas production, a state-owned tanker fleet, continental pipeline rail supplies, long-term contracts, strategic petroleum reserves, and domestic petroleum production with various measures in the domestic energy sector, such as boosting nuclear energy, gas production, energy efficiency, hydropower and alternative energy, and constraining demand, China is hedging against potential disruptions by developing a comprehensive risk management capacity.

98. Interviews with executives from shipping companies, a managing director from a shipping insurance company, an analyst in the tanker and petroleum market, and an analyst at one of the world's largest shipping banks: Beijing, May 2009; Hong Kong, August 2009; Oslo, October 2009; Beijing, March 2011; and Oslo, September 2011.

99. Erickson and Collins, "Beijing's Energy Security," 681–682.

100. Gabriel Collins, "China's Refining Expansion to Reshape Global Oil Trade," *Oil and Gas Journal* (February 18, 2008): 117; and Gabriel Collins and Michael Grubb, "Contemporary Chinese Shipbuilding Prowess," in *China Goes to Sea: Maritime Transformation in Comparative Historical Perspective*, ed. Andrew Erickson et al. (Annapolis, Md.: Naval Institute Press, 2009), 344–372.

101. Cole, *Sea Lanes and Pipelines*, 75.

102. Ibid., 78. New ships have lowered the average age of the global oil tanker fleet from over twenty years in 2002 to just over seventeen years in 2006. The tanker fleet still relies heavily on smaller ships: at the beginning of 2006, 7,635 ships—73.4 percent of all tankers—had a capacity of less than 40,000 dwt. The fleet included 471 VLCC tankers (200,000–320,000 dwt) and ten ULCC tankers (320,000 dwt and above). The VLCC and ULCC ships had an average age of 9.3 years, with six of the ULCC ships less than five years old. The trend toward larger tankers continues with thirty-one VLCC/ULCC tankers (above 200,000 dwt), totaling 9.5 million dwt, on order at the beginning of 2006. See Cole, *Sea Lanes and Pipelines*, 78.

103. Meng-di Gu and Shou-de Li, "China Seeks Oil Security Through Fleet Expansion," *Oil and Gas Journal* (December 8, 2008).

104. Thorsten Ludwig and Jochen Tholen, "Shipbuilding in China and Its Impact on European Shipbuilding Industry," part of the wider report *European Industries Shaken up by Industrial Growth in China: What Regulations Are Required for a Sustainable Economy?* (Brussels: European Metalworkers' Federation, 2006). The report comprises three sectors: shipbuilding, auto, and steel. See http://www.emf-fem.org/content/search?SearchText=China.

105. Ibid. It is uncertain whether Chinese shipping companies registered in Hong Kong or Macao are included in these numbers.

106. Gu and Li, "China Seeks Oil Security." In general, Unipec largely sorts out oil trade for Sinopec; China Oil operates mainly for CNPC (although jointly owned with Sinochem). CNOOC is not a major player in the downstream market.

107. "China's Largest Oil Tanker," *China Daily* (January 14, 2010). This contrasts with the estimate by Collins of closer to 10 percent.

108. Robert Wright et al., "Shipyards Struggle to Stay Above Waters," *Financial Times* (August 28, 2012), at http://www.ft.com/intl/cms/s/0/18ce5c5c-ee0c-11e1-b0e4-00144feab49a .html#axzz25P08Gl2x.

109. Interviews with an oil and shipping consultancy firm in Oslo and shipping industry experts in Hong Kong.
110. Collins, "China's Refining Expansion to Reshape Global Oil Trade," 117.
111. Moming Zhou, "Teekay First-Quarter Loss Widens as Tanker Rates Decline," *Bloomberg News* (May 12, 2011), http://www.bloomberg.com/news/2011-05-12/teekay-first-quarter -loss-widens-as-tanker-rates-decline-1-.html.
112. Collins, "China's Refining Expansion to Reshape Global Oil Trade," 114–115.
113. Erickson and Collins, "Beijing's Energy Security," 671, 673, 674.
114. Collins, "China's Refining Expansion to Reshape Global Oil Trade," 118–119.
115. Erickson and Collins, "Beijing's Energy Security," 683; Collins, "China's Refining Expansion to Reshape Global Oil Trade," 124.
116. Ibid., 680.
117. At the same time, the escalatory barrier created by operating state-owned tankers will not necessarily reduce the risk of piracy or other nontraditional or asymmetric security threats.
118. Liu and Jiang, "China's Energy Security Situation and Countermeasures," 85–86.
119. "Malacca Straits Oil Tanker Ablaze Following Collision," *OilVoice* (August 19, 2009), http://www.oilvoice.com/n/Malacca_Straits_Oil_Tanker_Ablaze_Following_Collision/736ec8734 .aspx.
120. "Taiwan Oil Tanker on Fire in Malacca Strait after Collision," *China Post* (August 20, 2009), http://www.chinapost.com.tw/asia/malaysia/2009/08/20/221213/Taiwan-oil.htm.
121. Neville Smith, "The Sun Rises on Safer Navigation," *BIMCO* (January 2008), https://www .bimco.org/Members%20Area/News/General_News/2008/01/03_Feature_Week_1.aspx.
122. Bo Kong, *China's International Petroleum Policy* (Santa Barbara, Calif.: Praeger Security International, 2010), 132.
123. Ibid.; Zheng Hong, "Confidence Building Measures and Nontraditional Security," in *Twenty-First-Century Seapower: Cooperation and Conflict at Sea*, ed. Robert Ross, Øystein Tunsjø, and Peter Dutton (London: Routledge, 2012).
124. See monthly and annual piracy and armed robbery reports by the International Maritime Organization at http://www.imo.org/ourwork/security/piracyarmedrobbery/pages/ piratereports.aspx; International Maritime Bureau, Piracy Reporting Centre (2009), 2, http://www.icc-ccs.org/piracy-reporting-centre. See also "Guidance Released on Arming Ship Personnel," *MaritimeSecurity.Asia* (June 8, 2011), http://maritimesecurity.asia/ free-2/piracy-update/guidance-released-on-arming-ship-personnel/; "Guidance Released on Arming Ship Personnel"; "ICS Changes Position on Using Armed Guard to Protect Ships Against Pirates," *MarineLog* (February 15, 2011), http://www.marinelog.com/ index.php?option=com_content&view=article&id=505:2011feb000152&catid=1:latest -news&Itemid=107.
125. Joshua Ho, "Security of Transportation Routes in Southeast Asia," presentation at the Energy Security in Asia conference, Beijing, May 20–21, 2009.
126. Joshua Ho, "Piracy in the Gulf of Aden: Lessons from the Malacca Strait," *RSIS Commentaries* (January 22, 2009), http://www.rsis.edu.sg/publications/Perspective/RSIS0092009.pdf.
127. Ho, "Security of Transportation Routes."

128. See http://www.recaap.org/.

129. This includes a forum for cooperation and discussions, a project-coordinating committee (for the removal of wrecks, capacity building, and the setting up of tide current and wind measurement systems), and aids to navigation funds.

130. Catherine Zara Raymond, "Countering Piracy and Armed Robbery in Asia: A Study of Two Areas," in *Twenty-First-Century Seapower: Cooperation and Conflict at Sea*, ed. Robert Ross, Øystein Tunsjø, and Peter Dutton (London: Routledge, 2012).

131. International Maritime Bureau, Piracy Reporting Centre (2009), 2, http://www.icc-ccs.org/piracy-reporting-centre.

132. Sam Bateman and Joshua Ho, "Somalia-Type Piracy: Why It Will Not Happen in Southeast Asia," *RSIS Commentaries* (November 24, 2008), http://www.rsis.edu.sg/publications/Perspective/RSIS1232008.pdf.

133. International Maritime Bureau, Piracy Reporting Centre (2009), 2, http://www.icc-ccs.org/piracy-reporting-centre.

134. Ibid., 26.

135. Bateman and Ho, "Somali-Type Piracy."

136. Ibid.

137. ICC International Maritime Bureau, "Piracy and Armed Robbery Against Ships," Annual Report, 1 January 2008–31 December 2008 (January 2009), http://ddata.over-blog.com/xxxyyy/0/50/29/09/Docs-Textes/Pirates2008RAP-BMI0901.pdf.

138. See Resolutions no. 1838 and no. 1851, http://www.un.org/News/Press/docs/2008/sc9467.doc.htm; http://www.un.org/News/Press/docs/2008/sc9541.doc.htm. See also Resolution no. 1814 (2008) and no. 1816 (2008).

139. "Two Fleets of Chinese Naval Escort Ships Meet in Gulf of Aden" (April 13, 2009), http://news.xinhuanet.com/english/2009-04/13/content_11180769.htm.

140. The second flotilla consisted of the destroyer DDG-167 *Shenzhen* and the frigate FFG-570 *Huangshan*, while the supply ship *Weishanhu*, part of the first flotilla, remained in the gulf. The third flotilla included the frigates FFG-529 *Zhoushan* and FFG-530 *Xuzhou*, while the supply ship *Qiandaohu* replaced the supply ship *Weishanhu*. By the end of November 2011, China had deployed ten different escort taskforces to the Gulf of Aden since January 2009 and had conducted 393 escort operations. See "Chinese Escort Taskforce Starts Independent Escort," *People's Daily* (November 23, 2011), http://eng.mod.gov.cn/DefenseNews/2011-11/23/content_4318638.htm.

141. On November 15, 2008, the VLCC *Sirius Star* was hijacked off the East African coast and taken into Somali territorial waters while pirates negotiated a ransom for the release of the ship and the crew.

142. Mark McDonald, "Chinese Warships Sail, Loaded for Pirates," *New York Times* (December 27, 2008).

143. "China to Deploy Ships off Somalia," *BBC News* (December 20, 2008), http://news.bbc.co.uk/2/hi/7793723.stm; McDonald, "Chinese Warships Sail, Loaded for Pirates."

144. Phillip C. Saunders, "Uncharted Waters: The Chinese Navy Sails to Somalia," *PacNet 3* (January 15, 2009).

145. "Chinese Flotilla Sails to Relieve Ships in Somalia Escort Mission," *Xinhua* (April 3, 2009), http://eng.mod.gov.cn/Database/MOOTW/2009-04/03/content_3100821.htm.

146. You Ji and Lim Chee Kia, "China's Naval Deployment to Somalia and Its Implications," *EAI Background Brief* 454 (May 29, 2009).

147. Saunders, "Uncharted Waters"; Daniel M. Hartnett, "The PLA's Domestic and Foreign Activities and Orientation," testimony before the U.S.-China Economic and Security Review Commission, March 4, 2009, 7.

148. Peter Dutton, Robert Ross, and Øystein Tunsjø, introduction to *Twenty-First-Century Seapower: Cooperation and Conflict at Sea*, ed. Robert Ross, Øystein Tunsjø, and Peter Dutton (London: Routledge, 2012).

149. Zheng Hong, "Confidence Building Measures," in *Twenty-First-Century Seapower: Cooperation and Conflict at Sea*, ed. Robert Ross, Øystein Tunsjø, and Peter Dutton (London: Routledge, 2012).

150. Ibid.

151. Former presidents Putin of Russia and Bush of the United States announced the creation of the GICNT during the G8 Summit in St. Petersburg, July 15, 2005. See http://www.nti .org/e_research/official_docs/inventory/pdfs/gicnt.pdf.

6. CHINA'S CONTINENTAL PETROLEUM STRATEGY

1. Stephen Blank, "Chinese Energy Policy in Central Asia and South Asia," testimony before the U.S.-China Economic and Security Review Commission (May 20, 2009), http://www .uscc.gov/hearings/2009hearings/transcripts/09_05_20_trans/09_05_20_trans.

2. Especially relevant in the case of the first stretch of the Kazakhstan-China pipeline (Atasu-Alashankou), when the China National Petroleum Company could not fill it with its own supplies. In addition, export commitments from Russian suppliers were either lacking or were held up by Transneft/GOR as a negotiating chip for its own pipeline.

3. The assumptions underlying this line of reasoning will be elaborated upon below. However, it is important to note that the Myanmar-China pipeline should be considered a transit pipeline because most of the oil for this pipeline will arrive by sea to the port of Kyaukphyu, located on Myanmar's west coast, and thus will face the same risks and threats as seaborne energy supplies. Indeed, this pipeline will face risks both at sea and on land, thereby providing limited energy security.

4. Again, it should be noted that the Myanmar pipeline is a transit pipeline for seaborne energy supplies and thus is also vulnerable to a blockade of Myanmar's coast or ports.

5. CNPC Web site; Bo Kong, *China's International Petroleum Policy* (Santa Barbara, Calif.: Praeger Security International, 2010), 82–83.

6. Kong, *China's International Petroleum Policy*, 83.

7. Xuanli Liao, "Central Asia and China's Energy Security," *China and Eurasia Forum Quarterly* 4, no. 4 (2006): 61–69.

8. Kong, *China's International Petroleum Policy*, 99–100.

9. Zha Daojiong, "Oiling the Wheels of Foreign Policy? Energy Security and China's International Relations," Asia Security Initiative Policy Series Working Paper 1 (March 2010).

10. Kong, *China's International Petroleum Policy*, 53–56.

11. Edward C. Chow and Leigh E. Hendrix, "Central Asia's Pipelines," *NBR Special Report* 2 (September 2010): 37; Liao, "Central Asia and China's Energy Security."

12. John Seaman, "Energy Security, Transnational Pipelines, and China's Role in Asia," *Asie Visions* 27 (Paris: Asia Centre, Institut française des relations internationals, April 2010), 25.

13. The Sino-Kazak crude pipeline was jointly developed by KazMunaiGaz and CNPC.

14. According to an agreement with the Kazakh Ministry of Energy and Mineral Resources, CNPC transferred 33 percent of its shares in PetroKazakhstan to KazMunaiGaz on July 5, 2006, retaining the remaining 67 percent stake in the company. See the CNPC website, http://www.cnpc.com.cn/eng/cnpcworldwide/euro-asia/Kazakhstan/.

15. Ryan Kennedy, "In the 'New Great Game,' Who Is Getting Played? Chinese Investment in Kazakhstan's Petroleum Sector," in *Caspian Energy Politics*, ed. Indra Overland et al. (London: Routledge, 2010), 123–125; Seaman, "Energy Security, Transnational Pipelines, and China's Role in Asia," 23–24.

16. CNPC website, http://www.cnpc.com.cn/eng/cnpcworldwide/euro-asia/Kazakhstan/.

17. Igor Danchenko, Erica Downs, and Fiona Hill, "One Step Forward, Two Steps Back? The Realities of a Rising China and Implications for Russia's Energy Ambitions," *Policy Paper* (Brookings Institution) 22 (August 2010): 6.

18. Based on research by Wood Mackenzie. Danchenko, Downs, and Hill, "One Step Forward, Two Steps Back?" 5.

19. "CNPC's Overseas Output Reaches Record on Acquisitions," *Bloomberg Businessweek* (January 20, 2010). CNPC had overseas oil production of 69.62 million tons in 2009: 18.57 million tons in Kazakhstan, 15 million tons in Sudan, and 10.43 million tons in Latin America.

20. "CNPC produced record oil, gas in Kazakhstan last year," *Reuters* (January 6, 2011), http://af.reuters.com/article/energyOilNews/idAFTOE70500K20110106.

21. Ibid.; Kennedy, "In the 'New Great Game,'" 123–125.

22. Kennedy, "In the 'New Great Game,'" 123.

23. Erica Downs, "Who's Afraid of China's Oil Companies?" in *Energy Security: Economics, Politics, Strategies, and Implications*, ed. Carlos Pascual and Jonathan Elkind (Washington, D.C.: Brookings Institution Press, 2010), 80; Kong, *China's International Petroleum Policy*. Of course, the construction of a pipeline also reflects the interests of the host country. The ability of Chinese NOCs to invest in expensive pipeline projects is highly valued in landlocked countries such as Kazakhstan and Turkmenistan. Hence, the difficulties that Chinese companies encountered in investing in Kazakhstan may have been related to the desire for additional commitments. The author thanks Jonas Graetz for pointing this out.

24. Kassymkhan Kapparov, Milena Ivanova-Venturini and Eka Gazadze, "China—Now Kazakhstan's Top Export Destination," *Renaissance Capital* (January 19, 2011); Richard Galpin, "Struggle for Central Asian Energy Riches," *BBC News* (June 2, 2010), http://www.bbc.co.uk/news/10131641; Erica Downs, presentation at the Energy Security in Asia conference, Beijing, May 2009; Andrew S. Erickson and Gabriel B. Collins, "China's Oil

Security Pipe Dream: The Reality, and Strategic Consequences of, Seaborne Imports," *Naval War College Review* 63, no. 2 (Spring 2010): 93; *Platts Oilgram* (July 3, 2009). Imports from Kazakhstan, accounting for roughly 8 percent of China's oil imports, are based on an estimate of roughly 5 mbd in 2010. In 2009, 400,000 b/d accounted for approximately 10 percent of China's total oil imports, which reportedly stood at 4.09 mbd in 2009. Seaman, "Energy Security, Transnational Pipelines, and China's Role in Asia," 7, 25.

25. "Central Asia to Provide 65 bcm per Year of Gas to China," *Central Asia Newswire* (June 16, 2011), http://www.universalnewswires.com/centralasia/viewstory.aspx?id=4272.

26. The author thanks Jonas Graetz for his comments.

27. Danchenko, Downs, and Hill, "One Step Forward," 5.

28. See the CNPC website, http://www.cnpc.com.cn/en/.

29. By June 2011, the Central Asian–China pipeline provided 10 percent of China's domestic gas consumption. See "Turkmenistan Provides 50 Percent of Chinese Gas Imports" (June 13, 2011), http://www.universalnewswires.com/centralasia/viewstory.aspx?id=4241.

30. Danchenko, Downs, and Hill, "One Step Forward," 6.

31. Downs, presentation at the Energy Security in Asia conference, Beijing, May 2009; Danchenko, Downs, and Hill, "One Step Forward," 6.

32. Russia wrongly believed it could prevent China from diversifying into the Central Asian gas market, and at the same time it was reluctant to supply China with resources from eastern Siberia, fearing a loss of control.

33. Chow and Hendrix, "Central Asia's Pipelines," 37.

34. International Energy Agency (IEA), *2007 World Energy Outlook*, 332, http://www.iea.org/weo/2007.asp; Gabe Collins, "China's Growing LNG Demand Will Shape Markets, Strategies," *Oil and Gas Journal* (October 15, 2007).

35. IEA, *2009 World Energy Outlook*, 366, 429, http://www.iea.org/weo/2009.asp.

36. Seaman, "Energy Security, Transnational Pipelines and China's Role in Asia," 8.

37. Shoichi Itoh, "The Geopolitics of Northeast Asia's Pipeline Development," *National Bureau of Research*, Special Report, no. 23 (September 2010): 21.

38. Leonty Eder, Philip Andrews-Speed, and Andrey Korzhubaev, "Russia's Evolving Energy Policy for Its Eastern Regions, and Implications for Oil and Gas Cooperation Between Russia and China," *Journal of World Energy Law and Business* 2, no. 3 (2009): 221–223; Seaman, "Energy Security, Transnational Pipelines, and China's Role in Asia," 17; Itoh, "The Geopolitics of Northeast Asia's Pipeline Development," 21–22.

39. Marshall Goldman, *Petrostate: Putin, Power, and the New Russia* (Oxford: Oxford University Press, 2008), 105–123.

40. Erica Downs, "Sino-Russian Energy Relations: An Uncertain Courtship," in *The Future of China-Russia Relations*, ed. James Bellacqua (Lexington: University Press of Kentucky, 2010), 156.

41. CNPC website, http://www.cnpc.com.cn/en/.

42. Vitaly Kozyrev, "China's Continental Energy Strategy: Russia and Central Asia," in *China's Energy Strategy: The Impact on Beijing's Maritime Policies*, ed. Gabriel B. Collins et al. (Annapolis, Md.: Naval Institute Press, 2008), 220.

43. Erickson and Collins, "China's Oil Security Pipe Dream," 96–97; Itoh, "The Geopolitics of Northeast Asia's Pipeline Development," 24.

44. Pavel Baev, *Russian Energy Policy and Military Power: Putin's Quest for Greatness* (London: Routledge 2008), 125; Goldman, *Petrostate*, 162.

45. See the report "The Energy Strategy of Russia for the Period up to 2030" (ESR-2030), approved by the Russian government in November 2009. Both former president Putin and President Medvedev, in addition to government figures such as Duma Chairman Boris Gryzlov and Deputy Prime Minister Igor Sechin, have supported increasing Russian energy exports from the Asia-Pacific region. See, among others, Mark A. Smith, "The Russo-Chinese Energy Relationship," *Russian Series* (Defence Academy of the United Kingdom) 10, no. 14 (October 21, 2010); Danchenko, Downs, and Hill, "One Step Forward," 13.

46. Danchenko, Downs, and Hill, "One Step Forward," 11.

47. "CNPC and Transneft Sign Agreement on Principle for the Construction and Operation of a Crude Pipeline," CNPC website (October 31, 2008), http://www.cnpc.com.cn/en/press/newsreleases/2008/10-31.htm.

48. Zhang Wen, "Russia-China Oil Pipeline Completed," *Global Times* (September 28, 2010), http://world.globaltimes.cn/europe/2010-09/577862.html.

49. "Russia-China Oil Pipeline Opens," *BBC News* (January 1, 2011), http://www.bbc.co.uk/news/world-asia-pacific-12103865.

50. "Russian Crude Oil Exports to the Far East—ESPO Starts Flowing," *Platts Special Report* (December 2009); Erickson and Collins, "China's Oil Security Pipe Dream," 97; *China Daily* (April 22, 2009).

51. See Erica Downs, *Inside China, Inc.: China Development Bank's Cross-Border Energy Deals* (Washington, D.C.: John L. Thornton China Center, Brookings Institution, 2011), for more details.

52. "Russian Crude Oil Exports to the Far East—ESPO Starts Flowing," *Platts Special Report* (December 2009).

53. Itoh, "The Geopolitics of Northeast Asia's Pipeline Development," 22.

54. "Russian Crude Oil Exports to the Far East—ESPO Starts Flowing," *Platts Special Report* (December 2009).

55. Seaman, "Energy Security, Transnational Pipelines, and China's Role in Asia," 20.

56. In addition, if Russia and China continue to transport crude oil from Russia to China by rail, then the available capacity will be substantial. If the capacity of the ESPO pipeline grows to 1 mbd and 600,000 b/d is piped to China through the spur pipeline, in addition to shipping more than 300,000 b/d by rail, then this could amount to somewhere between 10 to 15 percent of China's oil imports in the next several years. Of course, there is great uncertainty as to whether Russia can ship such large amounts of oil to China and the east.

57. Danchenko, Downs, and Hill, "One Step Forward," 7–8.

58. "Russian Crude Oil Exports to the Far East—ESPO Starts Flowing," *Platts Special Report* (December 2009).

59. Seaman, "Energy Security, Transnational Pipelines, and China's Role in Asia," 8.

60. See the data on world energy consumption in the BP *Statistical Review of World Energy 2011*, http://www.bp.com/sectionbodycopy.do?categoryId=7500&contentId=7068481. Many European countries import much more gas than the world average. For example, Germany relies on Russian gas exports for about 40 percent of its consumption of natural gas, a figure that is not likely to decrease as it seeks to phase out its nuclear power plants. China's gas imports are expected to rise to 90 bcm by 2015, and domestic production may grow to 170 bcm, reaching a total annual consumption of 260 bcm by 2015. These estimates imply an annual growth rate of 20 percent, with the share of gas in China's primary energy supply possibly reaching 8 percent by 2015. See Philip Andrews-Speed, "China's Booming Gas Sector: Threat or Promise?" *Oil, Gas, and Energy Law Intelligence* (February 2011).

61. Infrastructure is heavily geared toward the West, and the investment decision concerning infrastructure from Yamal, which holds Russia's largest reserves of natural gas, is also linked to the West.

62. The NEA reports that China's gas consumption increased by 20.4 percent to 110 bcm in 2010.

63. Keith Bradsher, "China Reportedly Plans Strict Goals to Save Energy," *New York Times* (March 4, 2011), http://www.nytimes.com/2011/03/05/business/energy-environment/05energy.html ?_r=1&pagewanted=print. See also Andrews-Speed, "China's Booming Gas Sector."

64. This only applies to oil production. For gas, Far Eastern and eastern Siberian production is not projected to increase substantially and will only constitute about one-sixth of overall production. The main increases in production will come from the Yamal peninsula and potentially Shtokman. Both point to increased LNG production rather than pipeline supplies. I thank Jonas Graetz for alerting me to this.

65. Goldman, *Petrostate*, 162.

66. Tetsuo Masuda, "Eastbound Flow of Russian Energy—Geopolitical Implications," presentation at the Norwegian Institute for Defense Studies, November 25, 2009; Itoh, "The Geopolitics of Northeast Asia's Pipeline Development," 19–20; Downs, "Sino-Russian Energy Relationship"; Danchenko, Downs, and Hill, "One Step Forward."

67. According to Downs, in 2007 CNPC offered Gazprom $5.28 per million btu. This was about 60 percent of the price at which Gazprom sold gas to Europe in mid-2008 ($13–$14 per million btu). Downs, "Sino-Russian Energy Relations," 156.

68. "Natural Gas Pricing Agreement with Russia Close," *China Daily* (January 20, 2011), http://www.china.org.cn/business/2011-01/20/content_21779889.htm; EIA, "Country Analysis Briefs: China," http://www.eia.gov/.

69. "Russia-China Oil Price Dispute Valued at $100M," *Reuters* (March 28, 2011), http://www.themoscowtimes.com/business/article/russia-china-oil-price-dispute-valued-at-100m/433814.html.

70. Sergei Blagov, "Russia Seeks to Increase Energy Supplies to China," *Eurasia Daily Monitor* (The Jamestown Foundation) 8, no. 122 (June 24, 2011), http://www.jamestown.org/programs/edm/single/?tx_ttnews%5Btt_news%5D=38092&cHash=34a7e086905462a4c8fe4adbc814be44.

71. Kozyrev, "China's Continental Energy," 221; Baev, *Russian Energy Policy*, 74; Goldman, *Petrostate*, 162.

72. Masuda, "Eastbound Flow"; Elena Shadrina, "Energy Cooperation in North East Asia: Impact on Regional Formation," presentation at the Norwegian Institute for Defense Studies, November 25, 2009.

73. "Rosneft May Up Oil Exports to China-Sechin," *Reuters* (June 16, 2011), http://uk.reuters.com/article/2011/06/16/rosneft-china-idUKLDE75F1752011o616.

74. CNPC website; Masuda, "Eastbound Flow."

75. Danchenko, Downs, and Hill, "One Step Forward," 12.

76. Goldman, *Petrostate*, 169.

77. Kozyrev, "China's Continental Energy," 221.

78. Baev, *Russian Energy Policy*, 74, 125.

79. Ibid., 125.

80. Catherine Belton, "Beijing Lends Russia $25bn for 20 Years of Oil Supplies," *Financial Times* (February 17, 2009).

81. Baev, *Russian Energy Policy*, 125.

82. See "Natural Gas Purchase and Sale Framework Agreement Signed Between China and Uzbekistan," CNPC website (June 10, 2010), http://www.cnpc.com.cn/en/press/newsreleases/Natural_gas_purchase_and_sale_framework_agreement_signed_between_China_and_Uzbekistan_.htm.

83. See CNPC website, http://www.cnpc.com.cn/en/cnpcworldwide/uzbekistan/Uzbekistan.htm.

84. "Myanmar President Visits China,"*Channelnewsasia.com* (May 26, 2011), http://www.channelnewsasia.com/stories/afp_asiapacific/view/1131338/1/.html. India and Thailand are other large trading partners.

85. The numbers are based on the SIPRI Arms Transfers Database. See also Jerker Hellström, "China's Foreign, Security and Defence Policy in a 10–20-Year Perspective," *FOI, Swedish Defence Research Agency* (2009), 24. See also Bo Kong, "The Geopolitics of the Myanmar-China Oil and Gas Pipelines," Special Report 23 (Seattle: National Bureau of Asian Research, 2010), 64.

86. Kong. "The Geopolitics of the Myanmar-China Oil and Gas Pipelines," 64.

87. Information about the contracts is taken from CNPC website and from a presentation by Jifeng Du, Chinese Academy of Social Sciences, Fairbank Center, Harvard University, February 25, 2010.

88. Kong, "The Geopolitics of the Myanmar-China Oil and Gas Pipelines," 57.

89. Erickson and Collins, "China's Oil Security Pipe Dream," 98.

90. Kong, "The Geopolitics of the Myanmar-China Oil and Gas Pipelines," 58.

91. Kong, *China's International Petroleum Policy*, 52; Kong. "The Geopolitics of the Myanmar-China Oil and Gas Pipelines," 60–61; Erickson and Collins, "China's Oil Security Pipe Dream," 98.

92. CNPC website and a presentation by Jifeng Du, Chinese Academy of Social Sciences, Fairbank Center, Harvard University, February 25, 2010.

93. Seaman, "Energy Security, Transnational Pipelines, and China's Role in Asia," 35.

94. Kong. "The Geopolitics of the Myanmar-China Oil and Gas Pipelines," 63.

95. Economist Intelligence Unit, "Deep Water Ahead? The Outlook for the Oil and Gas Industry in 2011," *The Economist* (February 2011): 8.

96. Richard E. Ericson, "Eurasian Natural Gas Pipeline: The Political Economy of Network Interdependence," *Eurasian Geography and Economics* 50, no. 1 (2009): 29–30.

97. Downs, "Inside China, Inc.," 41–45; Danchenko, Downs, and Hill, "One Step Forward," 12.

98. "Russian Crude Oil Exports to the Far East—ESPO Starts Flowing," 3.

99. Danchenko, Downs, and Hill, "One Step Forward," 12.

100. Ericson, "Eurasian Natural Gas Pipeline," 28–29.

101. Erickson and Collins, "China's Oil Security Pipe Dream," 104–105; Emmanuel Karagiannis, "China's Energy Security and Pipeline Diplomacy: Assessing the Threat of Low-Intensity Conflicts," *Harvard Asia Quarterly* (December 2010), http://asiaquarterly.com/2010/12/24/china%E2%80%99s-energy-security-and-pipeline-diplomacy-assessing-the-threat-of-low-intensity-conflicts/.

102. Kyaw Thein Kha, "Bomb Blast Rocks Dam Site," *Irrawaddy* (April 17, 2010); Kathrin Hille, "Burma Dam Disruption Concerns China," *Financial Times* (October 2, 2011), http://www.ft.com/intl/cms/s/0/bccaef18-ece7-11e0-be97-00144feab49a.html

103. Erickson and Collins, "China's Oil Security Pipe Dream," 105.

104. Danchenko, Downs, and Hill, "One Step Forward," 9.

105. Conversation with leading Norwegian energy experts and a Japanese energy expert and former IEA employee.

106. Carlos Pascual and Evie Zambetakis, "The Geopolitics of Energy: From Survival to Survival," in *Energy Security: Economics, Politics, Strategies, and Implications*, ed. Carlos Pascual and Jonathan Elkind (Washington, D.C.: Brookings Institution Press, 2010), 20.

107. Baev, *Russian Energy Policy*, 119–129; Goldman, *Petrostate*, 163.

108. A monopsony is a market in which only one buyer faces several sellers; a monopoly is a market in which only one seller faces many buyers. It is not strictly correct to say that Russia has a monopoly over the EU gas market since gas arrives to this market from the North Sea and by LNG and since the EU is not a single and coherent buyer. However, Russia is the leading actor in this market and does have a monopoly over some states, particularly in Eastern Europe and the transit states. While there are strong asymmetries in the disputes between former CIS countries and Russia, weak and small states can promote their interests when dealing with Russia. Central Asian states have obtained better deals with Russia through diversifying its petroleum export routes, Eastern European states were able to obtain subsidies and better deals on transfer by confronting Russia (with the support of the EU), and a small state like Norway managed to find an agreement with Russia over disputed maritime sovereignty claims in petroleum-rich areas in the Barents Sea.

109. Erickson and Collins, "China's Energy Security Pipe Dream," 100.

110. Saira H. Basit, "The Iran-Pakistan-India Pipeline Project, Fuelling Cooperation?" *Oslo Files* 4 (2008), http://www.ifspublications.com/index.php?page=shop.product_details&product_id=34&flypage=flypage.tpl&pop=0&option=com_virtuemart&Itemid=2&lang=nb.

111. Erickson and Collins, "China's Energy Security Pipe Dream," 104.

112. However, pipelines from landlocked resources in Russia and Kazakhstan not only bring security, reliability, and diversity to China's oil supply, but such overland pipeline projects are economically viable since the "fields filling these lines are so far from the sea that an overland line is the most effective way to transport their oil into the Chinese market." Erickson and Collins, "China's Energy Security Pipe Dream," 91, 107.

113. Chow and Hendrix, "Central Asia's Pipelines," 37.

114. Liao, "Central Asia and China's Energy Security"; James R. Holmes and Toshi Yoshihara, *China's Naval Strategy in the Twenty-First Century: The Turn to Mahan* (London: Routledge, 2008), 44.

115. Downs, "Who's Afraid of China's Oil Companies?" 92.

116. Xu, "China's Energy Security," 66.

117. Marc Lanteigne, "China, Energy Security, and Central Asian Diplomacy: Bilateral and Multilateral Approaches," in *Caspian Energy Politics*, ed. Indra Overland et al. (London: Routledge, 2010), 106.

118. Erica Downs, "China's 'New' Energy Administration," *China Business Review* 35, no. 6 (November/December 2008): 42.

119. Xu, "China's Energy Security," 66.

120. David Kerr, "Central Asian and Russian Perspectives on China's Strategic Emergence," *International Affairs* 86, no. 1 (2010): 141–143.

121. Liao, "Central Asia and China's Energy Security," 61.

122. Barry Naughton, "The Impact of the New Asia-Pacific Energy Competition on Russia and the Central Asian States," in *Energy Security in Asia*, ed. Michael Wesley (London: Routledge, 2007), 145.

123. Stephen J. Blank, "The Eurasian Energy Triangle: China, Russia, and the Central Asian States," *Brown Journal of World Affairs* 12, no. 2 (Winter 2005–Spring 2006): 55.

124. Xu Yi-chong, "China's Energy Security," in *Energy Security in Asia*, ed. Michael Wesley (London: Routledge, 2007), 66; Seaman, "Energy Security, Transnational Pipelines, and China's Role in Asia," 26–27.

125. Kapparov, Ivanova-Venturini, and Gazadze, "China—Now Kazakhstan's Top Export Destination." It should be noted that the distribution of revenue from the hydrocarbon sector and the economic development have only brought limited benefits to the ethnic Xinjiang Uyghurs, who remain hostile to the central and local governments. Nevertheless, the pipelines and growing infrastructure might achieve the goal of the central government to bring future development.

126. Kerr, "Central Asian and Russian Perspectives," 128.

127. Seaman, "Energy Security, Transnational Pipelines, and China's Role in Asia," 30.

128. Currently, southwestern China is the only region in the country that lacks medium- or large-scale oil refineries. See Kong, "The Geopolitics of the Myanmar-China Oil and Gas Pipelines," 59.

129. Ibid., 59–60.

130. Ibid., 60, 64.

131. Ibid., 60.

132. Bobo Lo, *Axis of Convenience: Moscow, Beijing, and the New Geopolitics* (Washington, D.C.: Brookings Institution Press, 2008).

133. Robert S. Ross, "The Rise of Russia, Sino-Russian Relations, and U.S. Security Policy," *Brief* (Royal Danish Defense College), June 2009.

134. U.S. Department of Defense, *Quadrennial Defense Review Report* (Washington, D.C., February 6, 2006).

135. The 2003 Iraq War, the Iranian nuclear issue, the North Korean nuclear question, controversy regarding U.S. missile defense plans, the 2008 war in Georgia, and other cases provided opportunities to challenge U.S. primacy, but neither China and Russia were willing to further consolidate their partnership to balance against the United States.

136. Øystein Tunsjø, "Geopolitical Shifts, Great-Power Relations, and Norway's Foreign Policy," *Cooperation and Conflict* 46, no. 1 (March 2011): 4–5.

137. Stephen M. Walt, "The Ties That Fray: Why Europe and America Are Drifting Apart," *National Interest* 54 (Winter 1998/1999); Richard Betts, "The Three Faces of NATO," *National Interest Online* (March/April 2009).

138. Tunsjø, "Geopolitical Shifts," 5.

139. Eugene Rumer, "The U.S. Interests and Role in Central Asia After K2," *Washington Quarterly* 29, no. 3 (2006): 141–154.

140. Danchenko, Downs, and Hill, "One Step Forward," 8.

141. Robert M. Cutler, "Does the ESPO Signal a New Sino-Russian Rapprochement?" *CACI Analyst* (March 25, 2009).

142. Naughton, "The Impact of the New Asia-Pacific Energy Competition," 140.

143. Rumer, "The U.S. Interests"; Jan Šír and Slavomir Horák, "China as an Emerging Superpower in Central Asia: The View from Ashkhabad," *China and Eurasian Forum Quarterly* 6, no. 2 (2008): 75–88; Matthew Oresman, "Reassessing the Fleeting Potential for U.S.-China Cooperation in Central Asia," *China and Eurasian Forum Quarterly* 6, no. 2 (2008): 10–11; Danchenko, Downs, and Hill, "One Step Forward."

144. Ericson, "Eurasian Natural Gas Pipeline," 35. However, Overland and Torjesen, drawing on BP's 2008 figures, estimate that "Russia produces a total of around 607 bcm of natural gas per year." Quoting the IEA, the same authors maintain that Turkmenistan exported 48 bcm to Russia in 2008 and Ukraine imported a total of 55 bcm from Russia, not including transit gas that Ukraine sent on to Europe. See Indra Overland and Stina Torjesen, "Kazakhstan's and Turkmenistan's Relations with Russia," in *Caspian Energy Politics*, ed. Indra Overland et al. (London: Routledge, 2010), 145–146.

145. Guliyev and Akhrarkhodjaeva note that Russia controls most of Central Asian exports, arguing that 80 percent of Kazakh crude oil exports was transported through Russia's pipeline network in 2007. But they also maintain that only 40 percent of Kazakhstan oil went through Russia in 2007. See Farid Guliyev and Nozima Akhrarkhodjaeva, "The Trans-Caspian Energy Route: Cronyism, Competition, and Cooperation in Kazakh Oil Exports," *Energy Policy* 37 (2009): 3173; Kennedy, "In the 'New Great Game,'" 119.

146. Blank, "The Eurasian Energy," 58–59.

147. Karagiannis, "China's Energy Security."

148. Baev, *Russian Energy Policy*, 62.

149. Goldman, *Petrostate*, 186–187.

150. Ericson, "Eurasian Natural Gas Pipeline," 40–41.

151. Quote taken from Baev, *Russian Energy Policy*, 73–74; Goldman, *Petrostate*.

152. Chow and Hendrix, "Central Asia's Pipelines," 38.

153. Kerr, "Central Asian and Russian Perspectives," 141–144.

154. However, it should be noted that it was the Russian construction firm Stroytransgas that built the 188-kilometer-long Turkmen section of the pipeline.

155. "Quotable Quotes from Chinese President's Central Asia Visit," *People's Daily* (December 15, 2009), http://english.peopledaily.com.cn/90001/90776/90883/6843114.html. See also http://www.cnpc.com.cn/en/press/Features/Flow_of_natural_gas_from_Central_Asia _.htm.

156. "Quotable Quotes from Chinese President's Central Asia Visit," *People's Daily* (December 15, 2009), http://english.peopledaily.com.cn/90001/90776/90883/6843114.html.

157. Lanteigne, "China, Energy Security, and Central Asian Diplomacy," 106. Nonetheless, it should be acknowledged that although China may have developed good elite relations, public opinion in a number of Central Asian states remains "wary if not hostile" to China's intrusion into the region. There are historical memories of earlier Chinese intrusions, and there is an underlying perception in the Central Asian states that "China is trying to recreate a system of tributary relations under modern conditions." See Kerr, "Central Asian and Russian Perspectives"; Kennedy, "In the 'New Great Game.'"

158. Karagiannis, "China's Energy Security"; Kong, "The Geopolitics of the Myanmar-China Oil and Gas Pipelines," 58.

159. Erickson and Collins, "China's Energy Security Pipe Dream," 92; Dennis Blair and Kenneth Lieberthal, "Smooth Sailing: The World's Shipping Lanes Are Safe," *Foreign Affairs* (May/June 2007).

160. Kha, "Bomb Blast Rocks Dam Site."

161. Erickson and Collins, "China's Energy Security Pipe Dream," 92.

162. Kong, "The Geopolitics of the Myanmar-China Oil and Gas Pipelines," 60.

163. Erickson and Collins, "China's Energy Security Pipe Dream"; Kong, "The Geopolitics of the Myanmar-China Oil and Gas Pipelines," 60.

164. Presentation by Jifeng Du, Chinese Academy of Social Sciences, Fairbank Center, Harvard University, February 25, 2010.

165. Karagiannis, "China's Energy Security."

166. "Myanmar's Neighbors Advance Pipeline Project," *Wall Street Journal* (November 2, 2009).

167. Kong, "The Geopolitics of the Myanmar-China Oil and Gas Pipelines," 58.

168. "China and Russia Veto US/UK-Backed Security Council Draft Resolution on Myanmar," *UN News Centre* (January 12, 2007), http://www.un.org/apps/news/story.asp?NewsID=2122 8&Cr=myanmar&Cr1. The multinational character of the gas pipeline, whereby CNPC obtained a majority stake of 51 percent, but the fact that Korea's Daewoo, India's Oil and Natural Gas Corporation Limited (ONGC), and the Gas Authority of India Limited (GAIL) are also significant stakeholders illustrates that the Myanmar government is seeking to play

the different actors against one another. Finally, recent events in Myanmar, whereby the country is opening up and turning closer to the United States and the West, may suggest that China's leverage over Myanmar should not be overstated.

169. Simultaneously, China's inroads into Myanmar have alarmed other countries, in particular India, which has launched several naval exercises and initiated cooperation through the Quadrennial Initiative—India, Japan, the United States, and Australia—plus Singapore, in response to the increasing presence of the PLAN in the Indian Ocean and the Bay of Bengal. Consequently, China's continental petroleum strategy and broader foreign policy objectives constitute both opportunities and risks that need to be managed through a hedging strategy.

7. GLOBAL, MARITIME, AND CONTINENTAL IMPLICATIONS

1. Traditional NOCs include National Iranian Oil Co., Libya National Oil Co., Iraq National Oil Co., and KazMunaiGaz; hybrid NOCs (hNOCs) include China National Offshore Oil Company (CNOOC), China National Petroleum Company (CNPC), Sinopec, ENI, Petronas, Rosneft, and Statoil; IOCs include BP, Chevron, Exxon Mobil, Total, and Royal Dutch Shell. hNOCs are characterized by the following characteristics: (1) varying levels of government ownership, (2) serving as stewards of national natural resources, and (3) operating both within and outside the home country. However, there is much variety within the new hNOCs category. For example, some are publicly listed and traded on either an international or local exchange. See Deloitte, "A New Era is Dawning: Rise of the Hybrid National Oil Company," 2010, http://www.deloitte.com/view/en_GX/global/industries/energy-resources/2e4745a204368210VgnVCM200000bb42fooaRCRD.htm.

2. Ibid., 4.

3. Erica Downs, "Who's Afraid of China's Oil Companies?" in *Energy Security: Economics, Politics, Strategies, and Implications*, ed. Carlos Pascual and Jonathan Elkind (Washington, D.C.: Brookings Institution Press, 2010), 80.

4. Julie Jiang and Jonathan Sinton, "Overseas Investments by Chinese National Oil Companies."

5. Oxford Analytica, "China: Energy Companies Change Strategic Tack" (April 26, 2010).

6. Philip Andrews-Speed and Roland Dannreuther, *China, Oil, and Global Politics* (London: Routledge, 2011), 74, 138; Xiao Wan, "CNPC, BP to Settle Iraq Oil Deal," *China Daily* (August 5, 2009), http://www.chinadaily.com.cn/business/2009-08/05/content_8522678.htm. These developments contradict both what the Chinese government feared when the United States started the war against Iraq in 2003 and popular assumptions suggesting that the United States waged war in Iraq to secure oil. China and its NOCs have capitalized on U.S. actions and have maneuvered into a stronger position than the one that existed before the war. Although the case may be different, a Chinese diplomat visited Libyan rebel leaders in Benghazi in June 2011, and representatives from the rebels visited Beijing several weeks later, indicating that China is lining up to take advantage of the conflict in Libya. "China Says Diplomat Meets Libyan Rebel Leaders in Benghazi," *Reuters* (June 7, 2011), http://af.reuters.com/article/libyaNews/idAFL3E7H706E20110607?pageNumber

=1&virtualBrandChannel=0; "China Boosts Support for Libya's Rebels," *Global Post* (June 23, 2011), http://www.globalpost.com/dispatch/news/regions/middle-east/110623/china -libya-rebel; "Libya 'Looking to Replace Eni,'" *UpstreamOnline* (July 7, 2011), http://www .upstreamonline.com/live/article265992.ece.

7. Not surprisingly, since the Chinese government has a stated aim of developing three world-class oil companies.

8. Downs, "Who's Afraid of China's Oil Companies?" 85–95.

9. Ibid., 82–89.

10. For instance, approximately fifteen companies are operating in Sudan. These companies are mainly from Asia, but the French oil company Total has a permanent representative in Khartoum. See Cindy Hurst, "China's Oil Rush in Africa," *IAGS Energy Security* (July 2006).

11. Esther Pan, "China, Africa, and Oil," Council on Foreign Relations, *Backgrounder* (January 26, 2007).

12. Satu P. Limaye, "The United States and Energy Security in the Asia-Pacific," in *Energy Security in Asia*, ed. Michael Wesley (London: Routledge, 2007), 19; Erica Downs, "The Chinese Energy Security Debate," *China Quarterly* 177 (2004): 36; Mikkal J. Herberg, "The Rise of Energy and Resource Nationalism in Asia," in *Strategic Asia 2010–11: Asia's Rising Power and America's Continued Purpose*, ed. Ashley J. Tellis et al. (Seattle: National Bureau of Asian Research, 2010). However, Kong argues that Chinese NOCs in large part have not been overpaying. Bo Kong, *China's International Petroleum Policy* (Santa Barbara, Calif.: Praeger Security International, 2010), and interview with Bo Kong, Washington, D.C., May 2010.

13. John Mitchell and Glada Lahn, "Oil for Asia," Briefing Paper, Chatham House, March 2007; Wang Haibin and Li Bin, "China's Choices About Energy Security Safeguards and Its New Security Concepts," *Contemporary Asia-Pacific Studies* 5 (2007): 21–30.

14. Interview with Professor Zha Dajiong, Beijing, November 2007.

15. Of course, Wang is also seeking to boost his political credentials and be promoted to higher positions in the government or the party. See Brian Spegele and Wayne Ma, "For China Boss, Deep-Water Rigs Are a 'Strategic Weapon,'" *Wall Street Journal* (August 29, 2012), http://online.wsj.com/article/SB10000872396390444233104577592890738740290.html.

16. Conversation with Arne Walther, Beijing, May 2009.

17. Yo Osumi, "The IEA in a Changing Energy World: What Is Its Value for China?" presentation at the Energy Security in Asia conference, Beijing, May 2009.

18. Carola Hoyos, "China Invited to Join IEA as Oil Demand Shifts," *Financial Times* (March 30, 2010), http://www.ft.com/intl/cms/s/0/0f973936-3beb-11df-9412-00144feabdco .html#axzz1f6Ps8wpK.

19. Yo Osumi, "The IEA in a Changing Energy World."

20. Ibid.

21. Gao Xiaohui, "Time Not Yet Ripe for China's IEA Membership: Analyst," *Global Times* (April 1, 2010), http://www.globaltimes.cn/business/comment/2010-04/518029.html.

22. See Flynt Leverett, "Consuming Energy: Rising Powers, the International Energy Agency, and the Global Energy Architecture," in *Rising States, Rising Institutions: Challenges for*

Global Governance, ed. Alen S. Alexandroff and Andrew F. Cooper (Washington, D.C.: Brookings Institution Press, 2010), 262.

23. Kenneth Waltz, *Theory of International Politics* (New York: Random House, 1979).

24. Arne Walther, "Global Energy Dialogue," keynote speech, Japan Cooperation Centre (January 26, 2011), http://www.jccp.or.jp/english/wp-content/uploads/k1_0126_mrwalther_royal-norwegian-embassy.pdf.

25. This can be linked to Ulrich Beck's definition of risk as a systematic way of dealing with hazards and insecurities induced and introduced by modernization itself. See Ulrich Beck, *Risk Society: Towards a New Modernity*, trans. Mark Ritter (Cambridge: Polity, 1992).

26. Robert Jervis, "Understanding the Bush Doctrine," *Political Science Quarterly* 118, no. 3 (2003): 381–382.

27. Yang Yi, "Engagement, Caution," *China Security* 3, no. 4 (Autumn 2007): 29–39.

28. Alex Perry, "China's New Focus on Africa," *Time* (June 24, 2010), http://www.time.com/time/specials/packages/article/0,28804,2000110_2000287_2000276,00.html.

29. Conversation with Zhao Hongto, Beijing, November 2011.

30. Grete De Lange, "Kina vil frakte oljen selv" (China will transport oil itself), *Aftenposten* (September 9, 2011), http://www.aftenposten.no/okonomi/utland/article4221367.ece; Magnus A. Koren, "Frontline i rødt" (Frontline in red), *E24* (August 26, 2011), http://e24.no/boers-og-finans/frontline-i-roedt/20093020.

31. De Lange, "Kina vil frakte oljen selv."

32. Interview with shipping analyst, Oslo, Norway, September 2011.

33. Randy Fabi and Jonathan Saul, "Global Economic Gloom to Sink Tanker Market Recovery Hopes," *Reuters* (August 18, 2011), http://in.reuters.com/article/2011/08/18/idINIndia-58841220110818.

34. Øystein Byberg, "Det verste markedet siden Svartedauen" (The worst market since the black death), *Hegnar Online* (May 25, 2011), http://www.hegnar.no/bors/shipping/article610650.ece.

35. Carsten Olav Coldevin, "Slutt på sagaen" (Slopes of the saga), *Stocklink* (October 19, 2011), http://www.stocklink.no/Article.aspx?id=84959.

36. Canadian Chamber of Shipping, "COS News—Week Ending 30 September 2011," no. 177 (October 7, 2011), http://www.chamber-of-shipping.com/index.php?option=com_content&view=article&id=827:cos-weekly-news-30-september-2011&catid=94:cos-weekly-news-archive&Itemid=74.

37. Interview with shipping analysts, Oslo, October 2011. However, Norwegian shipping and tanker experts argued in November 2012 that China controls about 60 VLCCs, which is roughly 10 percent of the world fleet of 608 vessels. Interviews in Oslo and Beijing, November 2012.

38. Robert S. Ross, "The Geography of Peace: East Asia in the Twenty-First Century," *International Security* 23, no. 4 (Spring 1999); Robert S. Ross, "China's Naval Nationalism," *International Security* 34, no. 2 (Fall 2009): 46–81.

39. On the debate about a more assertive China, see, among others, Michael D. Swaine and Taylor Fravel, "China's Assertive Behavior," *China Leadership Monitor* 35 (Summer 2011); Thomas J. Christensen, "The Advantages of an Assertive China," *Foreign Affairs* 90, no. 2 (2011): 54–67; Wang Jisi, "China's Search for a Grand Strategy," *Foreign Affairs* 90, no. 2

(2011): 68–79; June Teufel Dreyer, "China as Rumpelstiltskin," *The Diplomat* (February 12, 2011); Minxin Pei, "China's Bumpy Ride Ahead," *The Diplomat* (February 12, 2011); Kerry Brown and Loh Su Hsing, "Trying to Read the New 'Assertive' China," Asia Program Paper, Chatham House, January 2011.

40. Such behavior corresponds with hedging on a grand strategic level. See Øystein Tunsjø, "Geopolitical Shifts, Great Power Relations, and Norway's Foreign Policy," *Cooperation and Conflict* 46, no. 1 (March 2011): 60–77.

41. See Peter Dutton, Robert Ross, and Øystein Tunsjø, eds., *Twenty-First Century Seapower* (London: Routledge, 2012). This issue was also examined during the International Order at Sea workshop, Norwegian Institute for Defense Studies (IFS), August 26, 2011. See http://ifs.forsvaret.no/publikasjoner/IOS/Sider/default.aspx.

42. Michael McDevitt and Fredric Vellucci Jr., "The Evolution of the People's Liberation Army Navy: The Twin Missions of Area-Denial and Peacetime Operations," in *Sea Power and the Asia-Pacific: The Triumph of Neptune?*, ed. Geoffrey Till and Patrick C. Bratton (London: Routledge, 2012), 75–92.

43. This is, for example, illustrated by the challenges to find a balance between developing state-of-the-art space surveillance networks and aircraft carriers.

44. Linda Jakobson, "China Prepares for an Ice-Free Arctic," *SIPRI Insight on Peace and Security* 2 (March 2010). See also Hu Zhengyue, Assistant Minister of Foreign Affairs, PRC, "China's Arctic Policy," presentation at the High North Study Tour, Svalbard, July 2, 2009; David Curtis Wright, *The Dragon Eyes the Top of the World*, China Maritime Studies 8 (Newport, R.I.: Naval War College, August 2011).

45. Sven G. Holtsmark and Brooke A. Smith-Windsor, eds., *Security Prospects in the High North: Geostrategic Thaw or Freeze?* (Rome: NATO Defense College, May 2009), 14.

46. The USGS report should be interpreted with caution. It is not based on comprehensive geological and seismic surveys of the areas involved, and no drilling has been conducted. Nonetheless, from the perspective of the petroleum industry, even if it is only half right that one-fifth or one-fourth of the world's untapped reserves of oil and natural gas lie under the Arctic Ocean, "then it is worth exploring," and assessments about the commerciality of the region will grow. See Kristine Offerdal, "High North Energy: Myths and Realities," in ibid., 166, 171; ibid., 14.

47. "Kina får en del av Barents-kaka" (China may be part of the Barents-cake), *E24* (August 26, 2011), http://e24.no/olje-og-raavarer/kina-faar-en-del-av-barents-kaka/20093018.

48. Conversation with top officials from the Norwegian Ministry of Petroleum and Energy, Oslo, October 2011; ibid.

49. Jakobson, "China Prepares for an Ice-Free Arctic."

50. The water depth in the Bering Strait measures 100 to 160 feet, whereas a VLCC needs water depth of a minimum of 100 feet. Hence, the risk of allowing VLCCs to transit the Bering Strait is great.

51. Emphasized in interviews with representatives from a major shipping company and a leading shipping insurance firm, Hong Kong, August 2009. Combating oil spills in icy waters is almost impossible with current technology.

52. In particular, Wright, *The Dragon Eyes*, provides a summary of alarmist Chinese writers commentating on the Arctic. See also Jakobson, "China Prepares for an Ice-Free Arctic"; Joseph Spears, "A Snow Dragon in the Arctic," *Asia Times* (February 8, 2011), http://www.atimes.com/atimes/China/MB08Ad01.html. For some examples of more alarmist writings, see Gordon G. Chang, "China's Arctic Play," *The Diplomat* (March 9, 2010), http://the-diplomat.com/2010/03/09/china%E2%80%99s-arctic-play/; Margaret Blunden, "The New Problem of Arctic Stability," *Survival* 51, no. 5 (2009): 121–142; James R. Holmes, "The Arctic Sea—A New Wild West?" *The Diplomat* (April 20, 2010), http://the-diplomat.com/2011/04/20/the-arctic-sea%E2%80%94a-new-wild-west/.

53. Jakobson, "China Prepares for an Ice-Free Arctic"; Wright, *The Dragon Eyes*.

54. Hu, "China's Arctic Policy"; conversations with international law experts from CIMA, Beijing, China, November 2011.

55. Linda Jakobson et al., "China's Energy and Security Relations With Russia," SIPRI Policy Paper 29 (October 2011); Bobo Lo, *Axis of Convenience: Moscow, Beijing, and the New Geopolitics* (Washington, D.C.: Brookings Institution Press, 2008); Tunsjø, "Geopolitical Shifts."

56. Tunsjø, "Geopolitical Shifts."

8. CONCLUSION

1. IEA, *World Energy Outlook*, 2007, http://www.iea.org/weo/2007.asp.

2. "Can" and "should" refer mainly to alarmist Chinese analysts, whereas "has" refers mostly to alarmist international writers.

3. John W. Garver, "Is China Playing a Dual Game in Iran?" *Washington Quarterly* 34, no. 1 (Winter 2011): 75–88.

4. Mikkal E. Herberg, statement before the U.S.-China Economic and Security Review Commission hearing on "China's Energy Consumption and Opportunities for U.S.-China Cooperation to Address the Effects of China's Energy Use," June 14–15, 2007.

5. An even more detrimental example for a number of European countries, Japan, and to some extent the United States would be if a war or turmoil/uprising were to occur in Saudi Arabia, which is China's leading supplier. This would lead to a major disruption of oil supplies in the international market, which would be reinforced if China then decided to ship its equity oil or marketing rights back home. In addition, China's oil trade with Iran, a country shunned by the West, will insure that China will be relatively better off than the other major oil importers.

6. Robert Jervis, "Understanding the Bush Doctrine," *Political Science Quarterly* 118, no. 3 (2003): 381–382.

BIBLIOGRAPHY

Albert, Mathias. "From Defending Borders Towards Managing Geopolitical Risks? Security in a Globalized World." *Geopolitics* 5, no. 1 (2001): 57–80.

Alden, Chris. *China in Africa.* London: Zed, 2007.

Alterman, Jon B., and John W. Garver. *The Vital Triangle: China, the United States, and the Middle East.* Washington, D.C.: Center for Strategic and International Studies, 2008.

Amoore, Louise, and Mareke de Goede, eds. *Risk and the War on Terror.* London: Routledge, 2008.

Andrews-Speed, Philip. "China's Booming Gas Sector: Threat or Promise?" *Oil, Gas, and Energy Law Intelligence* (February 2011).

——. "China's Energy Policy and Its Contribution to International Stability." In "Facing China's Rise: Guidelines for an EU Strategy," ed. Marcin Zaborowski, *Chailot Paper* 94 (December 2006): 71–81.

——. "China's Nuclear Power Plans After Fukushima." *Oil, Gas, and Energy Law Intelligence* (April 2011).

——. "China's Ongoing Energy Efficiency Drive: Origins, Progress, and Prospects." *Energy Policy* 37, no. 4 (April 2009): 1331–1344.

——. "China-Russia Energy Relations: Which Party Holds the Stronger Hand?" *Oil, Gas, and Energy Law Intelligence* (July 2011).

——. "The Institutions of Energy Governance in China." *IFRI* (January 2010).

Andrews-Speed, Philip, and Roland Dannreuther. *China, Oil, and Global Politics.* London: Routledge, 2011.

Andrews-Speed, Philip, Xuanli Liao, and Roland Dannreuther. "The Strategic Implications of China's Energy Needs." *Adelphi Paper* 346. Oxford: Oxford University Press, 2002.

Art, Robert J. "Europe Hedges Its Security Bets." In *Balance of Power: Theory and Practice in the Twenty-First Century*, ed. T. V. Paul, James J. Wirtz, and Michel Fortman, 179–213. Stanford, Calif.: Stanford University Press, 2004.

Austin, Angie. "Energy and Power in China: Domestic Regulation and Foreign Policy." Foreign Policy Centre, April 2005.

Baev, Pavel. *Russian Energy Policy and Military Power: Putin's Quest for Greatness*. London: Routledge, 2008.

Baldwin, David, ed. *Neorealism and Neoliberalism: The Contemporary Debate*. New York: Columbia University Press, 1993.

Barnes, Joe, and Amy Myers Jaffe. "The Persian Gulf and the Geopolitics of Oil." *Survival* 48, no. 1 (Spring 2006): 143–162.

Basit, Saira H. *The Iran-Pakistan-India Pipeline Project: Fuelling Cooperation?* Oslo Files 4. Oslo: Norwegian Institute for Defense Studies, 2008.

Basit, Saira, and Øystein Tunsjø. *Emerging Naval Powers in Asia: China and India's Quest for Seapower*. Oslo Files 2. Oslo: Norwegian Institute for Defense Studies, 2012.

Bateman, Sam, and Joshua Ho. "Somalia-type Piracy: Why It Will Not Happen in Southeast Asia." *RSIS Commentaries* (November 24, 2008).

Beck, Ulrich. *Risk Society: Towards a New Modernity*, trans. Mark Ritter. Cambridge: Polity, 1992.

Bergsten, Fred C., et al. *China's Rise: Challenges and Opportunities*. Washington, D.C.: Peterson Institute for International Economics, October 2009.

Betts, Richard. "The Three Faces of NATO." *National Interest Online* (March/April 2009).

Beveridge, Neil, and Angus Chan. "Bernstein Asia-Pac Energy: Coal Bed Methane: Is It Time for China's Unconventional Gas Revolution?" *Bernstein Research* (May 18, 2010).

Blagov, Sergei. "Russia Seeks to Increase Energy Supplies to China." *Eurasia Daily Monitor* (The Jamestown Foundation) 8, no. 122 (June 24, 2011).

Blair, Dennis, and Kenneth Lieberthal. "Smooth Sailing: The World's Shipping Lanes Are Safe." *Foreign Affairs* (May/June 2007).

Blank, Stephen. "Chinese Energy Policy in Central Asia and South Asia." Testimony before the U.S.-China Economic and Security Review Commission, May 20, 2009.

——. "The Eurasian Energy Triangle: China, Russia, and the Central Asian States." *Brown Journal of World Affairs* 12, no. 2 (Winter 2005/Spring 2006): 53–67.

Blumenthal, Dan. "Concerns with Respect to China's Energy Policy." In *China's Energy Strategy: The Impact on Beijing's Maritime Policies*, ed. Gabriel B. Collins et al., 418–436. Annapolis, Md.: U.S. Naval Institute Press, 2008.

Blunden, Margaret. "The New Problem of Arctic Stability." *Survival* 51, no. 5 (2009): 121–142.

Booth, Ken, and Nicholas J. Wheeler. *The Security Dilemma: Fear, Cooperation, and Trust in World Politics*. New York: Palgrave Macmillan, 2008.

Boutilier, James. "Great Nations, Great Navies: Looking for Sea Room in Asia." Presentation at the conference NATO-Russia Council: From Vladivostok to Vancouver, Vladivostok, May 11, 2006.

Bray, Scott. "Seapower Questions on the Chinese Submarine Force." U.S. Navy, Office of Naval Intelligence, December 20, 2006.

Brealey, Richard A., and Stewart C. Myers. *Financing and Risk Management*. New York: McGraw-Hill, 2003.

Brown, Kerry, and Loh Su Hsing. "Trying to Read the New 'Assertive' China." Asia Program Paper, Chatham House, January 2011.

Canning, Cherie. "Pursuit of the Pariah: Iran, Sudan, and Myanmar in China's Energy Security Strategy." *Security Challenges* 3, no. 1 (February 2007): 47–63.

Chang, Gordon G. "China's Arctic Play." *The Diplomat* (March 9, 2010).

Cheung, Kat. "Integration of Renewables: Status and Challenges in China." Working Paper, International Energy Agency (IEA), 2011.

Chow, Edward C., and Leigh E. Hendrix. "Central Asia's Pipelines." *NBR Special Report* 2 (September 2010): 29–42.

Christensen, Thomas J. "The Advantages of an Assertive China." *Foreign Affairs* 90, no. 2 (2011): 54–67.

Clinton, Hillary. "America's Pacific Centrury." *Foreign Policy* (November 2011).

Coker, Christopher. *War in an Age of Risk*. Cambridge: Polity, 2009.

Cole, Bernard D. *The Great Wall at Sea: China's Navy in the Twenty-First Century*. 2nd ed. Annapolis, Md.: Naval Institute Press, 2010.

——. "Oil for the Lamps of China—Beijing's Search for Energy." *McNair Paper* 67. National Defence University, 2003.

——. "Right-Sizing the Navy: How Much Naval Force Will Beijing Deploy?" In *Right Sizing the People's Liberation Army: Exploring the Contours of China's Military*, ed. Roy Kamphausen and Andrew Scobell, 523–556. Carlisle Barracks, Penn.: U.S. Army War College, 2007.

——. *Sea Lanes and Pipelines: Energy Security in Asia*. London: Praeger Security International, 2008.

Collins, Gabriel. "China Fills First SPR Site, Faces Oil, Pipeline Issues." *Oil and Gas Journal* (August 20, 2007).

——. "China Seeks Oil Security with New Tanker Fleet." *Oil and Gas Journal* (October 9, 2006).

——. "China's Growing LNG Demand Will Shape Markets, Strategies." *Oil and Gas Journal* (October 15, 2007).

——. "China's Refining Expansion to Reshape Global Oil Trade." *Oil and Gas Journal* (February 18, 2008).

Collins, Gabriel, and Michael Grubb. "Contemporary Chinese Shipbuilding Prowess." In *China Goes to Sea: Maritime Transformation in Comparative Historical Perspective*, ed. Andrew Erickson, Lyle J. Goldstein, and Carnes Lord, 344–372. Annapolis, Md.: Naval Institute Press, 2009.

Collins, Gabriel, and William S. Murray. "No Oil for the Lamps of China?" *Naval War College Review* 61, no. 2 (Spring 2008): 79–95.

Constantin, Christian. "Understanding China's Energy Security." *World Political Science Review* 3, no. 3 (2007): 1–30.

Cornelius, Peter, and Jonathan Story. "China and Global Energy Markets." *Orbis* 51, no. 1 (Winter 2007): 5–20.

Cornish, Margaret. *Behaviour of Chinese SOEs: Implications for Investment and Cooperation in Canada*. Canadian Council of Chief Executives, February 2012.

Cunningham, Edward. "China's Energy Governance: Perception and Reality." In *Audits of the Conventional Wisdom*. MIT Center for International Studies, March 2007.

Cunningham, Graham, and Hanzhi Ding. "The Asia Petrolizer." *Citigroup Global Markets* (November 9, 2009).

Currier, Carrie Liu, and Manochehr Dorraj, eds. *China's Energy Relations with the Developing World*. New York: Continuum, 2011.

Cutler, Robert M. "Does the ESPO Signal a New Sino-Russian Rapprochement?" *CACI Analyst* (March 25, 2009).

Danchenko, Igor, Erica Downs, and Fiona Hill. "One Step Forward, Two Steps Back? The Realities of a Rising China and Implications for Russia's Energy Ambitions." *Policy Paper* (Brookings Institution) 22 (August 2010).

Deloitte. "A New Era is Dawning: Rise of the Hybrid National Oil Company." 2010.

Dennys, Kathrine. "'A Foreign Policy Migraine': Assessing China's Role in the Conflict Between Sudan and South Sudan." *Consultancy Africa Intelligence* (June 18, 2012).

Devonshire-Ellis, Chris. "China's String of Pearls Strategy." *China Briefing* (March 18, 2009).

Diamond, Andrew F. "Dying with Eyes Open or Closed: The Debate Over a Chinese Aircraft Carrier." *Korean Journal* 13, no. 1 (Spring 2006): 35–58.

Dittrick, Paula. "Chinese Oil Companies Invest Heavily Abroad." *Oil and Gas Journal* (February 8, 2010).

Dow, James. "Arbitrage, Hedging, and Financial Innovation." *Review of Financial Studies* 11, no. 4 (Winter 1998): 739–755.

Downs, Erica S. "Business Interest Groups in Chinese Politics: The Case of the Oil Companies." In *China's Changing Political Landscape: Prospects for Democracy*, ed. Cheng Li, 121–141. Washington, D.C.: Brookings Institution Press, 2008.

——. "China." In *Energy Security Series, The Brookings Foreign Policy Studies* (Washington, D.C.: Brookings Institution, December 2006).

——. "China-Gulf Energy Relations." In *China and the Gulf: Implications for the United States*, ed. Bryce Wakefield and Susan L. Levenstein. Washington, D.C.: Woodrow Wilson International Center for Scholars, 2011).

——. "China's Energy Security." Ph.D. diss. Princeton University, 2004.

——. "China's 'New' Energy Administration." *China Business Review* 35, no. 6 (November–December 2008): 42–45.

——. "The Chinese Energy Security Debate." *China Quarterly* 177 (2004): 21–44.

——. "The Facts and Fiction of Sino-Africa Energy Relations." *China Security* 3, no. 3 (Summer 2007).

——. *Inside China, Inc.: China Development Banks Cross-Border Energy Deals*. Washington, D.C.: John L. Thornton China Center, Brookings Institution, 2011).

——. "Sino-Russian Energy Relations: An Uncertain Courtship." In *The Future of China-Russia Relations*, ed. James Bellacqua, 146–175. Lexington: University Press of Kentucky, 2010.

——. "Who's Afraid of China's Oil Companies?" In *Energy Security: Economics, Politics, Strategies, and Implications*, ed. Carlos Pascual and Jonathan Elkind, 76–77. Washington, D.C.: Brookings Institution Press, 2010.

Dreyer, June Teufel. "China as Rumpelstiltskin." *The Diplomat* (February 12, 2011).

Duffie, Darrell. *Future Markets.* Englewood Cliffs, N.J.: Prentice Hall, 1989.

Dutton, Peter, Robert S. Ross, and Øystein Tunsjø, eds. *Twenty-First-Century Seapower: Cooperation and Conflict at Sea.* London: Routledge, 2012.

Economist Intelligence Unit. "Deep Water Ahead? The Outlook for the Oil and Gas Industry in 2011." *The Economist* (February 2011).

Eder, Leonty, Philip Andrews-Speed, and Andrey Korzhubaev. "Russia's Evolving Energy Policy for its Eastern Regions, and Implications for Oil and Gas Cooperation Between Russia and China." *Journal of World Energy Law and Business* 2, no. 3 (2009): 219–242.

Elster, Jon. *Nuts and Bolts for the Social Sciences.* Cambridge: Cambridge University Press, 1989.

Energy Information Administration. "US, China Oil." Country Analysis Briefs.

Engdahl, William. "Darfur? It's the Oil, Stupid." *Geopolitics-Geoeconomics* (May 20, 2007).

Erickson, Andrew. "Can China Become a Maritime Power?" In *Asia Looks Seaward: Power and Maritime Strategy*, ed. Toshi Yoshihara and James R. Holmes, 70–110. London: Praeger Security International, 2008.

Erickson, Andrew, and Gabriel Collins. "Beijing's Energy Security Strategy: The Significance of a Chinese State-Owned Tanker Fleet." *Orbis* 51, no. 4 (Fall 2007): 665–684.

——. "China's Oil Security Pipe Dream: The Reality, and Strategic Consequences, of Seaborne Imports." *Naval War College Review* 63, no. 2 (Spring 2010): 88–111.

Erickson, Andrew, and Lyle Goldstein. "Gunboats for China's New 'Grand Canals'? Probing the Intersection of Beijing's Naval and Oil Security Policies." *Naval War College Review* 62, no. 2 (Spring 2009): 43–76.

Ericson, Richard E. "Eurasian Natural Gas Pipeline: The Political Economy of Network Interdependence." *Eurasian Geography and Economics* 50, no. 1 (2009): 28–57.

Eurasia Group. "China's Overseas Investment in Oil and Gas Production." Report for the U.S.-China Economic and Security Review Commission, October 16, 2006.

Foot, Rosemary. "Chinese Strategies in a U.S.-Hegemonic Global Order: Accommodating and Hedging." *International Affairs* 82, no. 1 (2006): 77–94.

Foucault, Michael. "Governmentality." In *Essential Works of Michael Foucault, 1954–1984*, vol. 3: *Power*, ed. James D. Faubion. New York: The New Press, 2000.

Fravel, Taylor. *Strong Borders, Secure Nation: Cooperation and Conflict in China's Territorial Disputes.* Princeton, N.J.: Princeton University Press, 2008.

Fusaro, Peter C., ed. *Energy Risk Management: Hedging Strategies and Instruments for the International Energy Markets.* New York: McGraw-Hill, 1998.

Fusaro Peter C., and Tom James. *Energy Hedging in Asia: Market Structure and Trading Opportunities.* New York: Palgrave Macmillan, 2005.

Garver, John W. "Is China Playing a Dual Game in Iran?" *Washington Quarterly* 34, no. 1 (Winter 2011): 75–88.

Garver, John W., Flynt Leverett, and Hillary Mann Leverett. "Moving (Slightly) Closer to Iran: China's Shifting Calculus for Managing Its 'Persian Gulf Dilemma,'" Johns Hopkins School of Advanced International Studies (SAIS), October 2009.

Giddens, Anthony. *Modernity and Self-Identity: Self and Society in the Late Modern Age.* Cambridge: Polity, 1991.

Gill, Bates, Chin-hao Huang, and Stephen J. Morrison. "Assessing China's Growing Influence in Africa." *China Security* 3, no. 3 (Summer 2007): 3–21.

Godement, François, Francois Nicolas, and Taizo Yakushiji. "An Overview of Options and Challenges." In *Asia and Europe: Cooperating for Energy Security*, ed. François Godement, Françoise Nicolas, and Taizo Yakushiji, 9–28. Paris: Centre Asia, Institut français des relations internationales, 2004.

Goh, Evelyn. "Meeting the China Challenge: The U.S. in Southeast Asian Regional Security Strategies." *Policy Studies* 16 (Washington, D.C.: East-West Center, 2005): 1–57.

——. "Understanding 'Hedging' in Asia-Pacific Security." *Pacific Forum* (CSIS) (August 31, 2006).

Goldman, Marshall. *Petrostate: Putin, Power, and the New Russia*. Oxford: Oxford University Press, 2008.

Grieco, Joseph M. *Cooperation Among Nations: Europe, America, and Non-Tariff Barriers to Trade*. Ithaca, N.Y.: Cornell University Press, 1990.

Gross, Donald G. "Transforming the U.S. Relationship with China." *Global Asia* 2, no. 1 (Spring 2007).

Guliyev, Farid, and Nozima Akhrarkhodjaeva. "The Trans-Caspian Energy Route: Cronyism, Competition, and Cooperation in Kazakh Oil Exports." *Energy Policy* 37 (2009): 3171–3182.

Hartnett, Daniel M. "The PLA's Domestic and Foreign Activities and Orientation." Testimony before the US-China Economic and Security Review Commission, March 4, 2009.

Hatemi, Peter, and Andrew Wedeman. "Oil and Conflict in Sino-American Relations." *China Security* 3, no. 3 (Summer 2007): 95–118.

Hellström, Jerker. "China's Foreign, Security, and Defence Policy in a 10–20-Year Perspective." *FOI, Swedish Defence Research Agency* (2009).

Heng, Yee-Kuang. *War as Risk Management: Strategy and Conflict in an Age of Globalised Risks*. London: Routledge, 2006.

Herberg, Mikkal E. Statement before the U.S.-China Economic and Security Review Commission hearing on "China's Energy Consumption and Opportunities for U.S.-China Cooperation to Address the Effects of China's Energy Use." Washington D.C., June 14–15, 2007.

Herberg, Mikkal, and David Zweig. "China's 'Energy Rise': The U.S. and the New Geopolitics of Energy." Pacific Council on International Policy, April 2010. http://www.pacificcouncil.org/document.doc?id=159.

——. "The Rise of Energy and Resource Nationalism in Asia." In *Strategic Asia 2010–11: Asia's Rising Power and America's Continued Purpose*, ed. Ashley J. Tellis, Andrew Marble, and Travis Tanner, 113–142. Seattle: National Bureau of Asian Research, 2010.

Ho, Joshua. "Piracy in the Gulf of Aden: Lessons from the Malacca Strait." *RSIS Commentaries* (January 22, 2009).

——. "Security of Transportation Routes in Southeast Asia." Presentation at the conference on Energy Security in Asia, Beijing, May 20–21, 2009.

Holmes, James R. "The Arctic Sea – A New Wild West?" *The Diplomat* (April 20, 2010).

Holmes, James R., and Toshi Yoshihara. *China's Naval Strategy in the Twenty-First Century: The Turn to Mahan*. London: Routledge, 2008.

———. "China's 'Caribbean' in the South China Sea." *SAIS Review* 26, no. 1 (Winter/Spring 2006): 79–92.

Holtsmark, Sven G., and Brooke A. Smith-Windsor, eds. *Security Prospects in the High North: Geostrategic Thaw or Freeze?* Rome: NATO Defense College, May 2009.

Houser, Trevor, and Roy Levy. "Energy Security and China's UN Diplomacy." *China Security* 4, no. 3 (Summer 2008): 63–73.

Howarth, Peter. *China's Rising Sea Power: The PLA Navy's Submarine Challenge.* London: Routledge, 2006.

Howes, Stephen. "China's Energy Intensity Target: On Track or Off?" *East Asian Forum* (March 31, 2010).

Hu Jintao. "An Open Mind for Win-Win Cooperation." Speech at the Asia Pacific Economic Cooperation (APEC) CEO summit, Busan, Republic of Korea, November 17, 2005.

Hu Jintao. "G8 Written Statement." St. Petersburg, Russia, July 17, 2006.

Huang, Chin-Hao. "China's Evolving Perspective on Darfur: Significance and Policy Implications." *PacNet Newsletter* 40 (July 25, 2008).

Human Rights Watch. "China's Involvement in Sudan: Arms and Oil." November 24, 2003.

Hurst, Cindy. "China's Oil Rush in Africa." *IAGS Energy Security* (July 2006).

International Energy Agency (IEA). "Oil & Gas Security Emergency Response of IEA Countries: People's Republic of China." 2012.

———. "Oil Market Report." January 16, 2009.

———. *2007 World Energy Outlook: China and India Insights.*

———. *2009 World Energy Outlook.*

———. *2010 World Energy Outlook.*

International Maritime Bureau. "Piracy and Armed Robbery Against Ships." Annual Report, January 2009.

Information Office of the State Council of the People's Republic of China. *China's National Defense in 2006.* Beijing: Foreign Language Press, 2006.

International Crisis Group (ICG). "China's Thirst for Oil." *Asia Report* 153 (June 9, 2008).

———. "The Iran Nuclear Issue: The View from Beijing." *Asia Briefing* 100 (Beijing/Brussels, February 17, 2010).

Itoh, Shoichi. "The Geopolitics of Northeast Asia's Pipeline Development." *National Bureau of Asian Research*, Special Report 23 (September 2010): 17–28.

Jaffe, Amy Myers, and Steven W. Lewis. "Beijing's Oil Diplomacy." *Survival* 44, no. 1 (Spring 2002): 115–134.

Jakobson, Linda. "China Prepares for an Ice-Free Arctic." *SIPRI Insight on Peace and Security* 2 (March 2010).

Jakobson, Linda, and Dean Knox. "New Foreign Policy Actors." *SIPRI Policy Paper* 26.

Jakobson, Linda, Paul Holtom, Dean Knox, and Jingchao Peng. "China's Energy and Security Relations With Russia." *SIPRI Policy Paper* 29 (October 2011).

Jakobson, Linda, and Zha Daojiong. "China and the Worldwide Search for Oil Security." *Asia-Pacific Review* 13, no. 2 (2006): 60–73.

Jervis, Robert. "Understanding the Bush Doctrine." *Political Science Quarterly* 118, no. 3 (2003): 365–388.

Ji Xiaohua. "It Is Not Impossible to Send Troops Overseas to Fight Terrorism." *Sing Tao Jih Pao* (Hong Kong) (June 17, 2004), A27, in Foreign Broadcast Information Service, CPP20040617000054.

Jiang, Julie, and Jonathan Sinton. "Overseas Investments by Chinese National Oil Companies: Addressing the Drivers and Impacts." *Information Paper* (IEA) (February 2011).

Johnson-Reiser, Sabine. "China's Hydropower Miscalculation." *China Brief* 12, no. 11 (May 25, 2012).

Kang Wu and Liutong Zhang. "China Works to Double SPR Capacity by 2013." *Oil and Gas Journal* (October 4, 2010).

——. "China's Strategic Reserves Capacity to Double by 2011." *Oil and Gas Journal* (September 21, 2009).

Kaplan, Robert D. "Center Stage for the Twenty-First Century, Power Plays in the Indian Ocean." *Foreign Affairs* 88, no. 2 (March/April 2009): 16–32.

Kapparov, Kassymkhan, Milena Ivanova-Venturini, and Eka Gazadze. "China—Now Kazakhstan's Top." Renaissance Capital, Strategic Equity Research, January 19, 2011.

Karagiannis, Emmanuel. "China's Energy Security and Pipeline Diplomacy: Assessing the Threat of Low-Intensity Conflicts." *Harvard Asia Quarterly* (December 2010).

Kemenade, Willem van. "Iran's Relations with China and the West: Cooperation and Confrontation in Asia." Netherlands Institute of International Relations, Clingendael, November 2009.

Kennedy, Ryan. "In the 'New Great Game,' Who Is Getting Played? Chinese Investment in Kazakhstan's Petroleum Sector." In *Caspian Energy Politics*, ed. Indra Overland, Heidi Kjaernet, and Andrea Kendall-Taylor, 123–125. London: Routledge, 2010.

Keohane, Robert O. *Neorealism and Its Critics*. New York: Columbia University Press, 1986.

Kerr, David. "Central Asian and Russian Perspectives on China's Strategic Emergence." *International Affairs* 86, no. 1 (2010): 127–152.

Klare, Michael T. "China's Global Shopping Spree; and China Seals Oil Deals." *Center for American Progress* (April 27, 2010).

——. *Rising Powers, Shrinking Planet: The New Geopolitics of Energy*. New York: Metropolitan, 2008.

Klare, Michael T., and Daniel Volman. "The African 'Oil Rush' and U.S. National Security." *Third World Quarterly* 27, no. 4 (May 2006): 609–628.

Kong, Bo. "An Anatomy of China's Energy Insecurity and its Strategies." Pacific Northwest Center for Global Security, Seattle, December 2005.

——. "China's Energy Decision-Making—Becoming More Like the United States?" *Journal of Contemporary China* 18, no. 62 (November 2009): 798–812.

——. *China's International Petroleum Policy*. Santa Barbara, Calif.: Praeger Security International, 2010.

——. "The Geopolitics of the Myanmar-China Oil and Gas Pipelines." *National Bureau of Asian Research,* Special Report 23 (2010): 55–65.

——. "Institutional Insecurity." *China Security* 3 (Summer 2006): 64–88.

Kozyrev, Vitaly. "China's Continental Energy Strategy: Russia and Central Asia." In *China's Energy Strategy: The Impact on Beijing's Maritime Policies*, ed. Gabriel B. Collins, Andrew S. Erickson, Lyle J. Goldstein, and William S. Murray, 202–251. Annapolis, Md.: Naval Institute Press, 2008.

Lam, Willy. "The Rise of the Energy Faction in Chinese Politics." *China Brief* (April 22, 2011).

Lanteigne, Marc. "China, Energy Security, and Central Asian Diplomacy: Bilateral and Multilateral Approaches." In *Caspian Energy Politics: Azerbaijan, Kazakhstan, and Turkmenistan*, ed. Indra Overland, Heidi Kjaernet, and Andrea Kendall-Taylor, 101–115. London: Routledge, 2009.

Lee, Henry, and Dan A. Shalmon. "Searching for Oil: China's Oil Initiatives in the Middle East." BCSIA Discussion Paper. Cambridge, Mass.: Belfer Center for Science and International Affairs, Kennedy School of Government, Harvard University, January 2007.

Lestera, Richard, and Edward Steinfeld. "China's Real Energy Crisis." *Harvard Asia Review* 9, no. 1 (Winter 2007).

Leung, Guy C. K. "China's Oil Use, 1990–2008." *Energy Policy* 38, no. 2 (2010): 932–944.

Leverett, Flynt. "Consuming Energy: Rising Powers, the International Energy Agency, and the Global Energy Architecture." In *Rising States, Rising Institutions: Challenges for Global Governance*, ed. Alen S. Alexandroff and Andrew F. Cooper, 240–265. Washington, D.C.: Brookings Institution Press, 2010.

——. "The Geopolitics of Oil and America's International Standing." Statement before the Committee on Energy and Natural Resources, U.S. Senate, January 10, 2007.

Li Peng. "China's Policy on Energy Resources." *Xinhua* (May 28, 1997), appearing as "Li Peng on Energy Policy," *World News Connection* (June 23, 1997), doi: drchi119_n_97001.

Li, Justin. "Chinese Investment in Iran: One Step Forward and Two Steps Backward." *East Asian Forum* (November 3, 2010).

Lieberman, Joseph. "China-US Energy Policies: A Choice of Cooperation or Collision." Council on Foreign Relations, November 30, 2005.

Lieberthal, Kenneth, and Michel Oksenberg. *Policy Making in China: Leaders, Structures, and Processes*. Princeton, N.J.: Princeton University Press, 1988.

Lieberthal, Kenneth, and Mikkal Herberg. "China's Search for Energy Security: Implications for U.S. Policy." *NBR Analysis* 17, no. 1 (2006).

Limaye, Satu P. "The United States and Energy Security in the Asia-Pacific." In *Energy Security in Asia*, ed. Michael Wesley, 15–27. London: Routledge, 2007.

Lin, Christina. "The Caspian Sea: China's Silk Road Strategy Converges with Damascus." *Pitts Report* (The Jamestown Foundation) (September 7, 2010).

Linda Jakobson and Zha Daojiong. "China and the Worldwide Search for Oil Security." *Asia-Pacific Review* 13, no. 2 (2006): 60–73.

Liu, Currier, and Manochehr Dorraj, eds. *China's Energy Relations with the Developing World*. New York: Continuum, 2011.

Lo, Bobo. *Axis of Convenience: Moscow, Beijing, and the New Geopolitics*. Washington, D.C.: Brookings Institution Press, 2008.

Ludwig, Thorsten, and Jochen Tholen. "Shipbuilding in China and Its Impact on the European Shipbuilding Industry." In *European Industries Shaken up by Industrial Growth in China: What*

Regulations Are Required for a Sustainable Economy? Brussels: European Metalworkers' Federation, November 2006.

McDevitt, Michael. "The Strategic and Operational Context Driving PLA Navy Building." In *Right Sizing the People's Liberation Army: Exploring the Contours of China's Military*, ed. Roy Kamphausen and Andrew Scobell, 481–522. Carlisle Barracks, Penn.: U.S. Army War College, 2007.

McDevitt, Michael, and Fredric Vellucci Jr. "The Evolution of the People's Liberation Army–Navy: The Twin Missions of Area-Denial and Peacetime Operations." In *Sea Power and the Asia-Pacific: The Triumph of Neptune?*, ed. Geoffrey Till and Patrick C. Bratton, 75–92. London: Routledge, 2012.

McVadon, Eric A. "China's Navy Today: Looking Toward Blue Water." In *China Goes to Sea: Maritime Transformation in Comparative Historical Perspective*, ed. Andrew Erickson, Lyle J. Goldstein, and Carnes Lord, 373–400. Annapolis, Md.: Naval Institute Press, 2009.

Medeiros, Evan. "Strategic Hedging and the Future of Asia-Pacific Stability." *Washington Quarterly* 29, no. 1 (2005–2006): 145–167.

Meidan, Michal, ed. *Shaping China's Energy Security: The Inside Perspective*. Paris: Asia Centre, 2007.

Meidan, Michal, Philip Andrews-Speed, and Ma Xin. "Shaping China's Energy Policy: Actors and Processes." *Journal of Contemporary China* 18, no. 61 (September 2009): 591–616.

Meng-di Gu and Shou-de Li. "China Seeks Oil Security Through Fleet Expansion." *Oil and Gas Journal* (December 8, 2008).

Minxin Pei. "China's Bumpy Ride Ahead." *The Diplomat*, February 12, 2011.

Mitchell, John, and Glada Lahn. "Oil for Asia." Briefing Paper. London: Chatham House, March 2007.

Monaghan, Andrew. "Russia and the Security of Europe's Energy Supplies: Security in Diversity?" *Conflict Studies Research Centre*, Special Series, Defence Academy of the United Kingdom, January 2007.

Moran, Theodore H. *China's Strategy to Secure Natural Resources: Risks, Dangers, and Opportunities*. Policy Analyses in International Economics 92. Washington, D.C.: Peterson Institute for International Economics, July 2010.

Morse, Edward. "Energy 2020: North America, the New Middle East?" *CITI GPS* (March 20, 2012).

——. "Pincer Movement." *Fixed Income Research* (Credit Suisse) (April 19, 2010).

Myers Jaffe, Amy, and Steven W. Lewis. "Beijing's Oil Diplomacy." *Survival* 44, no. 1 (Spring 2002): 115–134.

Nan Li. "The Evolution of China's Naval Strategy and Capabilities: From 'Near Coast' and 'Near Seas' to Far Seas." *Asian Security* 5, no. 2 (2009): 144–169.

Naughton, Barry. "The Impact of the New Asia-Pacific Energy Competition on Russia and the Central Asian States." In *Energy Security in Asia*, ed. Michael Wesley, 128–157. London: Routledge, 2007.

——. "SASAC and Rising Corporate Power in China." *China Leadership Monitor* 24 (Spring 2008).

Nerlich, Uwe. "Energy Security or a New Globalization of Conflicts? Oil and Gas in Evolving New Power Structures." *Strategic Insight* 7, no. 1 (February 2008).

Newmyer, Jacqueline. "Oil, Arms, and Influence: The Indirect Strategy Behind Chinese Military Modernization." *Orbis* (Spring 2009): 205–219.

Niazi, Tarique. "The Ecology of Strategic Interests: China's Quest for Energy Security from the Indian Ocean and the South China Sea to the Caspian Sea Basin." *China and Eurasia Forum Quarterly* 4, no. 4 (2006): 97–116.

Nieh, Daniel Kang Wu, Lijuan Wang, and Shi Fu. "Study Examines Chinese SPR Growth Alternatives." *Oil and Gas Journal* (July 23, 2007).

Nye, Joseph S. "Neorealism and Neoliberalism." *World Politics* 40, no. 2 (1988): 235–251.

O'Rourke, Ronald. "China Naval Modernization: Implications for U.S. Navy Capabilities— Background and Issues for Congress." Congressional Research Service, April 22, 2011.

Obama, Barack. "Remarks by President Obama to the Australian Parliament."The White House, Office of the Press Secretary, November 17, 2011.

Offerdal, Kristine. "High North Energy: Myths and Realities." In *Security Prospects in the High North: Geostrategic Thaw or Freeze?*, ed. Sven G. Holtsmark and Brooke A. Smith-Windsor, 151–178. Rome: NATO Defense College, May 2009.

Office of the U.S. Secretary of Defense. *Annual Report to Congress: Military Power of the People's Republic of China*. Washington, D.C., 2009.

Oresman, Matthew. "Reassessing the Fleeting Potential for U.S.-China Cooperation in Central Asia." *China and Eurasia Forum Quarterly* 6, no. 2 (2008): 5–13.

Overland Indra and Stina Torjesen. "Kazakhstan's and Turkmenistan's Relations with Russia." In *Caspian Energy Politics: Azerbaijan, Kazakhstan, and Turkmenistan*, ed. Indra Overland, Heidi Kjaernet, and Andrea Kendall-Taylor, 136–149. London: Routledge, 2009.

Pan, Esther. "China, Africa, and Oil." Council on Foreign Relations, *Backgrounder* (January 26, 2007).

Pascual, Carlos, and Evie Zambetakis. "The Geopolitics of Energy: From Survival to Survival." In *Energy Security: Economics, Politics, Strategies, and Implications*, ed. Carlos Pascual and Jonathan Elkind, 9–36. Washington, D.C.: Brookings Institution Press, 2010.

Pehrson, Christopher J. "String of Pearls: Meeting the Challenge of China's Rising Power Across the Asian Littoral." *US Army War College* (July 2006).

People's Republic of China. *Defense White Paper*. 2010.

Pohlner, Huw. "Chinese Dam diplomacy: Leadership and Geopolitics in Continental Asia." *East Asia Forum* (August, 19, 2010).

Posen, Barry. "Command of the Commons: The Military Foundations of U.S. Hegemony." *International Security* 28, no. 1 (2008): 5–46.

Rasmussen, Mikkel Vedby. *The Risk Society at War: Technology and Strategy in the Twenty-First Century*. Cambridge: Cambridge University Press, 2006.

Raymond, Catherine Zara. "Countering Piracy and Armed Robbery in Asia: A Study of Two Areas." In *Twenty-First Century Seapower: Cooperation and Conflict at Sea*, ed. Peter Dutton, Robert S. Ross, Øystein Tunsjø, 213–236. London: Routledge, 2012.

Rosen, Daniel, and Trevor Houser. *China Energy: A Guide for the Perplexed*. Washington, D.C.: Peterson Institute for International Economics, 2007.

Ross, Robert S. "China's Naval Nationalism." *International Security* 34, no. 2 (Fall 2009): 46–81.

——. "The Geography of Peace: East Asia in the Twenty-first Century." *International Security* 23, no. 4 (Spring 1999): 81–118.

——. "The Rise of Russia, Sino-Russian Relations, and U.S. Security." In *Perspectives for a European Security Strategy Towards Asia*, ed. Gustaaf Geeraerts and Eva Gross, 169–188. Brussels: Brussels University Press, 2011.

Rumer, Eugene. "The U.S. Interests and Role in Central Asia After K2." *Washington Quarterly* 29, no. 3 (2006): 141–154.

Samuels, Richard J. *Tokyo's Grand Strategy and the Future of East Asia*. Ithaca, N.Y.: Cornell University Press, 2007.

Saunders, Phillip C. "Uncharted Waters: The Chinese Navy Sails to Somalia." *PacNet* 3 (January 15, 2009).

Seaman, John. "Energy Security, Transnational Pipelines, and China's Role in Asia." *Asie Visions* 27. Paris: Asia Centre, Institut française des relations internationals, April 2010.

Shambaugh, David. "Asia in Transition: The Evolving Regional Order." *Current History* (April 2006): 153–160.

Shi Hongtao. "China's 'Malacca Straits.'" *Qingnian bao* (June 15, 2004), in Foreign Broadcast Information Service, CPP20040615000042.

Shirk, Susan. *The Political Logic of Economic Reform in China*. Berkeley: University of California Press, 1993.

Shlapal, David A., et al. *A Question of Balance, Political Context, and Military Aspects of the China-Taiwan Dispute*. Santa Monica, Calif.: RAND Corporation, 2009.

Sír, Jan, and Slavomir Horák. "China as an Emerging Superpower in Central Asia: The View from Ashkhabad." *China and Eurasian Forum Quarterly* 6, no. 2 (2008): 75–88.

Smith, Mark A. "The Russo-Chinese Energy Relationship." *Russian Series* (Defence Academy of the United Kingdom) 10, no. 14 (October 21, 2010).

Steinfeld, Edward S., Richard K. Lester, and Edward A Cunningham. "Greener Plants, Greyer Skies? A Report from the Front Lines of China's Energy Sector." *Energy Policy* 37, no. 5 (2009): 1809–1824.

Storey, Ian. "China's 'Malacca Dilemma.'" *China Brief* (The Jamestown Foundation) 6, no. 8 (April 12, 2006).

Swaine, Michael D., and Taylor Fravel. "China's Assertive Behavior." *China Leadership Monitor* 35 (Summer 2011).

Tang, James. *With the Grain or Against the Grain? Energy Security and Chinese Foreign Policy in the Hu Jintao Era*. Washington, D.C.: The Brookings Institution, 2006.

Taylor, Ian. "Beijing's Arms and Oil Interests in Africa." *China Brief* 5, no. 21 (October 13, 2005).

——. "China's Oil Diplomacy in Africa." *International Affairs* 82, no. 5 (2006): 937–959.

Terjesen, Bjørn, and Øystein Tunsjø, eds. *The Rise of Naval Powers in Asia and Europe's Decline*. Oslo File 6. Oslo: Norwegian Institute for Defense Studies, 2012

Tessman, Brock, and Wojtek Wolfe. "Great Powers and Strategic Hedging: The Case of Chinese Energy Security Strategies." *International Studies Review* 13 (2011): 214–240.

Tow, William. "Strategic Dimensions of Energy Competition in Asia." In *Energy Security in Asia*, ed. Michael Wesley, 161–173. London: Routledge, 2007.

Tunsjø, Øystein. "Geopolitical Shifts, Great Power Relations and Norway's Foreign Policy." *Cooperation and Conflict* 46, no. 1 (March 2011): 60–77.

——. "Hedging Against Oil Dependency: New Security Perspectives on Chinese Energy Security Policy." *International Relations* 24, no. 1 (March 2010): 25–45.

——. *U.S. Taiwan Policy: Constructing the Triangle*. London: Routledge, 2008.

——. "Zhongguo nengyuan anquan de dui chong zhanlüe" (China hedges its energy security bets). *Shijie jingji yu zhengzhi* (World economics and politics) (August 2008): 42–51.

U.S. Department of Defense. *Annual Report to Congress on the Military Power of the People's Republic of China*. Washington, D.C., 2007.

——. *National Defense Strategy*. Washington, D.C., June 2008.

——. *Quadrennial Defense Review Report*. Washington, D.C., February 6, 2006.

U.S. Department of Energy (DOE). *Energy Policy Act 2005 Section 1837: National Security Review of International Energy Requirements*. February 28, 2006.

Vakil, Sanam. "Iran: Balancing East Against West." *Washington Quarterly* 29, no. 4 (2006): 51–65.

Victor, David, and Linda Yueh. "The New Energy Order: Managing Insecurities in the Twenty-First Century." *Foreign Affairs* 89, no. 1 (2010): 61–73.

Vivoda, Vlado. "Diversification of Oil Import Sources and Energy Security: A Key Strategy or an Elusive Objective?" *Energy Policy* 37, no. 11 (2009): 4615–4623.

Walt, Stephen M. "The Ties That Fray: Why Europe and America Are Drifting Apart." *National Interest* 54 (Winter 1998/1999).

Walther, Arne. "The Geopolitics of Oil and Global Energy Security." Paper presented at The Oil Era: Emerging Challenges, the Sixteenth Annual Energy Conference, Abu Dhabi, November 8–10, 2010.

——. "Global Energy Dialogue." Keynote speech, Japan Cooperation Centre, January 26, 2011.

Waltz, Kenneth. *Man, the State, and War: A Theoretical Analysis*. New York: Columbia University Press, 1959.

——. *Theory of International Politics*. New York: Random House, 1979.

Wang Haibin. "Profits Outweigh Self-Reliance as China's Nuclear Industry Expands." *Uranium Intelligence Weekly* 44, no. 4 (November 1, 2010).

Wang Haibin and Li Bin. "China's Choices About Energy Security Safeguards and Its New Security Concepts." *Contemporary Asia-Pacific Studies* 5 (2007): 21–30.

Wang Jisi. "China's Search for a Grand Strategy." *Foreign Affairs* 90, no. 2 (2011): 68–79.

Wang Qingyi. "Energy Conservation as Security." *China Security* 2, no. 2 (Summer 2006): 89–105.

Weitz, Richard. "Why U.S. Keeps Hedging Over China." *The Diplomat* (January 11, 2011).

Wesley, Michael. "The Geopolitics of Energy in Asia." In *Energy Security in Asia*, ed. Michael Wesley, 1–12. London: Routledge, 2007.

The White House. *The National Security Strategy of the United States of America*. Washington, D.C., March 2006.

White, Hugh. "Why War in Asia Remains Thinkable." *Survival* 50, no. 8 (2008): 85–104.

Wood, Mackenzie. "Chinese NOCs Step up International Expansion." *Corporate Service Insight* (June 2010).

Wright, David Curtis. *The Dragon Eyes the Top of the World*. China Maritime Studies 8. Newport, R.I.: Naval War College, 2011.

Wu Kang and Ian Storey. "Energy Security in China's Capitalist Transition: Import Dependence, Oil Diplomacy, and Security Imperatives." In *China's Emergent Political Economy: Capitalism in the Dragon's Lair*, ed. Christopher A. McNally. London: Routledge, 2008.

Wu Lei and Shen Qinyu. "Will China Go to War Over Oil?" *Far Eastern Economic Review* (April 2006): 38–40.

Xiaoli Liu and Xinmin Jiang. "China's Energy Security Situation and Countermeasures." *International Journal of Energy Sector Management* 3, no. 1 (2009): 83–92.

Xu, Xiaojie. *Petro-Dragon's Rise: What It Means for China and the World* (Florence: European Press Academic Publishing, 2002).

Xu Yi-Chong. "China's Energy Security." *Australian Journal of International Affairs* 60, no. 2 (June 2006): 265–286.

——. "China's Energy Security." In *Energy Security in Asia*, ed. Michael Wesley, 42–67. London: Routledge, 2007.

Xuanli Liao. "Central Asia and China's Energy Security." *China and Eurasia Forum Quarterly* 4, no. 4 (2006): 61–69.

Xuecheng Liu. "China's Energy Security and Its Grand Strategy." *Policy Analysis Brief* (The Stanley Foundation) (September 2006).

Yang Yi. "Engagement, Caution." *China Security* 3, no. 4 (Autumn 2007): 29–39.

Yergin, Daniel. "Ensuring Energy Security." *Foreign Affairs* 85, no. 2 (March–April 2006): 69–82.

——. *The Quest: Energy, Security, and the Remaking of the Modern World*. London: Allen Lane, 2011.

——. Statement before the Committee on Energy and Commerce U.S. House of Representatives, May 4, 2004.

You Ji and Lim Chee Kia. "China's Naval Deployment to Somalia and Its Implications." *EAI Background Brief* 454 (May 29, 2009).

You Yi. "Dealing with the Malacca Strait Dilemma: China's Efforts to Enhance Energy Transportation Security." *EAI Background Brief* 329 (April 12, 2007).

Yu, Silvia. "China Doubling Down on Domestic Uranium Exploration Efforts." *Uranium Intelligence Weekly* 44, no. 4 (November 1, 2010).

Zambelis, Chris. "Shifting Sands in the Gulf: The Iran Calculus in China-Saudi Arabia Relations." *China Brief* 10, no. 10 (May 13, 2010).

Zha Daojiong. "Energy Interdependence." *China Security* 2, no. 2 (Summer 2006): 2–16.

——. "Oiling the Wheels of Foreign Policy? Energy Security and China's International Relations." Asia Security Initiative Policy Series, Working Paper 1, March 2010.

Zha Daojiong and Hu Weixing. "Promoting Energy Partnership in Beijing and Washington." *Washington Quarterly* 30, no. 4 (Autumn 2007): 105–115.

Zhang Wenmu. "Sea Power and China's Strategic Choices." *China Security* 3 (Summer 2006): 17–31.

Zhen Li. "China's Utility Powerhouses Push Into Nuclear." *Uranium Intelligence Weekly* 44, no. 4 (November 1, 2010).

Zheng Hong. "Confidence Building Measures and Nontraditional Security." In *Twenty-First-Century Seapower: Cooperation and Conflict at Sea*, ed. Peter Dutton, Robert S. Ross, Øystein Tunsjø, 298–315. London: Routledge, 2012.

Zissis, Carin. "Crafting a U.S. Policy on Asia." Council on Foreign Relations (April 10, 2007).

Zweig, David, and Bi Jinhai. "China's Global Hunt for Energy." *Foreign Affairs* 84, no. 5 (September–October 2005): 25–38.

INDEX